VLSI

T0239673

Circuits for Emerging Applications

Devices, Circuits, and Systems

Series Editor

Krzysztof Iniewski
CMOS Emerging Technologies Research Inc.,
Vancouver, British Columbia, Canada

PUBLISHED TITLES:

Integrated Power Devices and TCAD Simulation
Yue Fu, Zhanming Li, Wai Tung Ng, and Johnny K.O. Sin

Internet Networks: Wired, Wireless, and Optical Technologies
Krzysztof Iniewski

Labs on Chip: Principles, Design, and Technology
Eugenio Iannone

Low Power Emerging Wireless Technologies
Reza Mahmoudi and Krzysztof Iniewski

Medical Imaging: Technology and Applications
Troy Farncombe and Krzysztof Iniewski

Metallic Spintronic Devices
Xiaobin Wang

MEMS: Fundamental Technology and Applications
Vikas Choudhary and Krzysztof Iniewski

Micro- and Nanoelectronics: Emerging Device Challenges and Solutions
Tomasz Brozek

Microfluidics and Nanotechnology: Biosensing to the Single Molecule Limit
Eric Lagally

MIMO Power Line Communications: Narrow and Broadband Standards, EMC, and Advanced Processing
Lars Torsten Berger, Andreas Schwager, Pascal Pagani, and Daniel Schneider

Mobile Point-of-Care Monitors and Diagnostic Device Design
Walter Karlen

Nano-Semiconductors: Devices and Technology
Krzysztof Iniewski

Nanoelectronic Device Applications Handbook
James E. Morris and Krzysztof Iniewski

Nanopatterning and Nanoscale Devices for Biological Applications
Šeila Selimović

Nanoplasmonics: Advanced Device Applications
James W. M. Chon and Krzysztof Iniewski

Nanoscale Semiconductor Memories: Technology and Applications
Santosh K. Kurinec and Krzysztof Iniewski

Novel Advances in Microsystems Technologies and Their Applications
Laurent A. Francis and Krzysztof Iniewski

Optical, Acoustic, Magnetic, and Mechanical Sensor Technologies
Krzysztof Iniewski

Optical Fiber Sensors: Advanced Techniques and Applications
Ginu Rajan

PUBLISHED TITLES:

Organic Solar Cells: Materials, Devices, Interfaces, and Modeling
Qiquan Qiao

Radiation Effects in Semiconductors
Krzysztof Iniewski

Semiconductor Radiation Detection Systems
Krzysztof Iniewski

Smart Grids: Clouds, Communications, Open Source, and Automation
David Bakken

Smart Sensors for Industrial Applications
Krzysztof Iniewski

Solid-State Radiation Detectors: Technology and Applications
Salah Awadalla

Technologies for Smart Sensors and Sensor Fusion
Kevin Yallup and Krzysztof Iniewski

Telecommunication Networks
Eugenio Iannone

Testing for Small-Delay Defects in Nanoscale CMOS Integrated Circuits
Sandeep K. Goel and Krishnendu Chakrabarty

VLSI: Circuits for Emerging Applications
Tomasz Wojcicki

Wireless Technologies: Circuits, Systems, and Devices
Krzysztof Iniewski

FORTHCOMING TITLES:

Analog Electronics for Radiation Detection
Renato Turchetta

**Cell and Material Interface: Advances in Tissue Engineering,
Biosensor, Implant, and Imaging Technologies**
Nihal Engin Vrana

Circuits and Systems for Security and Privacy
Farhana Sheikh and Leonel Sousa

CMOS: Front-End Electronics for Radiation Sensors
Angelo Rivetti

CMOS Time-Mode Circuits and Systems: Fundamentals and Applications
Fei Yuan

**Electrostatic Discharge Protection of Semiconductor Devices
and Integrated Circuits**
Juin J. Liou

FORTHCOMING TITLES:

Gallium Nitride (GaN): Physics, Devices, and Technology
Farid Medjdoub and Krzysztof Iniewski

Implantable Wireless Medical Devices: Design and Applications
Pietro Salvo

Laser-Based Optical Detection of Explosives
Paul M. Pellegrino, Ellen L. Holthoff, and Mikella E. Farrell

Mixed-Signal Circuits
Thomas Noulis and Mani Soma

Magnetic Sensors: Technologies and Applications
Simone Gambini and Kirill Poletkin

MRI: Physics, Image Reconstruction, and Analysis
Angshul Majumdar and Rabab Ward

Multisensor Data Fusion: From Algorithm and Architecture Design to Applications
Hassen Fourati

Nanoelectronics: Devices, Circuits, and Systems
Nikos Konofaos

Nanomaterials: A Guide to Fabrication and Applications
Gordon Harling, Krzysztof Iniewski, and Sivashankar Krishnamoorthy

Optical Imaging and Sensing: Technology, Devices, and Applications
Dongsoo Kim and Ajit Khosla

Physical Design for 3D Integrated Circuits
Aida Todri-Sanial and Chuan Seng Tan

Power Management Integrated Circuits and Technologies
Mona M. Hella and Patrick Mercier

Radiation Detectors for Medical Imaging
Jan S. Iwanczyk and Polad M. Shikhaliev

Radio Frequency Integrated Circuit Design
Sebastian Magierowski

Reconfigurable Logic: Architecture, Tools, and Applications
Pierre-Emmanuel Gaillardon

Soft Errors: From Particles to Circuits
Jean-Luc Autran and Daniela Munteanu

Wireless Transceiver Circuits: System Perspectives and Design Aspects
Woogeun Rhee

VLSI
Circuits for
Emerging
Applications

EDITED BY **TOMASZ WOJCICKI**
Sidense Corporation, Ottawa, Ontario, Canada

KRZYSZTOF INIEWSKI MANAGING EDITOR
CMOS Emerging Technologies Research Inc.
Vancouver, British Columbia, Canada

CRC Press
Taylor & Francis Group
Boca Raton   London   New York

CRC Press is an imprint of the
Taylor & Francis Group, an **informa** business

CRC Press
Taylor & Francis Group
6000 Broken Sound Parkway NW, Suite 300
Boca Raton, FL 33487-2742

First issued in paperback 2017

© 2015 by Taylor & Francis Group, LLC
CRC Press is an imprint of Taylor & Francis Group, an Informa business

No claim to original U.S. Government works
Version Date: 20140912

ISBN 13: 978-1-138-07605-1 (pbk)
ISBN 13: 978-1-4665-9909-3 (hbk)

Library of Congress Cataloging-in-Publication Data

VLSI : circuits for emerging applications / editor, Tomasz Wojcicki.
    pages cm -- (Devices, circuits, and systems ; 34)
  Includes bibliographical references and index.
  ISBN 978-1-4665-9909-3 (hardback)
  1. Integrated circuits, Very large scale integration. I. Wojcicki, Tomasz.

TK7874.75.V57155 2014
621.39'5--dc23

2014033184

Visit the Taylor & Francis Web site at
http://www.taylorandfrancis.com

and the CRC Press Web site at
http://www.crcpress.com

Contents

Preface

Only recently the world celebrated the 60th anniversary of the invention of the first transistor. The first integrated circuit was built a decade later, with the first microprocessor designed in the early 1970s. Today, integrated circuits are part of almost every aspect of our daily life. They help us to live longer and more comfortably, and to do more, and do it faster. And all that is possible because of the relentless search for new materials, new circuit designs, and new ideas happening on a daily basis at universities and within the industry around the globe.

Proliferation of integrated circuits in our daily lives does not mean making more of the same. It is actually the opposite. It is about making more of something completely different and customized for a particular application. And today's circuit designers cannot complain about the shortage of things to work with.

All leading semiconductor foundries are offering now at least six different process nodes, from 180 nm down to 16 nm, with each node having two, three, or even more flavors. There are at least three different input-output (IO) voltage standards—3.3 V, 2.5 V, and 1.8 V. And apart from the mainstream complementary metal oxide semiconductor (CMOS) process, each foundry offers more options such as GaAs, silicon on insulator (SOI), and GaN; new, even more exotic materials are not far behind. It all gives engineers an almost unlimited number of options and choices to make to achieve their objectives or their application.

There is no book that can provide a comprehensive view of what is happening in the world of micro- and nanoelectronics. And this book does not intend or pretend to be the one that does. The following 20 chapters give a snapshot of what researchers are working on now. Readers are encouraged to review the index and this preface before diving into a specific chapter or two. Maybe readers are going to find something related to their field of interest or maybe something completely unrelated but inspiring that will provoke new thoughts and ideas.

CHAPTER 1: INTEGRATION OF GRAPHICS PROCESSING CORES WITH MICROPROCESSORS
(Deepak C. Sekar and Chinnakrishnan Ballapuram)

Integration is at the forefront of practically any design activity. Putting more features in the same piece of silicon has been driving all segments of the industry, starting from manufacturing and ending with CAD tools. This chapter talks about the recent trend of combining multiple cores in a single chip and boost performance of the overall system this way, instead of just increasing the system clock speed. With more advanced processes, integration of many components such as peripheral control hubs, dynamic random access memory (DRAM) controllers, modems, and, more importantly, graphics processors have become possible. Single-chip integration of graphics processing units (GPUs) with central processing units (CPUs) has emerged and also brought many challenges that arise from integrating disparate devices or architectures, starting from overall system architecture, software tools, programming and memory models, interconnect design, power and performance, transistor requirements, and process-related constraints.

CHAPTER 2: ARITHMETIC IMPLEMENTED WITH SEMICONDUCTOR QUANTUM-DOT CELLULAR AUTOMATA (Earl E. Swartzlander Jr., Heumpil Cho, Inwook Kong, and Seong-Wan Kim)

The second chapter introduces us to quantum-dot cellular automata (QCA). Nanotechnology and QCA open new ways of performing what has been done for decades using the standard CMOS technology. What has been done for years at the transistor level, QCA attempt to do at the single electron level. Instead of voltage levels, QCA use electron position to save the information. And from there build more complicated functional elements like adders, multipliers and dividers, ripple carry and carry lookahead adders, and many more. As there is still some time before QCA elements can be physically implemented, this chapter investigates possible concept and architectures that can be implemented in various types of QCA technologies currently being developed.

CHAPTER 3: NOVEL CAPACITOR-LESS A2RAM MEMORY CELLS FOR BEYOND 22-NM NODES (Noel Rodríguez and Francisco Gamiz)

Each technology or material will reach its limits sooner or later. It can be power consumption, speed, minimum voltage operation, scaling, and so on. But even before that happens, researchers are trying to find a replacement. Chapter 3 shows us what is happening in the world of nonvolatile memories. A lot of energy is spent on such alternatives as resistive RAM (Re-RAM), magneto-resistive RAM (MRAM), or floating-body RAM (FB-RAM). The last alternative is the focal point of this chapter. According to the authors, floating-body DRAM seems to offer new ways of storing the charge and sensing it reliably and at the same time offering superior density and power performance.

CHAPTER 4: FOUR-STATE HYBRID SPINTRONICS–STRAINTRONICS: EXTREMELY LOW-POWER INFORMATION PROCESSING WITH MULTIFERROIC NANOMAGNETS POSSESSING BIAXIAL ANISOTROPY (Noel D'Souza, Jayasimha Atulasimha, and Supriyo Bandyopadhyay)

For the second time in this book we will be exploring nanotechnology; this time the focus is multiferroic nanomagnets. Multiferroic nanomagnets (consisting of piezoelectric and magnetostrictive layers) can be used as multistate switches to perform both Boolean and non-Boolean computing while dissipating only about a few hundreds of kT/bit of energy at clock rates of ~1GHz. At these power and speed levels, multiferroic nanomagnets can be used as memory elements for data storage, logic gates for Boolean computing, and associative memory for complex tasks such as ultrafast image reconstruction and pattern recognition.

CHAPTER 5: IMPROVEMENT AND APPLICATIONS OF LARGE-AREA FLEXIBLE ELECTRONICS WITH ORGANIC TRANSISTORS (Koichi Ishida, Hiroshi Fuketa, Tsuyoshi Sekitani, Makoto Takamiya, Hiroshi Toshiyoshi, Takao Someya, and Takayasu Sakurai)

Chapter 5 takes us into another popular and promising branch of electronics—organic transistors. Organic transistors offer completely new ways of building, integrating, and using electronic devices. With the advantage of the circuit integration on a plastic thin film with either printable or printing technology, the organic transistor is a promising candidate for realizing large-area flexible electronics. In this chapter, authors will present the most recent advancements in organic transistor technologies.

CHAPTER 6: SOFT-ERROR MITIGATION APPROACHES FOR HIGH-PERFORMANCE PROCESSOR MEMORIES (Lawrence T. Clark)

Chapter 6 brings up the well-known topic of soft errors. First DRAM, and then SRAM and logic designers have been dealing with soft errors for years. But as fabrication processes are reaching new scaling levels, the likelihood of radiation-induced bit upsets and logic transients in both terrestrial and aerospace environments is increasing and making their mitigation even more important. Several techniques have been developed to deal with soft errors, and this chapter investigates the efficiency of some of them, error detection and correction, redundancy, and error checking.

CHAPTER 7: DESIGN SPACE EXPLORATION OF WAVELENGTH-ROUTED OPTICAL NETWORKS-ON-CHIP TOPOLOGIES FOR 3D-STACKED MULTI- AND MANY-CORE PROCESSORS (Luca Ramini and Davide Bertozzi)

Chapter 7 introduces us to yet another new and very promising technology—photonics. Integration of many cores on the same chip has been discussed in Chapter 1. Solving interconnection problems has been identified as one of the many problems facing engineers working on the multicore, multichip designs. Photonic interconnect technology promises to lower the power and improve the bandwidth of multi- and many-core integrated systems. The presented material makes interesting insights into the latest developments in components, tools and technologies, and challenges preventing photonics from being widely adopted.

CHAPTER 8: QUEST FOR ENERGY EFFICIENCY IN DIGITAL SIGNAL PROCESSING: ARCHITECTURES, ALGORITHMS, AND SYSTEMS (Ramakrishnan Venkatasubramanian)

Cloud computing in a few short years turned from concept to reality. At the moment it is mostly associated with data storage, but imaging, video, and analytics processing in cloud-based computing and handheld devices is becoming increasingly common. Energy efficient digital signal processing (DSP) is key in delivering the best imaging and video experience at the lowest power budget. DSP architectures have evolved over the last three decades from simple vector co-processors to multicore DSP processor architecture-based systems to enable energy efficient DSP. This chapter explores the recent advances in industry and academia to solve this quest for DSP energy efficiency, holistically—in architecture, algorithms, and systems.

CHAPTER 9: NANOELECTROMECHANICAL RELAYS: AN ENERGY EFFICIENT ALTERNATIVE IN LOGIC DESIGN (Ramakrishnan Venkatasubramanian and Poras T. Balsara)

Lower power consumption is the ultimate goal of any design. And it does not apply to handheld, battery-powered devices only. Power consumption becomes important in all aspects of human life. The carbon dioxide footprint of large computer farms is something everyone is concerned with. The authors of this chapter introduce us to the concept of nanoelectromechanical (NEM) relays, which are a promising class of emerging devices that offer zero off-state leakage and behave like an ideal switch. The zero leakage operation of relays has renewed the interest in relay-based low-power logic design. This chapter covers the recent advances in relay-based logic design, memory design, and power electronics circuit design. NEM relays offer unprecedented 10X–30X energy efficiency improvement in logic design for low-frequency operation and has the potential to break the CMOS efficiency barrier in power electronic circuits as well.

CHAPTER 10: HIGH PERFORMANCE AND CUSTOMIZABLE BIOINFORMATIC AND BIOMEDICAL VERY-LARGE-SCALE-INTEGRATION ARCHITECTURES (Yao Xin, Benben Liu, Ray C.C. Cheung, and Chao Wang)

Bioinformatics is an application of computer science and information technology to the field of biology and medicine. It is a hybrid and emerging field in which computational tools, information systems, databases, and methodologies support-ing biology and genomic research are designed and developed. Challenges facing designers working in this field are the sheer amount of data and bandwidth needed to process high complexity algorithms. Off-the-shelf CPUs are operating close to their speed limits, therefore, new architectures are needed to supplement or replace them. This chapter presents two successful architectures designed for bioinformatic applications. The first one is called stochastic state point process filter (SSPPF), which is an effective tool for tracking coefficients in neural spiking activity research. And the second one is a parameterizable very-large-scale-integrated (VLSI) architecture for geometric biclustering developed for data mining operations.

CHAPTER 11: BASICS, APPLICATIONS, AND DESIGN OF REVERSIBLE CIRCUITS (Robert Wille)

Moore's Law has been shaping the electronics industry for the last several decades. Even though it was declared dead many times, and so far prematurely, everybody expects that the ongoing miniaturization will eventually come to a halt. When feature sizes of a single transistor approach the atomic scale, no further improvement can be expected. Therefore, the search for alternative technologies continues. This chapter looks at a promising new computing paradigm called reversible computation. In a standard computation, the output does not always define the state of the inputs—for example, if the output of a two input AND gate is 0, the state of the inputs is not 100% determined. In contrast, reversible computation uses operations where inputs can be derived from outputs and vice versa. This characteristic leads to some interesting and promising applications. The author tries to show how reversible circuits can help to solve many practically relevant problems much faster than conventional solutions.

CHAPTER 12: THREE-DIMENSIONAL SPINTRONICS (Dorothée Petit, Rhodri Mansell, Amalio Fernández-Pacheco, JiHyun Lee, and Russell P. Cowburn)

In this chapter, we are looking one more time at nonvolatile storage memories. Alternatives to the standard charge storage NVM memories have been researched for many years. Spintronics is one of them, although much less advanced than memories based on magnetism, for example. Spintronics uses the spin and the charge of the electron to store the information. In this chapter, authors discuss the possibility of using spintronics to design three-dimensional, highly scalable, low-power, and radiation-hardened storage devices.

CHAPTER 13: SOFT-ERROR-AWARE POWER OPTIMIZATION USING DYNAMIC THRESHOLD (Selahattin Sayil)

Power consumption has been mentioned in books a few times as the limiting factor of further integration of integrated circuits. In this chapter, a technique to lower the power consumption is looked at from the reliability perspective. One common technique to lower power consumption is to lower the supply voltage level. This trend has been observed for years across all recent process nodes. Main voltage supply level dropped from 1.8 to 0.8–0.9 V for the most advanced 20- to 28-nm processes. IO voltage dropped at the same time from 5 to 2.5 V and now to 1.8 V. Reduction in voltage levels is possible by the reduction of threshold voltages to maintain the circuit speed and high current drive. However, lower threshold voltages mean higher sensitivity to soft errors. In this chapter, the author presents an analysis of various dynamic threshold voltage reduction techniques from the perspective of the soft error tolerance.

CHAPTER 14: THE FUTURE OF ASYNCHRONOUS LOGIC
(Scott C. Smith and Jia Di)

This chapter talks about yet another way of overcoming the problem of scalability and increasing power consumption of the chips, which would allow even more complex and highly integrated applications—asynchronous logic.

Asynchronous logic is not a new idea; it has been discussed for many decades, but never became a mainstream design methodology. Synchronous circuits with well-established design flow and tools dominate the industry. But it is believed that asynchronous design will gain ground over the next 5–10 years. This chapter will provide the reader with more details about the state-of-the-art asynchronous logic, its current utilization in the industry, and its future.

CHAPTER 15: MEMRISTOR-CMOS-HYBRID SYNAPTIC DEVICES
EXHIBITING SPIKE-TIMING-DEPENDENT PLASTICITY (Tetsuya Asai)

The synapse is a structure that allows one neuron to communicate with another one by sending an electrical or chemical signal. In this chapter, the author introduces a concept of an electronic system mimicking functions of a neuronal interconnect system by using nonvolatile resistors. Such a device would have the ability to update the synaptic connections based on spike timing differences between pre- and postneuronal devices. The synaptic device consists of a Re-RAM, a capacitor, and *metal oxide semiconductor field effect transistors* (MOSFETs). Through extensive experimental results, the author shows that the memristor-based synaptic device offers a promising alternative for the implementation of nonvolatile analog synapses.

CHAPTER 16: VERY-LARGE-SCALE INTEGRATION
IMPLEMENTATIONS OF CRYPTOGRAPHIC
ALGORITHMS (Tony Thomas)

The growth of the Internet and mobile communication technologies has increased the importance of cryptography in our day-to-day life. In the present day communication systems, high-speed hardware implementations of cryptographic algorithms for data encryption, cryptographic hashing, and digital signatures are required. This chapter examines the VLSI implementations of major symmetric and asymmetric key cryptographic algorithms, hash functions, and digital signatures.

CHAPTER 17: DYNAMIC INTRINSIC CHIP ID FOR HARDWARE SECURITY (Toshiaki Kirihata and Sami Rosenblatt)

Counterfeit devices have been in circulation for many years, either recovered from recycled equipment or stolen from the supply chain, in both cases endangering the reliability of sometimes live supporting systems. To keep the military, avionic, or health supply chains as safe as possible, engineers came up with very sophisticated counterfeit hardware. Conventional chip authentication procedures utilize markers such as bar codes or nonvolatile memory, which are extrinsic to the actual product and can be copied or transferred with inexpensive resources. In this chapter, authors describe an emerging approach to enable physically unclonable functions (PUFs) using intrinsic features of a VLSI chip.

Authors propose to use embedded dynamic random access memory (eDRAM) to create a PUF based on a skewed and controllable random bit pattern generator. The concept is demonstrated using IBM 32-nm high-K/metal gate eDRAM.

CHAPTER 18: ULTRA-LOW-POWER AUDIO COMMUNICATION SYSTEM FOR FULL IMPLANTABLE COCHLEAR IMPLANT APPLICATION (Yannick Vaiarello and Jonathan Laudanski)

Health care is the area where new VLSI circuits can make a big difference. From monitoring and communicating vital signs to administering drugs, integrated circuits can help do it all. But chips and systems built for medical applications are facing severe acceptance criteria. Usually they have to be small, consume low power, and they always have to be extremely reliable.

This chapter talks about chips for cochlear implants, a medical device that restores the hearing sensation. There are hundreds of thousands of people experiencing the loss of hearing to a smaller or larger degree. The authors present a wireless microphone chip designed in a 130-nm CMOS process for transmitting an audio signal to an implanted cochlear neurostimulator.

CHAPTER 19: HETEROGENEOUS MEMORY DESIGN (Chengen Yang, Zihan Xu, Chaitali Chakrabarti, and Yu Cao)

In this chapter, the authors look one more time at various memory technologies and introduce a new memory system, which they believe will work the best in sub-10-nm technologies. The authors argue that there will be no single winner in the battle to be the next technology of choice. Every new memory, whether phase change memory or spin-transfer torque magnetic memory, has its advantages but none will be flexible enough to dominate the market. That is why the authors' heterogeneous memory system, which combines advantages of the different memory technologies to deliver superior system performance, seems to offer the best of all solutions. The authors try to show how the integration of multiple memory technologies at the same level can be used to improve system latency and energy, providing superior reliability at the same time.

CHAPTER 20: SOFT-ERROR RESILIENT CIRCUIT DESIGN
(Chia-Hsiang Chen, Phil Knag, and Zhengya Zhang)

In this final chapter the authors talk about soft errors and how designers are trying to build circuits that can be more resistant and work in environments where the probability of encountering soft errors is high. Research on soft errors started in the early 1970s when it was understood that the main source of particles that can create soft errors were actually integrated circuit packages. As packages evolved and contained fewer and fewer compounds creating alpha particles, the focus shifted toward cosmic rays. Techniques to combat soft errors were introduced at device and system levels. Authors are reviewing a few of them, such as a circuit-level hardening technique called dual interlocked storage cells, built-in soft error resilience elements, and more.

MATLAB ® is a registered trademark of The MathWorks, Inc. For product information, please contact:
The MathWorks, Inc.
3 Apple Hill Drive
Natick, MA 01760-2098 USA
Tel: 508 647 7000
Fax: 508-647-7001
E-mail: info@mathworks.com
Web: www.mathworks.com

Editors

Tomasz Wojcicki is a semiconductor industry veteran with over 30 years of experience in integrated circuit (IC) design and development, project management, and functional management.

Tomasz is with Sidense Corp.—the leading provider of One-Time Programmable (OTP) Nonvolatile Memory (NVM) Solutions—from its inception. He works there in various capacities including vice president of engineering and vice president of customer engineering support.

Previously he spent 14 years with MOSAID Technologies Inc., initially as the IC design engineer, and later as the technical and project leader developing high-end DRAM chips for first-tier DRAM manufacturers. He was instrumental in growing MOSAID's design services into a multimillion-dollar business, then serving as the director responsible for development of a family of classification products based on dynamic and static content addressable memories (CAMs).

Tomasz started his career in the early 1980s at the Institute of Electron Technology in Warsaw, Poland, in progressively more responsible roles. He graduated with an MSEE degree from Warsaw Technical University. He holds six patents, and has authored and coauthored several publications.

Krzysztof (Kris) Iniewski is manager of R&D at Redlen Technologies Inc., a start-up company in Vancouver, Canada. Redlen's revolutionary production process for advanced semiconductor materials enables a new generation of more accurate, all-digital, radiation-based imaging solutions. Kris is also a president of CMOS Emerging Technologies Research Inc. (www.cmosetr.com), an organization of high-tech events covering communications, microsystems, optoelectronics, and sensors. During his career, Dr. Iniewski has held numerous faculty and management positions at the University of Toronto, University of Alberta, Simon Fraser University, and PMC-Sierra Inc. He has published over 100 research papers in international journals and conferences. He holds 18 international patents granted in the United States, Canada, France, Germany, and Japan. He is a frequent invited speaker and has consulted for multiple organizations internationally. He has written and edited several books for CRC Press, Cambridge University Press, IEEE Press, Wiley, McGraw-Hill, Artech House, and Springer. His personal goal is to contribute to healthy living and sustainability through innovative engineering solutions. In his leisure time, Kris can be found hiking, sailing, skiing, or biking in beautiful British Columbia. He can be reached at kris.iniewski@gmail.com.

Contributors

Tetsuya Asai
Graduate School of Information Science
and Technology
Hokkaido University
Sapporo, Japan

Jayasimha Atulasimha
Mechanical and Nuclear Engineering
Virginia Commonwealth University
Richmond, Virginia, USA

Chinnakrishnan Ballapuram
Intel
Santa Clara, California, USA

Poras T. Balsara
University of Texas at Dallas
Richardson, Texas, USA

Supriyo Bandyopadhyay
Electrical and Computer Engineering
Virginia Commonwealth University
Richmond, Virginia, USA

Davide Bertozzi
Engineering Department (ENDIF)
University of Ferrara
Ferrara, Italy

Yu Cao
School of Electrical, Computer, and
Energy Engineering
Arizona State University
Tempe, Arizona, USA

Chaitali Chakrabarti
School of Electrical, Computer, and
Energy Engineering
Arizona State University
Tempe, Arizona, USA

Chia-Hsiang Chen
University of Michigan
Ann Arbor, Michigan, USA

Ray C.C. Cheung
Department of Electronic Engineering
City University of Hong Kong
Hong Kong

Heumpil Cho
Department of Electrical and Computer
Engineering
University of Texas at Austin
Austin, Texas, USA

Lawrence T. Clark
Ira A. Fulton Schools of Engineering
Arizona State University
Tempe, Arizona, USA

Russell P. Cowburn
Cavendish Laboratory
University of Cambridge
Cambridge, United Kingdom

Jia Di
Department of Computer Science &
Computer Engineering
University of Arkansas
Fayetteville, Arkansas, USA

Noel D'Souza
Mechanical and Nuclear Engineering
Virginia Commonwealth University
Richmond, Virginia, USA

Amalio Fernández-Pacheco
Cavendish Laboratory
University of Cambridge
Cambridge, United Kingdom

Hiroshi Fuketa
University of Tokyo
Tokyo, Japan

Francisco Gamiz
Department of Electronics
University of Granada
Granada, Spain

Koichi Ishida
Techniesche Universität Dresden
Dresden, Germany

Seong-Wan Kim
Department of Electrical and Computer
    Engineering
University of Texas at Austin
Austin, Texas, USA

Toshiaki Kirihata
IBM Systems & Technology Group
Microelectronics Division
Semiconductor Research &
    Development Center
Hopewell Junction, New York, USA

Phil Knag
University of Michigan
Ann Arbor, Michigan, USA

Inwook Kong
Department of Electrical and Computer
    Engineering
University of Texas at Austin
Austin, Texas, USA

Jonathan Laudanski
Neurelec/Oticon Medical
Vallauris, France

JiHyun Lee
Cavendish Laboratory
University of Cambridge
Cambridge, United Kingdom

Benben Liu
Department of Electronic Engineering
City University of Hong Kong
Hong Kong

Rhodri Mansell
Cavendish Laboratory
University of Cambridge
Cambridge, United Kingdom

Dorothée Petit
Cavendish Laboratory
University of Cambridge
Cambridge, United Kingdom

Luca Ramini
Engineering Department (ENDIF)
University of Ferrara
Ferrara, Italy

Noel Rodríguez
Department of Electronics
University of Granada
Granada, Spain

Sami Rosenblatt
IBM Systems & Technology Group
Microelectronics Division
Semiconductor Research &
    Development Center
Hopewell Junction, New York, USA

Takayasu Sakurai
University of Tokyo
Tokyo, Japan

Selahattin Sayil
Lamar University
Beaumont, Texas, USA

Deepak C. Sekar
Rambus
Sunnyvale, California, USA

Tsuyoshi Sekitani
University of Tokyo
Tokyo, Japan

Scott C. Smith
Department of Electrical and Computer
    Engineering
North Dakota State University
Fargo, North Dakota, USA

Takao Someya
University of Tokyo
Tokyo, Japan

Earl E. Swartzlander Jr.
Department of Electrical and Computer
    Engineering
University of Texas at Austin
Austin, Texas, USA

Makoto Takamiya
University of Tokyo
Tokyo, Japan

Tony Thomas
School of Computer Science and IT
Indian Institute of Information
    Technology and Management-Kerala
    (IIITM-K)
Thiruvananthapuram, Kerala, India

Hiroshi Toshiyoshi
University of Tokyo
Tokyo, Japan

Yannick Vaiarello
Neurelec/Oticon Medical
Vallauris, France

Ramakrishnan Venkatasubramanian
Texas Instruments, Inc.
Dallas, Texas, USA

Chao Wang
School of Computer Science and
    Technology
University of Science and Technology
    of China
Hefei, China

Robert Wille
Institute of Computer Science
University of Bremen
Bremen, Germany
and
Cyber-Physical Systems
DFKI GmbH
Bremen, Germany

Yao Xin
Department of Electronic Engineering
City University of Hong Kong
Hong Kong

Zihan Xu
School of Electrical, Computer, and
    Energy Engineering
Arizona State University
Tempe, Arizona, USA

Chengen Yang
School of Electrical, Computer, and
    Energy Engineering
Arizona State University
Tempe, Arizona, USA

Zhengya Zhang
University of Michigan
Ann Arbor, Michigan, USA

1 Integration of Graphics Processing Cores with Microprocessors

Deepak C. Sekar and Chinnakrishnan Ballapuram

CONTENTS

1.1 INTRODUCTION

Power and thermal constraints have caused a paradigm shift in the semiconductor industry over the past few years. All market segments, including phones, tablets, desktops, and servers, have now reduced their emphasis on clock frequency and shifted to multicore architectures for boosting performance. Figure 1.1 clearly shows this trend of saturating frequency and increasing core count in modern processors. With Moore's Law, on-die integration of many components such as peripheral control hubs, dynamic random-access memory (DRAM) controllers, modems, and more importantly graphics processors has become possible. Single-chip integration of graphics processing units (GPUs) with central processing units (CPUs) has emerged and also brought many challenges that arise from integrating disparate devices/architectures, starting from overall system architecture, software tools, programming and memory models, interconnect design, power and performance, transistor requirements, and process-related constraints. This chapter provides insight into the implementation, benefits and problems, current solutions, and future challenges of systems having CPUs and GPUs on the same chip.

1

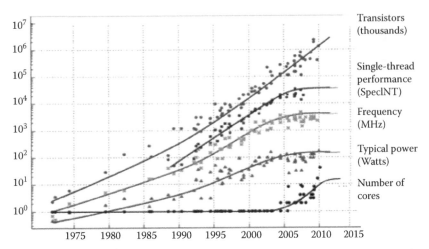

FIGURE 1.1 Microprocessor trends over the past 35 years. (Naffziger, S., Technology impacts from the new wave of architectures for media-rich workloads, *Symposium on VLSI Technology* © 2011 IEEE.)

1.2 WHY INTEGRATE CENTRAL PROCESSING UNITS AND GRAPHICS PROCESSING UNITS ON THE SAME CHIP?

CPU and GPU microarchitecture have evolved over time, though the CPU progressed at a much faster pace as graphics technology came into prominence a bit later than the CPU. Graphics is now getting more attention through games, content consumption from devices such as tablets, bigger sized phones, smart TVs, and other mobile devices. Also, as the performance of the CPU has matured, additional transistors from process shrink are used to enhance 3D graphics and media performance, and integrate more disparate devices on the same die. Figure 1.2 compares a system having a discrete graphics chip with another having a GPU integrated on the same die as the CPU. The benefits of having an integrated GPU are immediately apparent [1]:

- Bandwidth between the GPU and the DRAM is increased by almost three times. This improves performance quite significantly for bandwidth-hungry graphics functions.
- Power and latency of interconnects between the CPU chip and the GPU chip (of the multichip solution) are reduced.
- Data can be shared between the CPU and the GPU efficiently through better programming and memory models.
- Many workloads stress the GPU or the CPU and not both simultaneously. For GPU-intensive workloads, part of the CPU power budget can be transferred to the GPU and vice versa. This allows better performance–power trade-offs for the system.

Besides these benefits, the trend of integrating GPUs with CPUs has an important scalability advantage. GPUs are inherently parallel and are known to benefit linearly

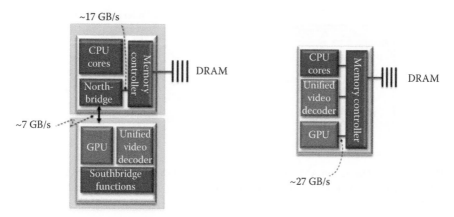

FIGURE 1.2 A multichip central processing unit–graphics processing unit solution (Left). A single-chip central processing unit–graphics processing unit solution (Right). (Naffziger, S., Technology impacts from the new wave of architectures for media-rich workloads, *Symposium on VLSI Technology* © 2011 IEEE.)

with density improvements. Moore's Law is excellent at providing density improvements, even though many argue that the performance and power improvements it used to provide have run out of steam. By integrating GPUs, the scalability of computing systems is expected to be better.

1.3 CASE STUDY OF INTEGRATED CENTRAL PROCESSING UNIT–GRAPHICS PROCESSING UNIT CORES

In this section, we describe two modern processors, AMD Llano (Advanced Micro Devices, Sunnyvale, CA) and Intel Ivy Bridge (Intel, Santa Clara, CA), which have both integrated CPUs and GPUs on the same die. These chips are often referred to as accelerated processing units (APUs).

1.3.1 AMD LLANO

The AMD Llano chip was constructed in a 32-nm high-k metal gate silicon-on-insulator technology [2]. Figure 1.3 shows the integrated die that includes four CPU cores, a graphics core, a unified video decoder, and memory and input/output (I/O) controllers. The total die area is 227 mm^2. CPU cores were x86 based, with 1 MB of L2 cache allocated per core. Each CPU core was 17.7 mm^2 including the L2 cache. Power gating was aggressively applied to both the core and the L2 cache to minimize power consumption. A dynamic voltage and frequency system (DVFS) was used that tuned supply voltage as a function of clock frequency to minimize power. Clock frequency was tuned for each core based on power consumption and activity of other CPU cores and the GPU. This was one of the key advantages of chip-level CPU and GPU integration—the power budget could be flexibly shared between these components based on workload and activity.

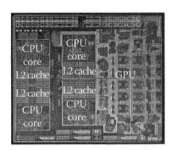
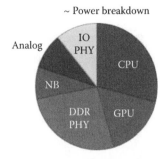

FIGURE 1.3 The 32-nm AMD Llano chip and a breakdown of its power consumption. IO PHY and DDR PHY denote interface circuits for input/outputs and dynamic random-access memory, respectively, and NB denotes the Northbridge.

The GPU used a very long instruction word (VLIW) core as a basic building block, which included four stream cores, one special function stream core, one branch unit, and some general purpose registers. Each stream core could coissue a 32-bit multiply and dependent ADD in a single clock. Sixteen of these VLIW cores were combined to form a single instruction, multiple data (SIMD) processing unit. The GPU consisted of five such SIMDs, leading to a combined throughput of 480 billion floating point operations per second. Power gating was implemented in the GPU core as well, to save power. The GPU core occupied approximately 80 mm^2, which was nearly 35% of the die area. Power consumption of the GPU was comparable to that of the CPU for many workloads, as shown in Figure 1.3. The CPU cores and the GPU shared a common memory in Llano systems, and a portion of this memory could be graphics frame buffer memory. Graphics, multimedia, and display memory traffic were routed through the graphics memory controller, which arbitrated between the requestors and issued a stream of memory requests over the Radeon Memory Bus to the Northbridge (Figure 1.4). Graphics memory controller accesses to frame buffer memory were noncoherent and did not snoop processor caches. Graphics or multimedia coherent accesses to memory were directed over the Fusion Control Link, which was also the path for processor access to I/O devices. The memory controller arbitrated between coherent and noncoherent accesses to memory.

1.3.2 Intel Ivy Bridge

Ivy Bridge was a 22-nm product from Intel that integrated CPU and GPU cores on the same die [3]. The four x86 CPU cores and graphics core were connected through a ring interconnect and shared the memory controller. Ivy Bridge had 1.4 billion transistors and a die size of about 160 mm^2. It was the first product that used a trigate transistor technology.

Figure 1.5 shows the system architecture of Ivy Bridge, where a graphics core occupied a significant portion of the total die. All coherent and noncoherent requests from both CPU and GPU were passed through the shared interconnect. The shared ring interconnect provided hundreds of GB/s bandwidth to the CPU and GPU cores. The last level cache is logically one, but physically distributed to independently deliver data.

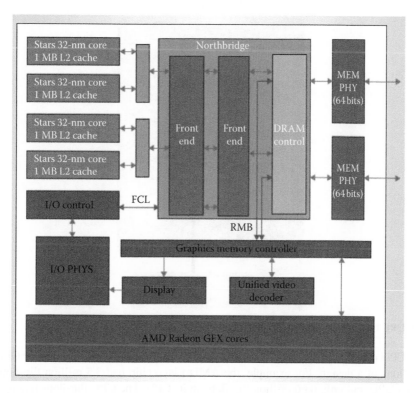

FIGURE 1.4 Block diagram of the AMD Llano chip. FCL denotes Fusion Control Link, MEM denotes memory, PHY denotes physical layers, and RMB denotes Radeon Memory Bus.

In Llano, coherent requests from the GPU went through a coherent queue and the noncoherent requests directly went to the memory. In Ivy Bridge, the CPU and GPU could share data in the bigger L3 cache, for example. The CPU could write commands to the GPU through the L3 cache, and in turn the GPU could flush data back to the L3 cache for the CPU to access. Also, the bigger L3 cache reduced memory bandwidth requirements and hence led to overall lower power consumption. Two different varieties of GPU cores were developed to serve different market segments. The graphics performance is mainly determined by the number of shader cores. The lower end segment had eight shader cores in one slice, whereas the next level segment had two slices. Different components of the processor were on different power planes to dynamically turn on/off the segments based on demands to save power. The CPU, GPU, and system agent were on different power planes to dynamically perform DVFS.

1.4   TECHNOLOGY CONSIDERATIONS

The fundamentally different nature of CPU and GPU computations places interesting requirements on process and device technology [1]. CPUs rely on using high-performance components, whereas GPUs require high-density, low-power components. This leads to the use of performance-optimized standard cell libraries

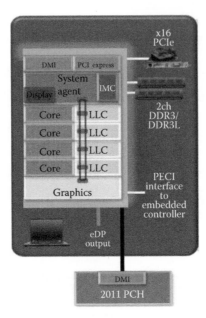

FIGURE 1.5   Block diagram of the Intel Ivy Bridge chip.

for CPU portions of a design and density-optimized standard cell libraries for GPU portions of a design. For example, the AMD Llano chip had 3.5 million flip-flops in its GPU, but only 0.66 million flip-flops in its CPU. The CPU flip-flops required higher performance and so were optimized differently. The flip-flop used for CPU cores occupied 50% more area than the flip-flop used for GPUs. The need for higher performance in CPU blocks led to the use of lower threshold voltages and channel lengths in CPU standard cell libraries compared to GPU ones.

The need for higher density in GPUs also leads to the requirement for smaller size wires than a pure-CPU process technology. Smaller size wiring causes more wire delay issues because wire resistivity increases exponentially at smaller dimensions. This is because of scattering at sidewalls and grain boundaries of wires as well as the fact that the diffusion barrier of copper occupies a bigger percentage of wire area.

In the long term, the differences in technology requirements for CPU and GPU cores could lead to 3D integration solutions. This would be particularly relevant for mobile applications where heat is less of a constraint. CPU cores could be stacked on a layer built with a high-performance process technology, whereas GPU cores could be stacked on a different layer built with a density-optimized process technology. DRAM could be stacked above these layers to provide the high memory bandwidth and low latency required for these systems. Figure 1.6 shows a schematic of such a system.

1.5   POWER MANAGEMENT

Most workloads emphasize either the serial CPU or the GPU and do not heavily utilize both simultaneously. By dynamically monitoring the power consumption in each CPU and GPU, and tracking the thermal characteristics of the die, watts that

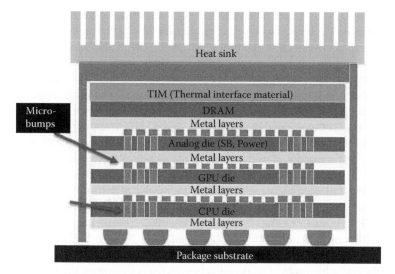

FIGURE 1.6 Three-dimensional integration of central processing unit and graphics processing unit cores. (Naffziger, S., Technology impacts from the new wave of architectures for media-rich workloads, *Symposium on VLSI Technology* © 2011 IEEE.)

go unused by one compute element can be utilized by others. This transfer of power, however, is a complex function of locality on the die and the thermal characteristics of the cooling solution. The efficiency of sharing is a function of where the hot spot is and will vary across the spectrum of power levels. While the CPU is the hot spot on the die, for example, a 1 W reduction in CPU power could allow the GPU to consume an additional 1.6 W before the lateral heat conduction from CPU to GPU heats the CPU enough to be the hot spot again [1]. As the GPU consumes more power, it finally becomes the hot spot on the die, and the reverse situation occurs. A power-management system that maintains repeatable performance must have sophisticated power-tracking capability and thermal modeling to ensure maximum compute capability is extracted from a given thermal solution. Once that is in place, a chip with a CPU and a GPU can deliver far more computation within a thermal envelope than either design in isolation.

1.6　SYSTEM ARCHITECTURE

With the integration of CPU and GPU, there are a few possible system architectures, ranging from separate CPU/GPU memory to a unified one. Until recently, before CPU and GPU were on the same die, the CPU and GPU had their own memory, and data was copied between these two memories for operation. AMD's road map shows progression from separate physical memory to hardware context switching for GPUs, wherein the chip will be able to decide which of its heterogeneous cores would best fit needs of a particular application [4]. In 2011, AMD partitioned the physical memory into two regions, one each for CPU and GPU. The CPU paging mechanism was handled by hardware and operating system (OS), whereas the GPU paging was handled by the driver. AMD plans to add unified memory address space for CPU and GPU and then add hardware context switching for the GPU. Intel in 2011 used an on-die unified memory controller for CPU/GPU, including sharing the last level cache between CPU and GPU. The on-die GPUs also have different levels of caches similar to CPU for texture, color, and other data.

The data transfer speed and bandwidth between CPU and GPU are critical for performance, and hence a scalable interconnect between these disparate cores is an important factor in APU design. Intel uses a ring interconnect to communicate between many CPU cores and the GPU. In addition to the interconnect design, other considerations in system architecture design include cache organization in both CPUs and GPUs, number of cache levels in the hierarchy, cache sizes at each level, cache policies, and sharing the last level cache. AMD APU (Llano) uses two levels of CPU cache (L1 data cache is 64 KB, L2 cache is 1 MB 16-way) in each of four cores. There is no L3 cache and the memory controller is shared between the CPU and the GPU. In contrast, Intel's Sandy Bridge has L1 data cache of 32 KB, L2 of 256 KB in each of four CPU cores, and inclusive L3 of 8 MB shared between four CPU cores and GPU using ring interconnect with common memory controller behind the cache hierarchy. The decision to selectively determine the type of data to be cacheable/uncacheable and coherent/noncoherent between the CPU and the GPU can improve performance and bandwidth between cores in both types of system architecture design. Also, general purpose GPU (GPGPU) programming can take advantage of both the integrated cache and the memory controller design to tap into computing power of GPUs. The bandwidth between key units in the system dictates the overall system performance.

1.7   PROGRAMMING AND MEMORY MODELS

Hardware manufacturers and software companies have been providing and support-ing many tools and new languages to help parallel programming. These tools have evolved since the days when CPU and GPU were separate. First, let us look at a few tools provided by hardware manufacturers. We will then compare them with language extensions and standards proposed from Microsoft to write programs that support multicore CPUs, APUs, GPGPUs, and heterogeneous devices.

Compute Unified Device Architecture (CUDA) (Nvidia Corporation, Santa Clara, CA): In early 2000, a few computer research labs built GPGPU applica-tion programming interfaces (APIs) on top of graphics APIs to enable and sup-port GPGPU programming, and two such languages were BrookGPU and Lib Sh. CUDA is Nvidia's approach to the GPGPU programming problem that lets programmers easily off-load data processing to GPUs [5]. The language is C with Nvidia extensions that provide interfaces to allocate GPU memory, copy data from host to device memory and back, and declare global and shared scope variables, to name a few. First the data has to be copied to the GPU memory before GPU computation is invoked by the CPU, and the results are copied back to main memory after GPU computation. The speedup is based on how efficient the programmers code the parallelism. The CUDA architecture has evolved and it currently supports many high level languages and device-level APIs such as Open Computing Language (OpenCL) and DirectX (Microsoft, Redmond, Washington).

The integration of CPU and GPU on the same die will help ease memory bandwidth constraints.

OpenCL: This is a standard that provides a framework to parallelize programs for heterogeneous systems [6]. Programs written using OpenCL can not only take advantage of multiple CPU cores and GPU cores but also can use other heterogeneous processors in the system. OpenCL's main goal is to use all resources in the system and offer superior portability. It uses a data and task parallel computational model and abstracts the underlying hardware. Data management is similar to CUDA, where the application has to explicitly manage the data transfer between main memory and device memory.

Hardware Multicore Parallel Programming (HMPP): CUDA and OpenCL require programmers to rewrite their code in new language. HMPP, on the other hand, provides a set of compiler directives that supports multicore parallel programming in C [7]. It is a flexible and portable interface for developing parallel applications that can use GPU and other hardware accelerators in the system. The HMPP directives divide the program into multiple codelets that can be run on multiple hardware accelerators. The HMPP runtime handles the parallel execution of the codelets, which have been translated earlier into vendor programming model either by hand or with an available code generator. When the same code is run on a different system, the runtime HMPP will try to find the specified hardware accelerator and run it. If the specified accelerator is not available, the codelet will run on the host core. The HMPP directives also handle transfer of data between host memory and hardware accelerator memory. The HMPP model can also take advantage of shared memory systems like APUs and use fully synchronous execution between CPU and GPU. HMPP also supports Message Passing Interfaces and can run the codelet on a remote host with automatic generation of memory transfers.

Accelerated Massive Parallelism (AMP): Microsoft's C++ AMP extension allows programmers to express data parallelism as part of the C++ language and lets compilers and DirectX tools create one binary and run on any heterogeneous hardware that supports it to improve performance [8,9]. The C++ AMP understands if the data has to be copied to the GPU based on the memory model and is transparent to the programmer.

The complexity of programming CPU, GPU, and other accelerators is that memory models are different ranging from weak, nonuniform memory access, and so on, from a hardware perspective to a software-managed memory model and recently proposed unified memory model for CPU/GPU and heterogeneous processors. The C++ AMP programming model is progressive looking and works well with AMD's Heterogeneous System Architecture (HSA) road map that plans to support programming and memory models to support efficient programming of APUs.

Graphics Core Next (GCN): AMD's GCN is a new design to make GPUs capable of doing compute tasks equally well. AMD is moving from VLIW graphics machine to non-VLIW or SIMD machine as a basic block for GPU. In the new SIMD compute-unit-based GPU, AMD plans to support high level language features such as pointers, virtual functions, and support for GPU to directly call system services and I/O. Also, the new graphics architecture can serve all segments. In the client segment, the new architecture provides similar power/performance as a VLIW machine. And in the server segments, the SIMD basic block with the right software tools can be used as an efficient compute engine and also take advantage of integrated CPU and GPU. The APUs will also provide a view of unified memory between both CPU and GPU to make the data communication efficient by eliminating copying of data between host and accelerator memory that is required in CUDA or OpenCL.

HSA: As the demand to use GPU for purposes other than graphics has been increasing, the GPU architecture is also evolving to support both graphics and compute engine. Also, there is an increased need of software support to use GPUs as compute engines in parallel with other asymmetric cores in the system. HSA provides an ecosystem to build a powerful system from combining simple, efficient, unique, and disparate processors. AMD plans to support coherent and unified memory for CPU and GPU and then provide GPU context switching in future products. The HSA road map is to move from physical integration, architectural integration, and finally to system integration. The architectural integration supports unified address space, pageable system memory for GPU and then fully coherent memory between CPU and GPU. The system integration provides preemption, context switching, and quality of service. AMD's plan is to treat GPU as a first-class core and give equal privileges as CPU cores with HSA. Currently, AMD is positioned well with their commitment and support to hardware, architecture, and OS tools, and applications for HSA framework and ecosystem.

In summary, there are three main vectors in heterogeneous computing: Instruction Set Architecture (ISA), memory model, and programming model, which covers different memory models and ISAs. We can see that CUDA, OpenCL, HMPP, and C++ AMP target the programming model vector; Graphics Compute Next targets the ISA vector, whereas APUs themselves define the underlying memory model. Another important secondary vector and effect of memory model and programming model is data synchronization. The data synchronization between asymmetric cores depends on the memory model supported by hardware and the one used by the application. As more parallelism is exploited by different accelerators and hardware that support many core counts, the cost of synchronization increases probably to a point where the cost may exceed the benefit of result from parallelism. We know that graphics cores use high memory bandwidth. In a system with APUs, if this high bandwidth traffic from GPU is coherent with the CPU cores, then the snoop bandwidth to all the CPU cores will be high, which will not only increase power and reduce performance of CPU cores but also increase the latency of GPU core. So, an efficient synchronization mechanism becomes important as we move toward programming APUs and future heterogeneous systems. As the potential for parallelism is increased, memory consistency restrictions on hardware may limit the performance.

1.8   AREA AND POWER IMPLICATIONS IN ACCELERATED PROCESSING UNITS

The area dedicated for graphics is increasing as mobile and desktop devices are demanding more media and graphics performance. In AMD's Llano APU, GPU occupies around 35% of the die area, whereas Intel's Sandy Bridge GPU occupies 20% of the die area. Also, the frequency at which the GPU operates is increasing and is taking more percentage of the total power budget, and hence efficient power management is required to increase the battery life. There may come an inflection point, where bigger control units like thread scheduler, dispatcher, scoreboard logic, register file management, and other control units will become a bottleneck and adding more parallel execution units in the GPU may not provide the expected benefit. Similar to the multicore CPUs, instantiation of multiple GPU units is a possible solution. But, the management of two GPU cores has to be supported by drivers and this opens up new challenges for driver management.

1.9 GRAPHICAL PROCESSING UNITS AS FIRST-CLASS PROCESSORS

GPUs have matured over time and have powerful graphics and compute engine capabilities. Also, programmers have many options to program APUs, and the tools range from language extensions and declarations/annotations in the programming language to aid compilers and low-level APIs with good debugging tools. Recent trends are to use onboard GPU clusters, networked GPU clusters, and GPU virtualization along with CPU clusters for many applications and creation of clouds and personal cloud networks. All these techniques still use CPU and device driver handshakes to communicate data between host and device memory. Though the GPUs are used for diverse purposes and applications, complete integration of CPUs and GPUs is missing due to lack of framework, standards, and challenges in making this happen. To extend GPU support to the next level, the OS needs to treat GPUs as first-class hardware for complete integration of CPUs and GPUs in the APUs.

To understand the requirements and challenges in making the GPU a first-class processor from the OS perspective, let us briefly look at the current method of interaction between the OS, applications, and CPU and GPU hardware. The main application runs on the CPU and fills up system memory with required information for graphics processing. The graphics part of the main application uses OpenCL, CUDA, or any other previously described programming model API to communicate with the device driver. The device driver in turn will send GPU commands to a ring buffer that acts as a queue to the underlying GPU hardware. Current GPU schedulers mostly use First In, First Out process scheduling that assumes a run-to-completion model and does not preempt tasks for intelligent scheduling. Some of the basic requirements for an OS like preemption and task migration are not supported in the current generation of GPUs.

Figure 1.7 captures the previously described flow between the main application running on CPU that uses system memory and device drivers to run the graphics application on graphics hardware. Recent architectures allow the CPU part and the graphics part of the main application to share information directly in system memory that is transparent to the OS. The drivers, CPU, and GPU part of the application take care of data transfer between cores and coherency.

The number of shader cores and other necessary blocks are increased to handle more data and make the GPU more powerful. There comes a point where just increasing the number of cores may not be easily achieved; instead, adding another

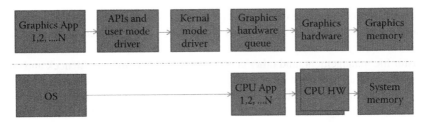

FIGURE 1.7 Interaction between operating system, central processing unit, and graphics processing unit to run graphics application.

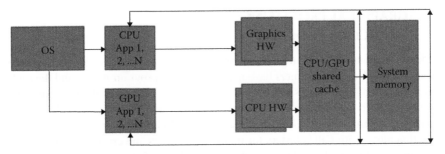

FIGURE 1.8 The operating system treats both central processing units and graphics processing units as first-class processors.

GPU core would be easier and complexity-effective. When the number of GPUs has to be increased, the system architecture has to be redesigned. One solution is to add intelligence at the device driver level. Another solution is to add preemption for GPUs and provide control to the OS and treat GPUs as first-class processors [10–12]. With the addition of GPUs, the OS needs to take GPU execution times, resources, migration of GPU tasks, and other information into consideration to come up with new scheduling strategies accordingly. Also, the OS needs to consider different underlying architectures and binaries that will affect the way programs are loaded and dynamically linked. Another challenge is the migration of an already started task from CPU core to GPU core for acceleration will require saving the states from CPU cores and migrating it to GPU cores. Current GPUs provide support for indirect branching to make calls and save state information during exceptions. These primitives and support for shared virtual memory and unified address space proposed in future GPUs are steps in the right direction to make GPUs first-class processors from both a programmer and an OS perspective.

Figure 1.8 shows a future ecosystem, where CPUs and GPUs can communicate through a shared cache and shared memory system. The OS schedules the workloads to both CPUs and GPUs and treats them as equals.

1.10 SUMMARY

Moore's Law has made the integration of CPUs and GPUs on the same chip possible. This has several important implications for process technology, circuit design, architecture, programming, and memory models, as well as software tools. In terms of process technology, GPUs prefer slower but higher density libraries compared to CPUs, and require smaller size wires as well. This leads to separately optimized process technologies for CPU and GPU portions of a chip, and could eventually lead to 3D integration solutions where CPU and GPU portions of a chip can be stacked atop each other. Power budgets can be efficiently shared between CPU and GPU portions of a design based on thermal and power delivery considerations. Different memory sharing models and interconnect networks are possible for chips that integrate CPUs and GPUs, and performance can be quite sensitive to these decisions. Currently, there are different programming models based on memory models, ISAs, and synchronization mechanisms with high overhead. These constructs will ease the programming model and also help define simpler memory models that help portability and programmability of APU programs. Unified address spaces, preemption, task migration, and other constructs will enable the OS to treat GPUs as CPUs in the long term.

REFERENCES

1. Naffziger, S. Technology impacts from the new wave of architectures for media-rich workloads, *Symposium on VLSI Technology*, pp. 6–10, IEEE, Honolulu, HI, June 2011.
2. Branover, A., D. Foley, and M. Steinman. AMD fusion APU: Llano, *Micro*, pp. 28–37, IEEE, 32(2), 2012.
3. Damaraju, S., V. George, S. Jahagirdar, T. Khondker, R. Milstrey, S. Sarkar, S. Siers, I. Stolero, and A. Subbiah. A 22 nm IA multi-CPU and GPU system-on-chip, *Proceedings of International Solid State Circuits Conference*, IEEE, San Francisco, CA, 2012.
4. Rogers, P. The Programmer's Guide to the APU Galaxy, AMD Fusion Developer Summit (AFDS) Keynote, 2011.
5. Nvidia. What is CUDA, http://www.nvidia.com/object/cuda_home_new.html.
6. Khronos Group. Developer Overview, http://www.khronos.org/developers.
7. Wikipedia. OpenHMPP, accessed April 4, 2012, http://en.wikipedia.org/wiki/OpenHMPP.
8. Microsoft Developer Network. C++ AMP (C++ Accelerated Massive Parallelism), accessed April 4, 2012, http://msdn.microsoft.com/en-us/library/hh265137(v = vs.110).aspx.
9. Kanter, D. Adaptive clocking in AMD's steamroller, Real World Technologies, May 6, 2014, http://realworldtech.com/page.cfm?ArticleID = RWT062711124854.
10. Kato, S., K. Lakshmanan, Y. Ishikawa, and R. Rajkumar. Resource sharing in GPU-accelerated windowing systems, *Proceedings of the IEEE Real-Time and Embedded Technology and Applications Symposium*, pp. 191–200, IEEE, Chicago, IL, 2011.
11. Kato, S., S. Brandt, Y. Ishikawa, and R. R. Rajkumar. Operating systems challenges for GPU resource management, *Proceedings of the Seventh International Workshop on OSPERT 2011*, Porto, Portugal, pp. 23–32, 2011.
12. Beisel, T., T. Wiersema, C. Plessl, and A. Brinkmann. Cooperative multitasking for heterogeneous accelerators in the Linux Completely Fair Scheduler, *Proceedings IEEE International Conference on Application-Specific Systems, Architectures, and Processors (ASAP)*, pp. 223–26, IEEE Computer Society, Santa Monica, CA, September 2011.

2 Arithmetic Implemented with Semiconductor Quantum-Dot Cellular Automata

Earl E. Swartzlander Jr., Heumpil Cho, Inwook Kong, and Seong-Wan Kim

CONTENTS

2.1   INTRODUCTION

It is expected that the role of complementary metal–oxide–semiconductor (CMOS) as the dominant technology for very-large-scale integrated circuits will encounter serious problems in the near future due to limitations such as short channel effects, doping fluctuations, and increasingly difficult and expensive lithography at nanoscale. The projected expectations of diminished device density and performance and increased power consumption encourage investigation of radically different technologies. Nanotechnology, especially quantum-dot cellular automata (QCA) provides new possibilities for computing owing to its unique properties [1]. QCA relies on a fresh physical phenomena (Coulombic interaction), and its logic states are not stored as voltage levels, but rather as the position of individual electrons. Even though the physical implementation of devices is still being developed, it is appropriate to investigate QCA circuit architecture.

There are several types of QCA technologies including semiconductor QCA, metal island QCA, molecular QCA, and magnetic QCA. This chapter is based on semiconductor QCA, but the concepts and the design concepts apply to the other forms as well. In fact, as one of the unique aspects of QCA technology is that communication on the chip is slow whereas the logic operations are fast, it is expected that many of the QCA design concepts will apply to CMOS as the feature sizes continue to shrink.

Section 2.2 describes the fundamentals of semiconductor QCA circuits. Section 2.3 shows the design of adders in QCA. Three standard types of adders, ripple carry, carry lookahead, and conditional sum, are designed. An interesting result is that by optimizing the design of a full adder (FA) to take advantage of QCA characteristics, the ripple carry adder achieves better performance than the carry lookahead and conditional sum adders (CSAs), even for large word sizes.

Small array and "fast" Wallace multipliers are compared. Finally, a Goldschmidt divider is implemented. All of the designs have been simulated with QCADesigner, a simulator that simplifies the estimation of area and delay.

2.2 INTRODUCTION TO QCA TECHNOLOGY

A QCA cell can be viewed as a set of four quantum dots that are positioned at the corners of a square and charged with two free electrons that can tunnel through to neighboring dots. Potential barriers between cells make the two free electrons locate on individual dots at opposite corners to maximize their separation by Coulomb repulsion. With four corners there are two valid configurations allowing binary logic. Two bistable states result in polarizations of P = +1 and P = −1 as shown in Figure 2.1. Figure 2.1a shows regular cells whereas Figure 2.1b shows rotated cells that can be used to make coplanar signal crossings.

2.2.1 QCA Building Blocks

The QCA wire, inverter, and majority gate are the basic QCA elements [2,3]. The binary value propagates from input to output by Coulombic interactions between cells. The wire made by cascading cells could be a horizontal row or a vertical columns of cells. Regular cells can be placed next to off center cells as shown in Figure 2.2. As shown in Figure 2.3, inverters can be built by placing cells off center to produce opposite polarization. Figure 2.4 shows a three input majority gate, which is the fundamental QCA logic function. The logic equation for a majority gate is

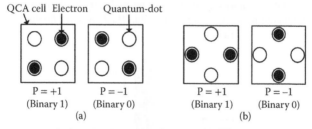

FIGURE 2.1 Basic QCA cells with two possible polarizations. (a) Regular cells. (b) Rotated cells. (I. Kong et al, Design of Goldschmidt Dividers with Quantum-Dot Cellular Automata, *IEEE Transactions on Computers* © 2013 IEEE.)

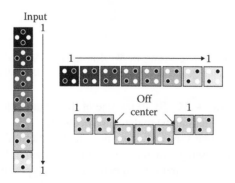

FIGURE 2.2 QCA wires. (S.-W. Kim and E. E. Swartzlander, Jr., Parallel Multipliers for Quantum-Dot Cellular Automata, *IEEE Nanotechnology Materials and Devices Conference* © 2009 IEEE.)

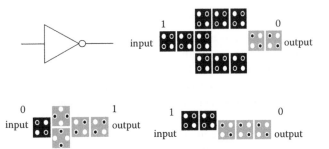

FIGURE 2.3 Various QCA inverters. (S.-W. Kim and E. E. Swartzlander, Jr., Parallel Multipliers for Quantum-Dot Cellular Automata, *IEEE Nanotechnology Materials and Devices Conference* © 2009 IEEE.)

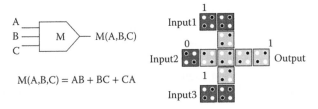

$$M(A,B,C) = AB + BC + CA$$

FIGURE 2.4 A QCA majority gate (S.-W. Kim and Earl E. Swartzlander, Jr., Parallel Multipliers for Quantum-Dot Cellular Automata, *IEEE Nanotechnology Materials and Devices Conference* © 2009 IEEE.)

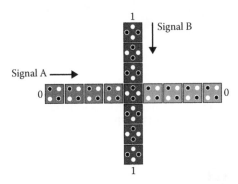

FIGURE 2.5 Coplanar "wire" crossing. (S.-W. Kim and E. E. Swartzlander, Jr., Multipliers with Coplanar Crossings for Quantum-Dot Cellular Automata, *IEEE NANO* © 2010 IEEE.)

M(A, B, C) = AB + BC + AC. By fixing one of the inputs to logic 0 or to logic 1, 2-input AND and OR gates can be realized, respectively.

In QCA, there are two types of crossings: coplanar crossings and multilayer crossovers. Coplanar crossings, such as the example shown in Figure 2.5, use only one layer but require regular and rotated cells. The two types of cells do not interact with each other when they are properly aligned, so they can be used for coplanar crossings. Published information suggests that coplanar crossings may be very

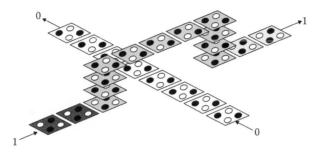

FIGURE 2.6 Multilayer "wire" crossover. (H. Cho and E. E. Swartzlander, Jr., Modular design of conditional sum adders using quantum-dot cellular automata, *Proceedings 6th IEEE Conference on Nanotechnology* © 2006 IEEE.)

sensitive to misalignment [4]. In a coplanar crossing, there is a possibility of a loose binding of the signal with the possibility of back-propagation from the far side. Preventing this requires more clock zones between the regular cells across the rotated cells.

An example of a multilayer crossover is shown in Figure 2.6. They use more than one layer of cells similar to the multiple metal layers in a conventional integrated circuit. The multilayer crossover is conceptually simple although there are questions about its realization, as it requires two overlapping active layers with vertical via connections. Previous work has examined the possibility of the multilayer QCA, but there has been no reported implementation yet [5]. Both types of crossovers are used in this chapter, multilayer for the adders and coplanar for the multipliers and divider.

2.2.2 QCA CLOCKING

With adiabatic switching accomplished by modulating the interdot tunneling barrier of the QCA cell, a multiple-phase clock is used in QCA. The cells are grouped together into pipelined zones [1,6,7]. This clocking scheme allows one zone of cells to perform a calculation and have its state held by raising its interdot barriers. The four clock phases shown in Figure 2.7 are used as switch, hold, release, and relax. During the *switch* phase, interdot potential barriers are lowered by applying an input signal, and data propagation occurs via electron tunneling. By gradually raising the barriers, the cells become polarized. For the *hold* phase the barriers are held high so the cell maintains its polarization and the output of the subarray can be used as inputs to the next stage. During the *release* and *relax* phases, the QCA cells start to lose their polarization by lowering the barriers and then they remain in an unpolarized state. This nature of clocking leads to inherent self-latching in QCA.

Generally, QCA circuits have very significant wiring delays. For a fast design in QCA, complexity constraints are very critical issues and the design needs to use architectural techniques to boost the speed considering these limitations.

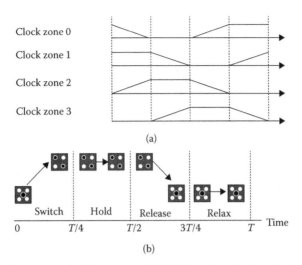

(a)

(b)

FIGURE 2.7   Four phase clocking scheme. (a) Clocking for a clock cycle with each lagging by 90°. (b) Interdot barriers in a clocking zone.

2.2.3   QCA DESIGN RULES

The QCA cells are 18 nm by 18 nm with a 2 nm space on all four sides, with 5 nm diameter quantum dots. The cells are placed on a grid with a cell center-to-center distance of 20 nm. Because there are propagation delays between cell-to-cell reactions, there should be a limit on the maximum cell count in a clock zone. This insures proper propagation and reliable signal transmission. In this chapter, a maximum length of 16 cells is used. The minimum separation between two different signal wires is the width of two cells. Multilayer crossovers are used here for wire crossings. They use more than one layer of cells like a bridge. The multilayer crossover design is conceptually simple although there are questions about its realization, as it requires two overlapping active layers with vertical via connections. Alternatively, coplanar "crossovers" that may be easier to realize can be used with some modification to the basic designs. For circuit layout and functionality checking, a simulation tool for QCA circuits, QCADesigner, is used [8]. This tool allows users to do a custom layout and then verify QCA circuit functionality by simulations.

2.3   QCA ADDERS

Section 2.2.1 showed that interconnections incur significant complexity and wire delay when implemented with QCAs, so transistor circuit designs that assume wires have negligible complexity and delay need to be reexamined. In QCA, as the complexity increases, the delay may increase because of the increased cell counts and wire connections.

2.3.1 CARRY-FLOW RIPPLE CARRY ADDER*

In this subsection, the adder design is that of a conventional ripple carry adder, but with a FA whose layout is optimized for QCA technology [9]. The proposed adder design shows that a very low delay can be obtained with an optimized layout. To avoid confusion with conventional ripple carry adders, the new layout is referred to as the carry-flow ripple carry adder (CFA).

2.3.1.1 Basic Design Approach

Equations for a FA realized with majority gates and inverters are shown below. In a ripple carry adder most of the delays come from carry propagation. For faster calculation, reducing the carry propagation delay (i.e., carry-in to carry-out delay of a FA) is most important. The usual approach for fast carry propagation is to add additional logic elements. In this design, simplification is used instead.

$$s_i = a_i b_i c_i + a_i \overline{b_i} \overline{c_i} + \overline{a_i} b_i \overline{c_i} + \overline{a_i} \overline{b_i} c_i \tag{2.1}$$

$$s_i = M\left(\overline{M}\left(a_i,b_i,c_i\right), M\left(a_i,b_i,\overline{c_i}\right), c_i\right) \tag{2.2}$$

$$s_i = M\left(\overline{c}_{i+1}, M\left(a_i,b_i,\overline{c_i}\right), c_i\right) \tag{2.3}$$

$$c_{i+1} = a_i b_i + b_i c_i + a_i c_i \tag{2.4}$$

$$c_{i+1} = M(a_i,b_i,c_i) \tag{2.5}$$

In QCA, the path from carry-in to carry-out only uses one majority gate. The majority gate always adds one more clock zone (one-quarter clock delay). Thus, each bit in the words to be added requires at least one clock zone, which sets the minimum delay.

2.3.1.2 Carry-Flow Full Adder Design

Figures 2.8a and 2.8b show the schematic and the layout of the carry-flow FA. The schematic and layout are optimized to minimize the delay and area. The carry propagation delay for 1 bit is a quarter clock and the delay from data inputs to the sum output is three-quarter clocks.

The wiring channels for the input/output synchronization should be minimized as wire channels add significantly to the circuit area. The carry-flow FA shown in Figure 2.8b requires only a one cell vertical offset between the carry-in and carry-out.

Figures 2.9 and 2.10 show 4- and 32-bit ripple carry adders, respectively, realized with carry-flow FAs. From the layouts, it is clear that for large adders, much of the area is devoted to skewing the input data and deskewing the outputs.

* Subsection 2.3.1 is based on [9].

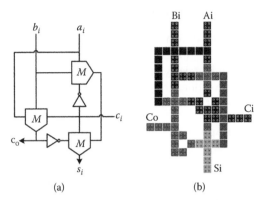

(a)   (b)

FIGURE 2.8 Carry-flow full adder. (a) Schematic. (b) Layout. (H. Cho and E. E. Swartzlander, Jr., Adder and multiplier design in quantum-dot cellular automata, *IEEE Transactions on Computers* © 2009 IEEE.)

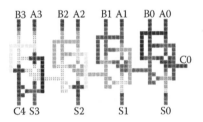

FIGURE 2.9 Layout of a 4-bit carry-flow adder. (H. Cho and E. E. Swartzlander, Jr., Adder and multiplier design in quantum-dot cellular automata, *IEEE Transactions on Computers* © 2009 IEEE.)

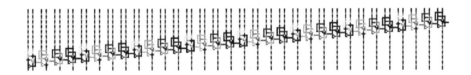

FIGURE 2.10 Layout of a 32-bit carry-flow adder. (H. Cho and E. E. Swartzlander, Jr., Adder and multiplier design in quantum-dot cellular automata, *IEEE Transactions on Computers* © 2009 IEEE.)

2.3.1.3 Simulation Results

The input and output waveforms for an 8-bit CFA are shown in Figure 2.11. For clarity, only 8-bit CFA simulation results are shown. The first meaningful output appears in the third clock period after 2.5 clock delays. First and last input/output pairs are highlighted.

2.3.2 Carry Lookahead Adder[†]

2.3.2.1 Architectural Design

The carry lookahead adder (CLA) has a regular structure. In CMOS implementations, it achieves high speed with a moderate complexity. For this section, 4-, 16-, and 64-bit CLA designs were developed following basic CMOS pipelined adder designs [10]. The pipelined designs avoid feedback signals (used in regular CMOS CLAs) that are difficult to implement with QCAs. Figures 2.12 through 2.14 show block diagrams of the designs for 4-, 16-, and 64-bit CLAs, respectively. The designs use 4-bit slices for the lookahead logic, so each factor of 4 increase in word size requires an additional level of lookahead logic.

The PG block has a generate output, $g_i = a_i b_i$, which indicates that a carry is "generated" at bit position i, and a propagate output $p_i = a_i + b_i$, which indicates that a

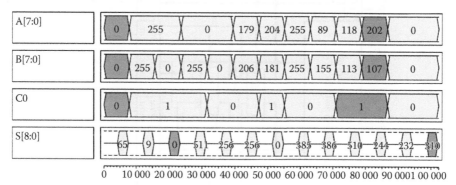

FIGURE 2.11 Simulation results for 8-bit carry-flow adder. (H. Cho and E. E. Swartzlander, Jr., Adder and multiplier design in quantum-dot cellular automata, *IEEE Transactions on Computers* © 2009 IEEE.)

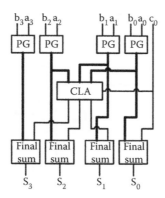

FIGURE 2.12 4-bit carry lookahead adder block diagram. (H. Cho and E. E. Swartzlander, Jr., Adder designs and analyses for quantum-dot cellular automata, *IEEE Transactions on Nanotechnology* © 2007 IEEE.)

[†] Subsection 2.3.2 is based on [11].

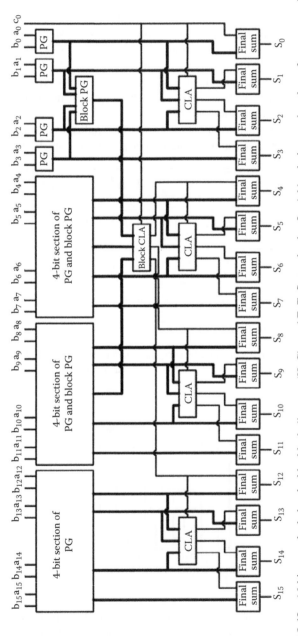

FIGURE 2.13 16-bit carry lookahead adder block diagram. (H. Cho and E. E. Swartzlander, Jr., Adder designs and analyses for quantum-dot cellular automata, *IEEE Transactions on Nanotechnology* © 2007 IEEE.)

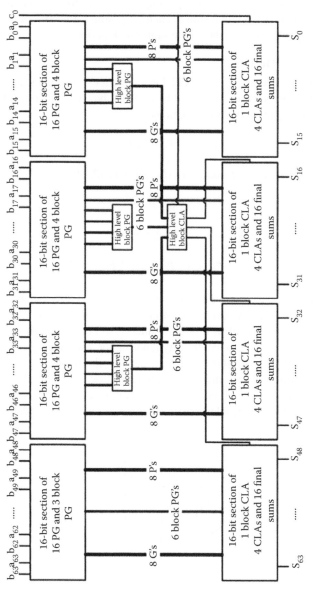

FIGURE 2.14 64-bit carry lookahead adder block diagram. (H. Cho and E. E. Swartzlander, Jr., Adder designs and analyses for quantum-dot cellular automata, *IEEE Transactions on Nanotechnology* © 2007 IEEE.)

carry entering bit position i will "propagate" to the next bit position. They are used to produce all the carries in parallel at the successive blocks. The block PG section produces and transfers block generate/propagate signals to the next higher level. The CLA and block CLA sections are virtually identical except for the different hierarchy of their positions and additional bypassing signals. Their outputs and PG outputs are used to calculate the final sum at each bit position. Because of the pipeline design, all sum signals are available at the same clock period.

2.3.2.2 Schematic Design

Using the block PG equations, AND/OR logic functions are mapped to majority gates to build the 4-bit section of PG and block PG. The CLA and block CLA sections are described by the following equations:

$$P_b = P_{i+3}P_{i+2}P_{i+1}P_i \tag{2.6}$$

$$g_b = g_{i+3} + P_{i+3}g_{i+2} + P_{i+3}P_{i+2}g_{i+1} + P_{i+3}P_{i+2}P_{i+1}g_i \tag{2.7}$$

$$c_{i+1} = g_i + p_i c_i \tag{2.8}$$

$$c_{i+2} = g_{i+1} + p_{i+1}g_i + p_{i+1}p_i c_i \tag{2.9}$$

$$c_{i+3} = g_{i+2} + p_{i+2}g_{i+1} + p_{i+2}p_{i+1}g_i + p_{i+2}p_{i+1}p_i c_i \tag{2.10}$$

Because of the characteristics of majority gates in QCA, a half adder and a FA have the same complexity. The FA design shown in Figure 2.8 uses three majority gates. An exclusive OR gate (equivalent to a half adder) also needs three majority gates. Thus, both adders have the same complexity ignoring the wire routing complexity.

The final sum adder is similar to a FA except that the inputs are p_i, g_i, and c_i rather than a_i, b_i, and c_i. Using these inputs, wire routing is simplified.

After the optimization, the final sum adder is implemented with three majority gates. Figure 2.15 shows the gate level diagram of the given final sum adder equation.

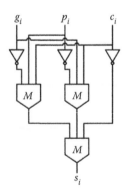

FIGURE 2.15 Final sum adder schematic. (H. Cho and E. E. Swartzlander, Jr., Adder designs and analyses for quantum-dot cellular automata, *IEEE Transactions on Nanotechnology* © 2007 IEEE.)

$$s_i = p_i g_i c_i + p_i \overline{g}_i \overline{c}_i + \overline{p}_i g_i \overline{c}_i + \overline{p}_i \overline{g}_i c_i \qquad (2.11)$$

$$s_i = M\left(\overline{M}\left(\overline{p}_i g_i c_i\right), \overline{M}\left(p_i \overline{g}_i c_i\right), \overline{c}_i\right) \qquad (2.12)$$

2.3.2.3 Layout Design

Figures 2.16 and 2.17 show the layouts of 4- and 16-bit CLAs from QCADesigner.

2.3.2.4 Simulation Results

With QCADesigner version 2.0.3, the circuit functionality of the CLA is verified. The following parameters are used for a bistable approximation: cell size = 20 nm, number of samples = 102,400, convergence tolerance = 0.00001, radius of effect = 41 nm, relative permittivity = 12.9, clock high = $9.8e^{-22}J$, clock low = $3.8e^{-23}J$, clock amplitude factor = 2, layer separation = 11.5 nm, maximum iterations per sample = 10,000 [8].

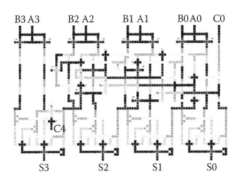

FIGURE 2.16 4-bit carry lookahead adder layout. (H. Cho and E. E. Swartzlander, Jr., Adder designs and analyses for quantum-dot cellular automata, *IEEE Transactions on Nanotechnology* © 2007 IEEE.)

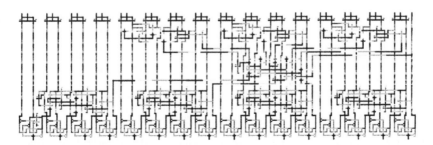

FIGURE 2.17 16-bit carry lookahead adder layout. (H. Cho and E. E. Swartzlander, Jr., Adder designs and analyses for quantum-dot cellular automata, *IEEE Transactions on Nanotechnology* © 2007 IEEE.)

For clarity, only the 4-bit CLA simulation results are shown. The input and output waveforms are shown in Figure 2.18. The first meaningful output appears in the fourth clock tick after 3.5 clock delays. The first and last input/output pairs are highlighted.

2.3.3 CONDITIONAL SUM ADDER[‡]

2.3.3.1 Architectural Design

In CMOS, the CSA is frequently used when the highest speed is required. CSAs of 4, 8, 16, 32, and 64 bits were designed and simulated [12]. The structures are based on the recursive relations shown in Figure 2.19.

This design can be divided into two half-size calculations. The upper half calculation is duplicated (one assuming a carry-in of 0 and one assuming a carry-in of 1).

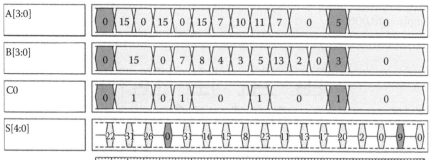

FIGURE 2.18 Simulation results for 4-bit CLA using QCADesigner version 2.0.3. (H. Cho and E. E. Swartzlander, Jr., Adder designs and analyses for quantum-dot cellular automata, *IEEE Transactions on Nanotechnology* © 2007 IEEE.)

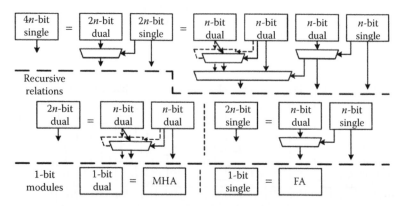

FIGURE 2.19 Recursive structure of a conditional sum adder. (H. Cho and E. E. Swartzlander, Jr., Modular design of conditional sum adders using quantum-dot cellular automata, *Proceedings 6th IEEE Conference on Nanotechnology* © 2006 IEEE.)

[‡] Subsection 2.3.3 is based on [12].

The carry output from the lower half is used to select the correct upper half output. This process is continued recursively down to the bit level.

These recursive relations produce modular designs. Figure 2.20 shows the block diagram of an 8-bit CSA. The blocks just below the modified half adders (MHAs) are referred to as level 1. Successive lower blocks are called level 2, 3, and so on.

2.3.3.2 Schematic Design

The following equations are used for CSAs. a_i and b_i denote inputs at bit position i. s_i represents the sum output at bit position i and c_{i+1} represents the carry output generated from bit position i. s_i^p means the sum of bit position when the carry input value is p and c_{i+1}^q means the carry output of bit position i when the carry input is q. At bit position i, the definitions of sum and carry output are $s_i^0 = a_i \oplus b_i$, $s_i^1 = a_i \overline{\oplus} b_i = \overline{a_i \oplus b_i}$, $c_{i+1}^0 = a_i b_i$ and $c_{i+1}^1 = a_i + b_i$.

$$s_0 = a_0 \oplus b_0 \oplus c_0 \tag{2.13}$$

$$c_1 = M(a_0 b_0 c_0) \tag{2.14}$$

$$s_1 = s_1^0 \overline{c_1} + s_1^1 c_1 \tag{2.15}$$

$$c_2 = c_2^0 \overline{c_1} + c_2^1 c_1 \tag{2.16}$$

$$s_i = s_i^0 \overline{c_i} + s_i^1 c_i \tag{2.17}$$

$$c_{i+1} = c_{i+1}^0 \overline{c_i} + c_{i+1}^1 c_i \tag{2.18}$$

As shown in Figure 2.20, the circuits are composed of a FA, MHAs, and multiplexers. The schematics of the MHA circuits are shown in Figure 2.21. Figure 2.21a and b shows two options for half-adder modules. In transistor circuits, the duplicated half adder (DHA) has the same delay and more area than the MHA.

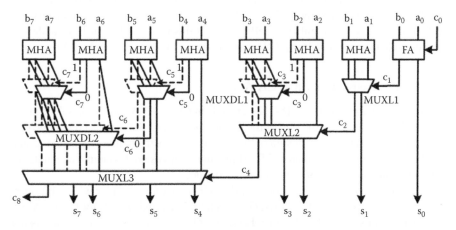

FIGURE 2.20 8-bit conditional sum adder block diagram. (H. Cho and E. E. Swartzlander, Jr., Modular design of conditional sum adders using quantum-dot cellular automata, *Proceedings 6th IEEE Conference on Nanotechnology* © 2006 IEEE.)

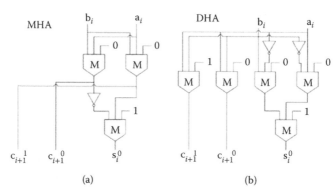

FIGURE 2.21 Half adder schematics. (a) Modified half adder. (b) Duplicated half adder. (H. Cho and E. E. Swartzlander, Jr., Modular design of conditional sum adders using quantum-dot cellular automata, *Proceedings 6th IEEE Conference on Nanotechnology* © 2006 IEEE.)

In QCA, the area of the DHA is still large but the delay is slightly less than that of the MHA by one-quarter of a clock. This trade-off governs the circuit choice. The general MHA module for CSA has four outputs, but QCA design only has three outputs. The sum for a carry-in of one is the complement of the sum for a carry-in of zero. The inverter is realized at the destination, which eliminates a wire. Thus, two-wire channels are reduced to one.

The required multiplexers are similar to transistor circuits. The inverters for complementing the sums are implemented just before the multiplexers. Successive levels of multiplexers are implemented in the same manner.

2.3.3.3 Layout Design

Figure 2.22a and b shows the layouts for the two types of half adder modules. A DHA module has larger area and more cells, but it has more delay margin. This shows a difference from transistor circuits. DHA modules are used in the implementations for the timing margin. The width of the design is dominated by the multiplexers. The heights of the MHA and the DHA are the same. The only disadvantage of DHA is a small increase in the number of cells used, but this is negligible for large adders.

The layouts for 4- and 16-bit CSAs are shown in Figures 2.23 and 2.24.

2.3.4 COMPARISON OF THE QCA ADDERS

Table 2.1 compares the 4-, 8-, 16-, 32-, and 64-bit CFAs, CLAs, and CSAs. The areas reported in Table 2.1 are the size of the bounding box (i.e., the smallest rectangle that contains the layout). As the design of the CSA is roughly triangular in shape, the size can be reduced by approximately 25% if the unused areas are taken into account, but it will still be the largest of the three types of adders.

From the values shown in Table 2.1, the cell counts for an adder with n-bit operands are roughly $O(n^{1.21})$ for CFAs, $O(n^{1.32})$ for CLAs, and $O(n^{1.5})$ for CSAs. Areas are $O(n^{1.42})$ for CFAs, $O(n^{1.47})$ for CLAs, and $O(n^{1.73})$ for CSAs. Delays are $n+0.5$ for RCAs, $O(n^{0.8})$ for CLAs, and $O(n^{0.91})$ for CSAs. These results show that the design overheads are indeed significant.

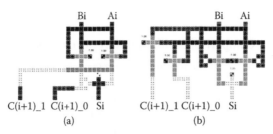

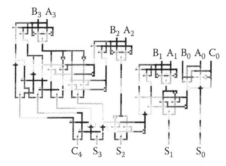

FIGURE 2.22 Half adder layouts. (a) Modified half adder. (b) Duplicated half adder. (H. Cho and E. E. Swartzlander, Jr., Modular design of conditional sum adders using quantum-dot cellular automata, *Proceedings 6th IEEE Conference on Nanotechnology*) © 2006 IEEE.)

FIGURE 2.23 4-bit conditional sum adder layout. (H. Cho and E. E. Swartzlander, Jr., Modular design of conditional sum adders using quantum-dot cellular automata, *Proceedings 6th IEEE Conference on Nanotechnology* © 2006 IEEE.)

FIGURE 2.24 16-bit conditional sum adder layout. (H. Cho and E. E. Swartzlander, Jr., Modular design of conditional sum adders using quantum-dot cellular automata, *Proceedings 6th IEEE Conference on Nanotechnology* © 2006 IEEE.)

The CFA is the best design in QCA. The complexity is significantly lower than that of the CLA or the CSA. Also, rather surprisingly, the delay is less, even for large adders.

The delays of the CLAs are less than that of the CSAs. Comparing QCA circuits with CMOS circuits, the main differences are observed in the CSAs. In transistor circuits, the CSA shows similar speed to the CLA, even though the size of the CSA

TABLE 2.1

Adder Comparisons with Multilayer Crossovers

Type	Complexity	Area (μm^2)	Delay (clocks)
4-bit CFA	371 cells	0.9×0.45	1.5
8-bit CFA	789 cells	1.8×0.53	2.5
16-bit CFA	1,769 cells	3.6×0.69	4.5
32-bit CFA	4,305 cells	7.1×1.0	8.5
64-bit CFA	11,681 cells	14.2×1.7	16.5
4-bit CLA	1,575 cells	1.7×1.1	3.5
8-bit CLA	3,988 cells	3.5×1.6	6.5
16-bit CLA	10,217 cells	7.0×2.2	10.25
32-bit CLA	25,308 cells	14.1×3.0	19
64-bit CLA	59,030 cells	28.2×3.7	31.5
4-bit CSA	1,999 cells	2.7×1.6	3.75
8-bit CSA	62,169 cells	5.9×2.6	7.75
16-bit CSA	16,866 cells	12.5×3.9	14
32-bit CSA	45,254 cells	25.7×6.2	25
64-bit CSA	129,611 cells	55.6×10.3	45

is larger than that of the CLA. But in QCA, the CSA is slower and the complexity and area are much greater than that of the CLA.

2.4   QCA MULTIPLIERS

This chapter explores the implementation of two types of parallel multipliers in QCA technology. Array multipliers that are well suited to QCA are constructed and formed by a regular lattice of identical functional units so that the structure is conformable to QCA technology without extra wire delay are considered in Section 2.4.2. Column compression multipliers, such as Wallace multipliers are implemented with several different operand sizes in Section 2.4.3.3. Section 2.4.4 gives a summary of multiplier design in QCA.

2.4.1   INTRODUCTION

Figure 2.25 shows the multiplication of two n-bit unsigned binary numbers that yields a $2n$-bit product. The basic equation is as follows:

$$P = A \times B \tag{2.19}$$

$$P = \sum_{j=0}^{n-1} a_j 2^j \times \sum_{i=0}^{n-1} b_i 2^i \tag{2.20}$$

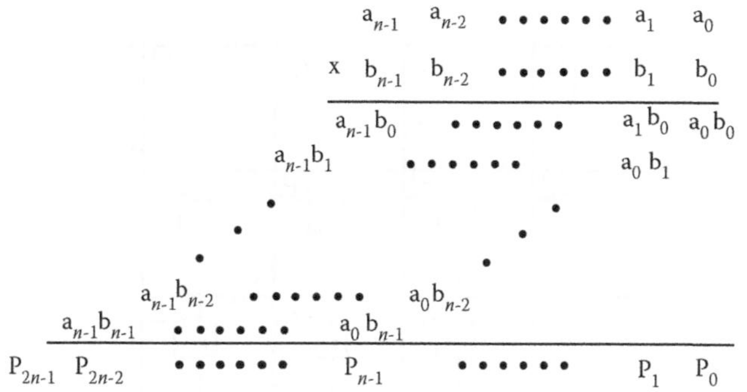

$$
\begin{array}{cccccccc}
 & a_{n-1} & a_{n-2} & \cdots\cdots & a_1 & a_0 \\
x & b_{n-1} & b_{n-2} & \cdots\cdots & b_1 & b_0 \\
\end{array}
$$

FIGURE 2.25 Multiplication of two n-bit binary numbers.

$$P = \sum_{k=0}^{2n-1} p_k 2^k \tag{2.21}$$

where the multiplicand is A ($= a_{n-1},\ldots, a_0$), the multiplier B ($= b_{n-1},\ldots, b_0$), and the product P ($= p_{2n-1},\ldots, p_{n-1},\ldots, p_0$).

2.4.2   QCA ARRAY MULTIPLIERS§

Array multipliers that are well suited to QCA are studied and analyzed in this section. An array multiplier is formed by a regular lattice of identical functional units so that the structure conforms to QCA technology without extra wire delay.

2.4.2.1   Schematic Design

A 4-bit by 4-bit array multiplier is shown in Figure 2.26. It has 5 rows and an irregular lattice [13]. The 4-bit by 4-bit array multiplier consists of 16 AND gates, 3 half adders, and 9 FAs. The carry outputs from each FA go to the next row. One operand propagates from left to right in the array multiplier. In general, the array multipliers have a latency of $4N - 2$. Table 2.2 shows the required hardware for the array multipliers.

2.4.2.2   Implementation of Array Multipliers with QCAs

The 4-bit by 4-bit array multiplier is implemented as shown in Figure 2.27. It can be extended to make a larger 8-bit by 8-bit multiplier as shown in Figure 2.28. The area of the array multiplier gets 4.35 times larger as the operand size is doubled because of its irregular lattice and one more last row.

2.4.3   WALLACE MULTIPLIERS FOR QCA**

2.4.3.1   Introduction

The Wallace strategy for fast multiplication is to form the N by N array of partial product bits, then reduce them to a two-row matrix with an equivalent numerical value, and then to add the two rows with a carry propagating adder [15]. To handle

§ Subsection 2.4.2 is based on [13].
** Subsection 2.4.3 is based on [14].

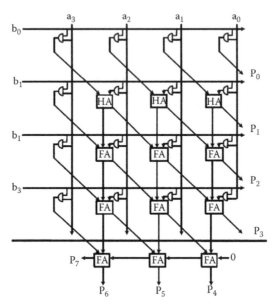

FIGURE 2.26 Schematic of an array multiplier. (S.-W. Kim and E. E. Swartzlander, Jr., Multipliers with Coplanar Crossings for Quantum-Dot Cellular Automata, *IEEE NANO* © 2010 IEEE.)

TABLE 2.2
Required Components for Array Multipliers

Type	Full Adders	Half Adders	AND Gates
Array: 4-bit by 4-bit	9	3	16
Array: 8-bit by 8-bit	49	7	64
Array: N-bit by N-bit	$(N-1)2$	$N-1$	$N2$

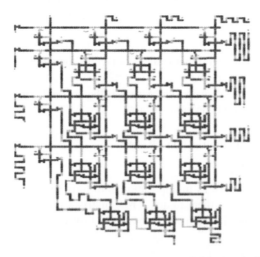

FIGURE 2.27 Layout of a 4-bit by 4-bit array multiplier. (S.-W. Kim and E. E. Swartzlander, Jr., Multipliers with Coplanar Crossings for Quantum-Dot Cellular Automata, *IEEE NANO* © 2010 IEEE.)

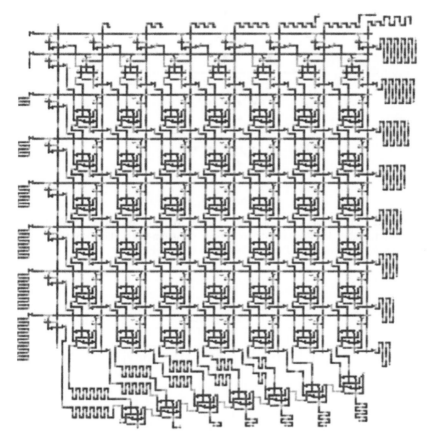

FIGURE 2.28 Layout of an 8-bit by 8-bit array multiplier. (S.-W. Kim and E. E. Swartzlander, Jr., Multipliers with Coplanar Crossings for Quantum-Dot Cellular Automata, *IEEE NANO* © 2010 IEEE.)

the reduction stage, Wallace's strategy is to combine partial product bits with the use of full and half adders at the earliest opportunity. Figure 2.29 shows a dot diagram of a 4-bit by 4-bit Wallace multiplier that has two reduction stages. A dot indicates a partial product of the multiplication. Plain and crossed diagonal lines indicate the outputs of a FA and a half adder, respectively. The block diagram of a 4-bit by 4-bit Wallace multiplier is shown in Figure 2.30.

2.4.3.2 Schematic Design

In each stage of the reduction, the Wallace multiplier conducts a preliminary grouping of rows into sets of three. Within each three-row set, FAs and half adders are used to reduce the three rows to two. Rows that are not part of a three-row set are transferred to the next stage without modification. The bits of these rows are considered in the later stages. An 8-bit by 8-bit Wallace multiplier has four reduction stages and intermediate matrix heights of 2, 3, 4, and 6. A dot diagram of

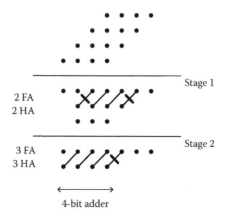

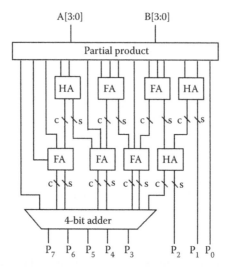

FIGURE 2.29 Dot diagram of a 4-bit by 4-bit Wallace reductions. (S.-W. Kim and E. E. Swartzlander, Jr., Parallel Multipliers for Quantum-Dot Cellular Automata, *IEEE Nanotechnology Materials and Devices Conference* © 2009 IEEE.)

FIGURE 2.30 Block diagram of a 4-bit by 4-bit Wallace multiplier. (S.-W. Kim and E. E. Swartzlander, Jr., Parallel Multipliers for Quantum-Dot Cellular Automata, *IEEE Nanotechnology Materials and Devices Conference* © 2009 IEEE.)

an 8-bit by 8-bit Wallace multiplier is shown in Figure 2.31. Table 2.3 shows the required hardware.

2.4.3.3 Implementation of Wallace Multipliers with QCAs

The layout of a 4-bit by 4-bit Wallace multiplier is shown in Figure 2.32. In these multipliers stair-like ripple carry adders are used for synchronization and to make each stage pipelined. A total of 3295 cells are used to make 4-bit by 4-bit Wallace

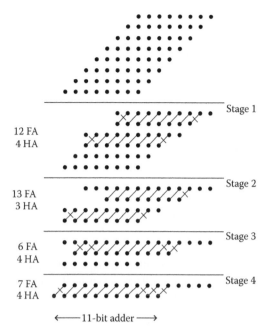

FIGURE 2.31 Dot diagram of 8-bit by 8-bit Wallace multiplier. (S.-W. Kim and E. E. Swartzlander, Jr., Multipliers with Coplanar Crossings for Quantum-Dot Cellular Automata, *IEEE NANO* © 2010 IEEE.)

TABLE 2.3
Required Adders for Wallace Multipliers

Type	Full Adders	Half Adders	Final Adder Size
Wallace: 4-bit by 4-bit	5	3	4
Wallace: 8-bit by 8-bit	38	15	11

multiplier with an area of 7.39 μm^2 [14]. An 8-bit by 8-bit Wallace multiplier is shown in Figure 2.33. The 8-bit by 8-bit multiplier has about 10 times as many cells as the 4-bit by 4-bit multiplier.

2.4.3.4　Simulation Results

Simulations were done with QCADesigner [8] assuming coplanar wire "crossings" and a maximum of 15 cells per clock zone. The size of the basic quantum cell was set at 18 nm by 18 nm with 5-nm diameter quantum dots. The center-to-center distance is set at 20 nm for adjacent cells. The following parameters are used for a bistable approximation: 51,200 samples, 0.001 convergence tolerance, 65-nm radius effect, 12.9 relative permittivity, $9.8e^{-22}J$ clock high, $3.8e^{-23}J$ clock low, 2-clock amplitude factor, 11.5 layer separation, 100 maximum iterations per sample.

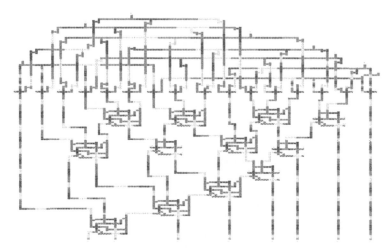

FIGURE 2.32   Layout of a 4-bit by 4-bit Wallace multiplier. (S.-W. Kim and E. E. Swartzlander, Jr., Parallel Multipliers for Quantum-Dot Cellular Automata, *IEEE Nanotechnology Materials and Devices Conference* © 2009 IEEE.)

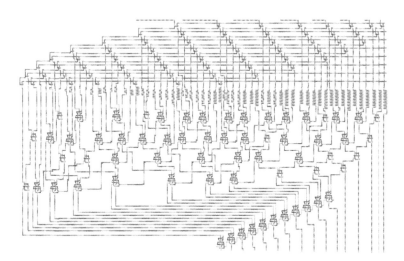

FIGURE 2.33   Layout of an 8-bit by 8-bit Wallace multiplier. (S.-W. Kim and E. E. Swartzlander, Jr., Multipliers with Coplanar Crossings for Quantum-Dot Cellular Automata, *IEEE NANO* © 2010 IEEE.)

The 4-bit by 4-bit Wallace multiplier has 10-clock latency, and 3295 cells with area that is comparable to the array multiplier. The 8-bit by 8-bit Wallace multiplier has more than 4 times the latency and 11 times the area of the 4-bit by 4-bit multiplier. These results show that simple and dense structures are needed. In addition, if they can be realized, multilayer wire crossings might mitigate the wire burden.

2.4.4 COMPARISON OF QCA MULTIPLIERS

Table 2.4 shows a comparison of the simulation results for 4 by 4 and 8 by 8 bit array and Wallace multipliers. The various 4-bit by 4-bit multipliers have 10- to 14-clock latency, and 3295–3738 cells. The 8-bit by 8-bit multipliers have roughly 2 to 3 times the latency and 4 to 8 times the cell count of the 4-bit by 4-bit multipliers. The 8-bit by 8-bit multipliers are much slower and larger than would be expected from CMOS multipliers, where the Wallace multiplier latency scales as the logarithm of the operand size. These results show that the most significant factor in the performance is the wiring. At least part of the explanation for this is that many of the wires are quite long. The long wires significantly affect the timing, 33.8% of the latency is due to the wiring.

Thus it seems that array multipliers are the best choice for QCA implementation. The latency is least (for all but the smallest multipliers) and the area is much less than Wallace multipliers.

2.5 QCA GOLDSCHMIDT DIVIDER

Convergent dividers where an initial estimate of the quotient in refined iteratively are used where high speed is required and when one or more multipliers are available. Section 2.5.3 presents the design of a Goldschmidt convergent divider in QCA [16].

With CMOS, large iterative computational circuits such as convergent dividers are often built with state machines to control the various computational elements. Because of QCA wire delays, state machines have problems due to long delays between the state machine and the units to be controlled. Even a simple 4-bit microprocessor that has been implemented with QCA [16] was done without using a state machine. Because of the difficulty of designing sequential circuits, there has been little research into using QCAs to realize large iterative computational units.

This section presents a design for a convergent divider using the Goldschmidt algorithm implemented with a data tag architecture to solve the difficulty in designing iterative computation units. In Section 2.5.1, the Goldschmidt iterative division algorithm is described. In Section 2.5.2, the data tag method is presented. In Section 2.5.3, an implementation of the Goldschmidt divider using the proposed method is reviewed in detail. Finally, simulation results are presented in Section 2.5.4.

TABLE 2.4
Comparison of QCA Multipliers

Type	Cell Count	Area (μm^2)	Latency
Array: 4-bit by 4-bit	3,738	6.02	14
Array: 8-bit by 8-bit	15,106	21.5	30
Wallace: 4-bit by 4-bit	3,295	7.39	10
Wallace: 8-bit by 8-bit	26,499	82.2	36

2.5.1   GOLDSCHMIDT DIVISION ALGORITHM[tt]

In Goldschmidt division [17,18], an approximate quotient converges toward the true quotient by multiple iterations. The division operation can be viewed as the manipulation of a fraction. The numerator (N) and the denominator (D) are each multiplied by a sequence of numbers so that the value of D approaches 1 and the value of N approaches the quotient. In the first step, both N and D are multiplied by F_0, an approximation to the reciprocal of D. Often F_0 is produced by a reciprocal table with very limited precision, thus, the product of D times F_0 is not exact, but has an error, ε. Therefore, the first approximation of the quotient is as follows:

$$Q = \frac{N \times F_0}{D \times F_0} = \frac{N_0}{D_0} = \frac{N_0}{1 - \varepsilon} \tag{2.22}$$

At the next iteration, N_0 and D_0 are multiplied by F_1, which is given by:

$$F_1 = (2 - D_0) = 2 - (1 - \varepsilon) = 1 + \varepsilon \tag{2.23}$$

Note that often the one's complement of D_i is used for F_{i+1} as that avoids the need to do a subtraction. The result is a slight (1 least significant bit [LSB]) increase in the error, which very slightly reduces the rate of convergence.

$$Q = \frac{N_0 \times F_1}{D_0 \times F_1} = \frac{N_0(1 + \varepsilon)}{(1 - \varepsilon)(1 + \varepsilon)} = \frac{N_0(1 + \varepsilon)}{(1 - \varepsilon^2)} = \frac{N_1}{D_1} \tag{2.24}$$

At the i+1-st iteration, F_i is as follows:

$$F_i = (2 - D_{i-1}) = 1 + \varepsilon^{2^{i-1}} \quad \text{for} \quad i > 0 \tag{2.25}$$

$$Q = \frac{N_i}{D_i} = \frac{N_{i-1} F_i}{D_{i-1} F_i} = \frac{N_{i-1}(1 + \varepsilon^{2^{i-1}})}{(1 - \varepsilon^{2^i})} \tag{2.26}$$

As the iterations continue, N_i converges toward Q with quadratic precision, which means that the number of correct digits doubles on each iteration.

To show the Goldschmidt division, consider the following example: $Q = 0.6/0.75$. From a look-up table, the approximate reciprocal of D (i.e., 0.75) is $F_0 = 1.3$:

$$Q = \frac{N \times F_0}{D \times F_0} = \frac{0.6 \times 1.3}{0.75 \times 1.3} = \frac{0.78}{0.975} = \frac{N_0}{D_0} \tag{2.27}$$

Then $F_1 = 2 - D_0 = 2 - 0.975 = 1.025$:

$$Q = \frac{N_0 \times F_1}{D_0 \times F_1} = \frac{0.78 \times 1.025}{0.975 \times 1.025} = \frac{0.7995}{0.999375} = \frac{N_1}{D_1} \tag{2.28}$$

Then $F_2 = 2 - D_1 = 2 - 0.999375 = 1.000625$:

$$Q = \frac{N_1 \times F_2}{D_1 \times F_2} = \frac{0.7995 \times 1.000625}{0.999375 \times 1.000625} = \frac{0.799999685}{0.999999609395} = \frac{N_2}{D_2} \tag{2.29}$$

[tt] Subsection 2.5.1 is based on [19,20].

The errors between N_0, N_1, and N_2 are 0.02, 0.0005, and 0.0000003125, respectively. This shows that the value of N_i converges quadratically to the value of Q.

2.5.2   Data Tag Method

To resolve the problem of communication commands from state machines to the computational units, a data tag method is used as shown in Figure 2.34. In this method, data tags travel with the data, and local tag decoders (TD in the Figure 2.34) generate control signals for the computational circuits (i.e., COMP1, COMP2). The tags travel with the data through the same number of pipeline stages as the corresponding computational circuits, and local tag decoders generate control signals appropriate to each datum. As the tags travel with the data and local tag decoders produce the control signals for the units, the synchronization issues that are a problem in state machines are significantly mitigated. In QCAs, the data tag method can be implemented very efficiently as the delays to keep the data tags synchronized to the data are generated inherently via the gates and wires in the QCAs.

An advantage of the data tag architecture is that each datum on a data path can be processed differently according to the tag information. For example, in typical Goldschmidt dividers controlled by a state machine, a new division cannot be started until the previous division is completed. There are many pipeline stages in QCA computational circuits, and most stages may be idle during iterations. With the data tag method, each datum on a data path can be processed by the operation that is required at that stage. As divisions at different stages are processed in a time-skewed manner, new divisions can be started while previous divisions are in progress as long as the initial pipeline stage of the data path is free. As a result, the throughput can be increased to a level that is much greater than that which is implied by the latency.

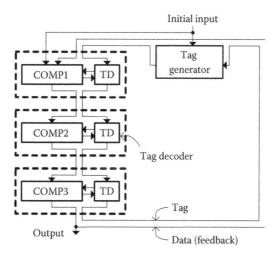

FIGURE 2.34   Computation unit implementation using data tags. (I. Kong et al., Design of a Goldschmidt iterative divider for quantum-dot cellular automata, *IEEE/ACM International Symposium on Nanoscale Architectures* © 2009 IEEE.)

A 12-bit Goldschmidt divider has been designed using the data tag method. The flowchart is shown in Figure 2.35. The divider uses a low-precision ROM that gives 4-bit values for F_0. The divider performs three iterations, the first with a value from the ROM followed by two iterations with the one's complement of D_i for F_{i+1}.

The flowchart realizes the steps from Section 2.5.1. To start a new division, the tag generator issues a new tag (DT = 1) for the data. On the first iteration, the factor is obtained from the ROM and used to multiply the denominator and numerator. For the second and third iterations, the factor is obtained by inverting the current value of the denominator. After the third iteration, the value of the numerator is output as the quotient. The local tag decoders control the multiplexers according to the tag associated with the data. Once a division has started, it progresses through the required iterations, irrespective of any other divisions that are being performed.

2.5.3 IMPLEMENTATION OF THE GOLDSCHMIDT DIVIDER

The Goldschmidt divider was designed using coplanar wire crossings with the design guidelines suggested by K. Kim et al. [21,22]. Coplanar wire crossings are used for this research, as a physical implementation of multilayer crossovers has not been demonstrated yet. If multilayer crossovers become available, the design will be slightly smaller

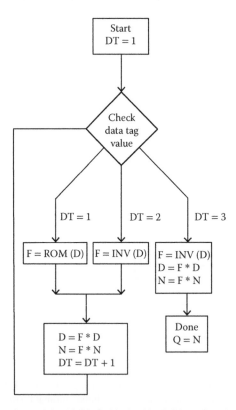

FIGURE 2.35   Flow chart of the 12-bit Goldschmidt divider using data tags.

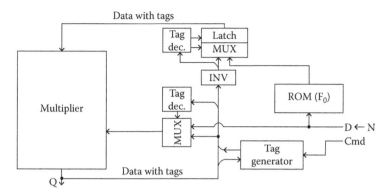

FIGURE 2.36 Block diagram of the Goldschmidt divider using the data tag method. (I. Kong et al., Design of a Goldschmidt iterative divider for quantum-dot cellular automata, *IEEE/ACM International Symposium on Nanoscale Architectures* © 2009 IEEE.)

and faster. The design guidelines [21] are kept except for the limitation on majority gate outputs. Robust operation of majority gates is attained by limiting the maximum number of cells that are driven by each output, which is verified using the coherence vector method. The maximum cell count for each circuit component in a clock zone is determined by simulations with sneak noise sources. For this work, the maximum length of a simple wire is 14 cells and the minimum length is 2 cells. As a result, each majority gate drives a line that is at least two cells long to insure proper operation.

A block diagram of the divider is shown in Figure 2.36. The main elements are the ROM and the multiplier. In addition, there are the tag generator, tag decoders, a word-wide inverter, and a few multiplexers. The CMD signal is asserted together with D, and a new tag is generated from the tag generator. Then N is entered. The tag decoders control the multiplexers and the latches using the tag that is associated with D. During the first iteration (used to normalize the denominator to a value that is close to 1), the multiplexers are set so that D and N are multiplied sequentially by F_0 from the reciprocal ROM. After the first denominator normalization step is completed, the tag is incremented by the tag generator. During the subsequent iterations, the multiplexers select D or N from the outputs of the multiplier and F_i that is computed by inverting the bits of D (as noted in Section 2.5.1; this one's complement operation approximates $2 - D$ with an error of 1 LSB). After three or four iterations, the final values of D and N have been computed, so N (the approximate value of Q) is output and the tag generator eliminates the tag.

2.5.3.1 2^3 by 3-Bit ROM Table

The 2^3 by 3-bit reciprocal ROM consists of a 3-bit decoder and an 8 by 3 ROM array as shown in Figure 2.37. All the ROM cells have the same access time, 7 clocks. The data are programmed by setting one input of the OR gate inside each ROM cell. Both the 12-bit and the 24-bit dividers use the same ROM. As the range of $D_i[0:11]$ is $0.5 \leq D < 1$ for the Goldschmidt division, $D_i[0:1]$ are always 01, so $D_i[2:4]$ is used as the input to the 3-bit ROM. Similarly, the ROM output is $F_0[1:3]$ as $F_0[0]$ is always 1. Thus, an 8 by 3 ROM implements a 32-word by 4-bit table.

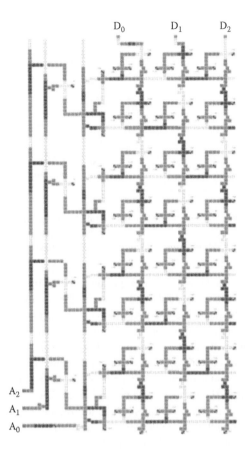

FIGURE 2.37 Layout of the 8-word by 3-bit reciprocal ROM. (I. Kong et al., Design of a Goldschmidt iterative divider for quantum-dot cellular automata, *IEEE/ACM International Symposium on Nanoscale Architectures* © 2009 IEEE.)

2.5.3.2 Array Multiplier

The divider uses a 12-bit by 12-bit array multiplier as array multipliers are attractive for QCA as shown in Section 2.4. The basic cell of the multiplier is a FA implemented with three majority gates. The cell has signal delays of 1 clock for the carry output, 2 clocks for the sum, and an area of 20 by 29 cells. The 12-bit by 12-bit multiplier has two inputs, A[11:0] and B[11:0], and a most significant output, M[11:0]. The latency of an N-bit by N-bit array multiplier is $4N - 2$, so the multiplier latency of 46 is the largest component of the latency to perform an iteration. The unused least significant outputs are not left unconnected as that would violate the QCA design guidelines. Additional dummy cells are attached to the unused outputs for robust transfers of the signals.

To realize Goldschmidt dividers of other sizes, much of the design remains the same as for the 12-bit divider. The multiplier size is changed to match the divider word size. If the ROM for F_0 is kept as 8 words of 3 bits, the number of iterations will change. If the ROM size is increased to 32 words of 5 bits, one fewer iteration is needed. The larger ROM may have slightly larger latency, but as one iteration (that includes a pair of high-latency multiplications) is saved, the total divider latency is reduced.

2.5.4 SIMULATION RESULTS

The layout of the 12-bit Goldschmidt divider is shown in Figure 2.38. The design has been implemented and simulated using QCADesigner version 2.0.3 [8]. Most default parameters for bistable approximation in QCADesigner version 2.0.3 are used except two parameters: the number of samples and the clock amplitude factor. As the recommended number of samples is 1000 times the number of clocks in a test vector, the number of samples is determined to be 226,000. Because adiabatic switching is effective to prevent a QCA system from relaxing to a wrong ground state [1], the clock amplitude factor is adjusted to 1.0 for more adiabatic switching. Other major parameters are as follows: size of QCA cell = 18 nm by 18 nm, center-to-center distance = 20 nm, radius of effect = 65 nm, and relative permittivity = 12.9.

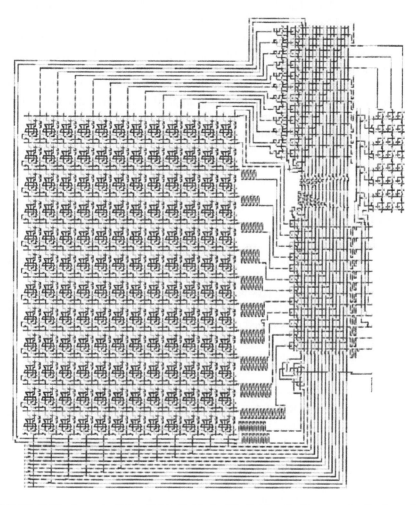

FIGURE 2.38 Layout of the 12-bit Goldschmidt divider. (I. Kong et al., Design of a Goldschmidt iterative divider for quantum-dot cellular automata, *IEEE/ACM International Symposium on Nanoscale Architectures* © 2009 IEEE.)

The area for the 12-bit Goldschmidt divider is 89.8 μm^2 (8.8 μm × 10.2 μm), and the total number of QCA cells is 55,562. The delays of the functional units are shown in Table 2.5. As a division requires three iterations, the total latency for a single isolated division is 219 clocks. Although this latency (in terms of the number of clocks) seems quite high, the clock rate for semiconductor QCA is on the order of 1 THz, so the time per isolated division can be less than 1/4 ns. Given the two clocks (one for D and the second for N) with a latency of 73 clocks per iteration, as many as 35 divisions can be started while the first one is progressing. Then successive quotients are available on every other clock.

The 12-bit Goldschmidt divider was tested using bottom-up verification as a full simulation for a case takes about 7 hours. Each unit block is verified exhaustively, and then the full integration is tested. A simulation of four consecutive divisions is shown in Figure 2.39. The first division computes 0.7080/0.7915. The inputs $D = 0.655_{16}$ and $N = 0.5aa_{16}$ are shown at the left side of the second row of Figure 2.24a. The results for this division ($Q = 0.8945_{10}$) are shown starting at clock 218 (sequence da_{16}) on the second row of Figure 2.28b, $D = 0.7ff_{16}$ and $Q = N = 0.728_{16}$. Three additional divisions are performed immediately after the first division to show that pipelining achieves a peak division throughput of one division for every two clock cycles.

Table 2.6 shows the results for the four example divisions. The first two columns give the numerator and denominator in hexadecimal and decimal, respectively. Column 3 gives the exact quotient. Columns 4 and 5 give the result computed by the Goldschmidt divider in hexadecimal and decimal. Finally, the last column gives the difference between the exact and the computed (QCA) quotients. In all four cases the computed quotients are accurate to within about 1 LSB.

2.6   CONCLUSION

A Goldschmidt divider (an iterative computational circuit) for QCA is implemented efficiently in a new architecture using data tags. The proposed data tag method avoids the synchronization problems that arise with conventional state machines in QCA because of the long delays between the state machines and the units to be controlled. In the proposed architecture, it is possible to start a new division at any iteration stage of a previous issued operation. As a result, the throughput is significantly increased because multiple division computations can be performed in a time-skewed manner using one iterative divider.

TABLE 2.5

Delays of the Functional Units

Functional Unit	Delay (Clocks)
Tag generator	3
Multiplexer and tag decoder	19
8 by 3 ROM	7
12-Bit array multiplier	46
Data bus	5

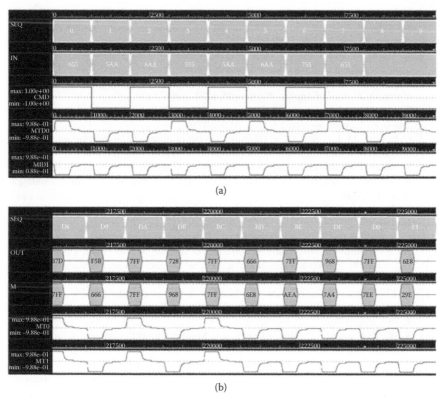

(a)

(b)

FIGURE 2.39 Simulation results. (a) Input vectors for four consecutive divisions. (b) Output waveforms for the four quotients. (I. Kong et al., Design of Goldschmidt dividers with quantum-dot cellular automata, *IEEE Transactions on Computers* © 2013 IEEE.)

TABLE 2.6
Example Divisions

Input		Quotient			Difference
			QCA		
Hex	Decimal	Exact	Hex	Decimal	Decimal
5aa/655	0.7080/0.7915	0.8945	728	0.8945	−0.00002
555/6aa	0.6665/0.8330	0.8001	666	0.7998	0.0003
6aa/5aa	0.8330/0.7080	1.1765	968	1.1757	0.0008
655/755	0.7915/0.9165	0.8636	6e8	0.8632	0.0003

REFERENCES

1. C. Lent and P. Tougaw, A device architecture for computing with quantum dots, *Proceedings of the IEEE*, 85, 541–557, 1997.
2. P. Tougaw and C. Lent, Logical devices implemented using quantum cellular automata, *Journal of Applied Physics*, 75, 1818–1825, 1994.

3. G. Snider, A. Orlov, I. Amlani, G. Bernstein, C. Lent, J. Merz, and W. Porod, Quantum-dot cellular automata: Line and majority logic gate, *Japanese Journal of Applied Physics*, 38, 7227–7229, 1999.

4. K. Walus, G. Schulhof, and G. A. Jullien, A method of majority logic reduction for quantum cellular automata, *IEEE Transactions on Nanotechnology*, 3, 443–450, 2004.

5. A. Gin, P. D. Tougaw, and S. Williams, An alternative geometry for quantum-dot cellular automata, *Journal of Applied Physics*, 85, 8281–8286, 1999.

6. K. Hennessy and C. S. Lent, Clocking of molecular quantum-dot cellular automata, *American Vacuum Society*, 19, 1752–1755, 2001.

7. C. S. Lent, M. Liu, and Y. Lu, Bennett clocking of quantum-dot cellular automata and the limits to binary logic scaling, *Nanotechnology*, 17, 4240–4251, 2006.

8. K. Walus, T. Dysart, G. Jullien, and R. Budiman, QCADesigner: A rapid design and simulation tool for quantum-dot cellular automata, *IEEE Transactions on Nanotechnology*, 3, 26–31, 2004.

9. H. Cho and E. E. Swartzlander, Jr., Adder and multiplier design in quantum-dot cellular automata, *IEEE Transactions on Computers*, 58(3), 721–727, 2009.

10. I. H. Unwala and E. E. Swartzlander, Jr., Superpipelined adder designs, *Proceedings IEEE International Symposium on Circuits and Systems*, 3, 1841–1844, 1993.

11. H. Cho and E. E. Swartzlander, Jr., Adder designs and analyses for quantum-dot cellular automata, *IEEE Transactions on Nanotechnology*, 6(3), 374–383, 2007.

12. H. Cho and E. E. Swartzlander, Jr., Modular design of conditional sum adders using quantum-dot cellular automata, *Proceedings 6th IEEE Conference on Nanotechnology*, 1, 363–366, 2006.

13. S.-W. Kim and E. E. Swartzlander, Jr., Multipliers with Coplanar Crossings for Quantum-Dot Cellular Automata, *IEEE NANO*, 953–957, Seoul, Korea, 2010.

14. S.-W. Kim and E. E. Swartzlander, Jr., Parallel Multipliers for Quantum-Dot Cellular Automata, *IEEE Nanotechnology Materials and Devices Conference*, 68–72, Traverse City, MI, 2009.

15. C. Wallace. A suggestion for a fast multiplier, *IEEE Transactions on Electronic Computers*, EC-13, 14–17, 1964.

16. K. Walus, M. Mazur, G. Schulhof, and G. A. Jullien, Simple 4-bit processor based on quantum-dot cellular automata (QCA), *16th International Conference on Application-Specific Systems, Architecture and Processors*, 288–293, 2005.

17. R. E. Goldschmidt, *Applications of Division by Convergence*, Master's thesis, Massachusetts Institute of Technology, Cambridge, MA, 1964.

18. S. F. Oberman and M. J. Flynn, Division algorithms and implementations, *IEEE Transactions on Computers*, 46, 833–854, 1997.

19. I. Kong, E. E. Swartzlander, Jr., and S.-W. Kim, Design of a Goldschmidt iterative divider for quantum-dot cellular automata, *IEEE/ACM International Symposium on Nanoscale Architectures*, 47–50, 2009.

20. I. Kong, S.-W. Kim, and E. E. Swartzlander, Jr., Design of Goldschmidt Dividers with Quantum-Dot Cellular Automata, IEEE Transactions on Computers, 2014 (in press).

21. K. Kim, K. Wu, and R. Karri, Towards designing robust QCA architectures in the presence of sneak noise paths, *Proceedings of the Conference on Design, Automation and Test in Europe*, 2, 1214–1219, 2005.

22. K. Kim, K. Wu, and R. Karri, The robust QCA adder designs using composable QCA building blocks, *IEEE Transactions on Computer-Aided Design of Integrated Circuits and Systems*, 26, 176–183, 2007.

3 Novel Capacitor-Less A2RAM Memory Cells for Beyond 22-nm Nodes

Noel Rodríguez and Francisco Gamiz

CONTENTS

3.1   INTRODUCTION

The semiconductor industry is facing a period where the survival of mature technologies is being questioned. Short-channel effects, leakage, and variability are suited as insurmountable obstacles that standard metal–oxide–semiconductor field-effect transistor (MOSFET) would not be able to overcome in the ultimate nodes. In the case of the memory field, the situation is analogous. Mainstream technologies (Flash, static random-access memory [SRAM], dynamic random-access memory [DRAM]) are threatened by particular issues related to their scalability.[1] In parallel new emerging alternatives, that is, resistive random-access memory, magnetoresistive random-access memory, floating-body DRAMs (FB-DRAMs), and so on, are claiming to solve the issues of the established technologies.

FB-DRAMs are one of the candidates positioned for SRAM/DRAM replacement. In this approach, the isolated body of a silicon-on-insulator (SOI) transistor is used to store charge that changes the electrical potential of the transistor and therefore the way it drives the current. This approach has a long way back since the first works suggesting the application of SOI-MOSFET as single-transistor memory cells were made back in the 1990s when Kuo et al.[2] anticipated the current flow modulation in fully depleted SOI transistor by modifying the charge at the opposite interface.[2] However, it was at the beginning of the new century when the floating-body memory cell became a subject of intense research. Since then, there have been multiple proposals combining different methods to store the charge in the body of the transistor and different ways to sense the state of the cell from simple MOSFET operation to bipolar current breakdown.[3–7] Each particular approach has its own advantages and drawbacks, but none of them has risen yet as competitive opponent to the standard technologies. The search of a charge-based cell, substitute of DRAMs, has continued evolving during the last years, bearing encouraging fruits as novel thyristor-like structures or multibody cells giving another twist to the subject.[8,9] In this chapter we will focus in one of the particular approaches pointed out as fast storage (SRAM, DRAM) substitute: the recently introduced floating-body A2RAM cell.

3.2 MULTIBODY FLOATING-BODY-1T-DYNAMIC RANDOM-ACCESS MEMORY

One of the trends followed by the FB-DRAM researches lies in the modification of the body of the transistor to create multibody structures defining dedicated regions for the charge storage and the current flow. The basics for introducing this complexity in the device arise from scalability constrains.[10] FB-DRAM cells are intrinsically very dependent on SOI technology. The scaling of the gate length of an SOI transistor requires for the simultaneous scaling of the silicon film thickness.[11] This condition introduces a paramount challenge: how to allow the simultaneous coexistence of holes (which represent the information in terms of charge) and electrons (which are the carriers involved in the current flow for discriminating the state in the device) in the same ultrathin silicon film. The solution to surmount this limitation implies innovative solutions enhancing the electrostatic potential differences that the body of the transistor can sustain. One possibility is to create specific regions (or bodies) to store the holes. Figure 3.1 shows three variants of multibody floating-body cells. In Figure 3.1a, a schematic of the single-transistor quantum well (QW) 1T-DRAM is presented.[12] This device is based on an engineered-body material that integrates, within the Si film, a thin layer of a material with a lower bandgap (i.e., SiGe). This last layer creates an electrostatic-potential well for storage of holes. It was theoretically demonstrated that

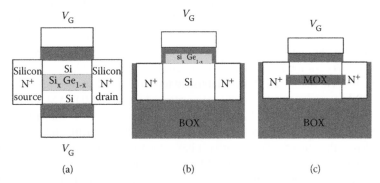

(a) (b) (c)

FIGURE 3.1 Schematic examples of multibody floating-body dynamic random-access memory cells. (a) Quantum well floating-body dynamic random-access memory: the holes are stored in a potential well created by the Si–Ge layer.[11] (b) Convex channel 1T-dynamic random-access memory: the holes are stored in a Si–Ge layer created into the gate stack.[12] (c) A-RAM memory cell: the body of the transistor is partitioned in two dedicated regions for hole storage and electron current sensing.[13] The bodies are separated by a low-k dielectric layer. This layer provides the electrostatic potential difference needed for electron–hole coexistence.

this structure improves the current sensing margin and scalability characteristics. As the QW devices are more scalable, thanks to the introduction of a "more efficient storage room," the effect of the volume reduction with the channel length is mitigated.

The second example of multibody FB-DRAM cell is the convex channel 1T-DRAM structure (Figure 3.1b), which using the bipolar junction transistor (BJT) programming technique, was proposed to improve the retention time.[13] Holes are stored beneath a raised gate oxide that may be filled by a smaller bandgap material (e.g., SiGe) (Figure 3.1b). As the holes stored during the "1" state programming reduce the body/source (drain) potential barrier, they easily diffuse through these junctions filling the SiGe region. The convex channel architecture provides a physical well for more effective storage of holes. Moreover, if a narrower bandgap material is used in the convex channel region, a deeper potential well is formed improving further the sensing margin and retention time.

Finally, the last example proposes the physical partitioning of the body of the transistor into two isolated regions (Figure 3.1c) by a middle oxide (MOX).[14] To take advantage of this structure, the MOX must have a dielectric constant smaller than silicon, for example, SiO_2. The resulting *semibodies* share the source and drain regions of the transistor. When this device is operated as a memory cell, the top semibody is used for majority carrier storage (holes) accommodated in a potential well created by the negative bias of the gate. The bottom semibody serves to sense the cell state via the minority carrier current. The low-k MOX constitutes the key advantage of this device: the electrostatic potential difference between the front and back interfaces is enlarged due to a higher electrostatic potential drop through the MOX.

3.3 A2RAM MEMORY CONCEPT AND FABRICATION

The second generation of the A-RAM cell or A2RAM, the main subject of this chapter, lays the foundations of a doping-based partitioning of the body of the transistor.[15] The device is a regular MOSFET with a retrograde N–P doping profile starting from the bottom of the body. The source and drain regions become electrically short-circuited due to the highly doped bottom region (named N-bridge) (Figure 3.2).

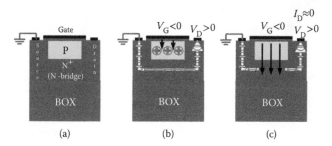

FIGURE 3.2 A2RAM memory cell: (a) The source and drain of a regular metal–oxide–semiconductor field-effect transistor are short-circuited by a buried N-type layer.[14] (b) In "1" state, the top body of the cell is charged with holes that screen the gate electric field allowing the current flow from source to drain through the highly doped N-bridge. (c) In "0" state, the top body is depleted of charge; the gate electric field is not screened; the N-bridge is fully depleted avoiding the current flow.

The memory states are defined by the conductivity level of the N-bridge. A negative gate bias is applied to create a potential well for the storage of the holes ("1" state) (Figure 3.2b). Under these circumstances, the electric field induced by the gate is screened by the positive charge of the P-type body (accumulated holes), and the electron concentration in the N-bridge (majority carriers) remains high. Biasing the drain, which is the way to easily test the state of the cell, makes an electron current to flow from source to drain through the N-bridge. In the "0" state, the holes are evacuated from the upper P-body: the gate electric field is no longer screened affecting the population of electrons of the N-bridge. If the device is well engineered, the N-bridge will become fully depleted during "0" state leading to a very low current flow, and therefore, a large margin between states (Figure 3.2c). One of the main advantages of this concept is that the use of an insulator substrate is not mandatory: the holes are stored in the potential well created at the top interface, and the electrons are confined by the N-bridge. There is no need for a physical insulator barrier at the back interface or a back-gate bias.

Prototypes of A2RAM cells have been fabricated in different technologies including 2.5-µm SOI, 2.5-µm bulk, and 22-nm SOI.[16] These devices were used as initial demonstrators for the viability of the concept, initially only supported by numerical simulations.[15] Figure 3.3a shows a transmission electron microscopy image of one of these prototypes fabricated in the more challenging 22-nm technology.[16] The fabrication adds two additional steps to the standard complementary metal–oxide–semiconductor (CMOS) process (Figure 3.3b). First, an arsenic implantation to introduce the N-type dopants for the N-bridge; second, a selective epitaxial growth (SEG) to generate the characteristic N–P retrograde doping profile establishing the body partitioning. Figure 3.3c shows results from ATHENA process simulations on the final vertical net doping profile of the device after the entire thermal load. It can be observed that the final profile achieves the bottom highly doped N-type layer under the low-doped P-type layer in a 36-nm thick body.

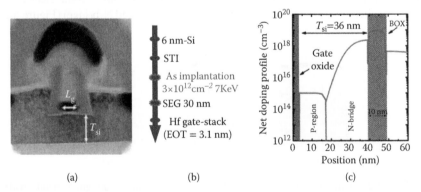

(a)   (b)   (c)

FIGURE 3.3   (a) Transmission electron microscopy image of an A2RAM memory cell fabricated on a 22-nm silicon-on-insulator process at CEA-LETI facilities.[15] (b) Simplified process flow: the key steps allowing the formation of the N-bridge are the implantation of the arsenic ions followed by a selective epitaxial growth. (c) Simulated resulting doping profile of the fabricated device after the entire thermal load.

It is worth noting that if the required maximum thickness limit for the silicon film can be relaxed, the process flow can be further simplified avoiding the SEG step. This could be interesting for nonultimate embedded memory applications with longer channels and consequently thicker silicon films.

A proper calibration of the doping of the N-bridge is necessary to ensure a large margin between states. Figure 3.4 shows an example of the target doping level for the N-bridge considering a device with a gate length of $L = 100$ nm. For a given N-bridge doping, if the bridge is too thick, the device will have large leakage in the "0" state becoming very difficult to fully deplete the N-doped layer with reasonable gate bias. In contrast, if the bridge is too thin, it will be fully depleted even in the "1" state by the action of the negative gate bias during holding and reading, leading to a very low "1" state current. In our fabricated prototype cells, the N-bridge thickness is around 22 nm and the average doping approximately 2×10^{17} cm^{-3}, therefore they are in the target area.

3.4   EXPERIMENTAL ELECTRICAL RESULTS

The clearest evidence of the viability of the A2RAM concept as memory cell is manifested when the device is biased with a pulse pattern emulating the actual memory operation of the device in a matrix. The voltage waveform is generated with an arbitrary function generator with the capability to measure in real time the driven current.[17] The bias sequence (Figure 3.5a) starts with a positive drain-bias pulse overlapped with a negative gate-bias pulse. The large electric field created in the drain-to-gate overlap region allows the injection of holes by band-to-band (BTB) tunneling.[15] The large concentration of holes after the W"1" pulse screens the electric field induced by the negative bias of the gate taking the N-bridge to its

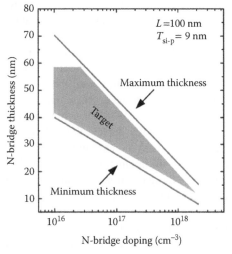

FIGURE 3.4   Maximum and minimum thickness of the N-bridge depending on its doping (assumed constant) to achieve a current ratio between the "1" and "0" state over 100 in an $L = 100$ nm with P-type body of 9 nm.

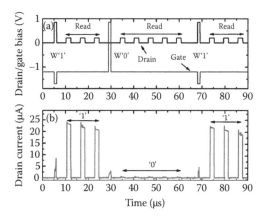

FIGURE 3.5 Experimental measurements of the drain current of an A2RAM memory cell with gate length $L = 100$ nm. (a) The voltage pattern for the test is applied to the drain and gate terminals. The sequence is as follows: write "1," read three times, write "0," read five times, write "1," and read three times (b) Drain current readout.

conductive state. Note that there is a peak of current during the W"1" event. This current is generated by the electron flow through the N-bridge due to the large drain bias during the writing process (see Figure 3.5b). When the hole concentration starts to be substantial, the N-bridge passes from a full-depletion situation to a partial-depletion situation presenting an increasing conductivity. A proper calibration of the width and level of the voltage pulses is necessary to avoid an excessive drain current during a W"1" event. This depends on the size of the cell.

After the W"1" event, the state of the cell is read three times. This is simply achieved by slightly increasing the drain bias to test whether the bridge is in a conductive condition or not. As observed, three pulses of current are immediately obtained in the drain corresponding to a conductive bridge situation. It is noticeable that the read current shows a decay with time. This is due to the overpopulation of holes in the body produced during the W"1" event. Actually, under the bias conditions used, the number of holes injected in the body is larger than the concentration that can be sustained by the gate bias corresponding to the thermodynamic equilibrium. The excess of holes gradually recombines leading to a slight current decrease until the equilibrium condition.

In the next step, a writing "0" event (W"0") is applied to the device (time ~30 μs). The bias conditions necessary to fully deplete the N-bridge are simply a positive voltage pulse in the gate. The sudden increase in the body potential forward biases the body-to-source and body-to-drain junctions evacuating the holes from the upper body of the transistor through these junctions. Because the body potential decreases quickly (as the gate bias return to a negative value), the body of the transistor enters in a deep-depletion condition.

The state of the cell is read five times by increasing the drain bias leading to negligible current in this axis scale. The difference with the previous current levels ("1" state) is enough to discriminate clearly the state of the device. More results on the current ratio between "1" and "0" states are presented in Figure 3.6. The ratio is shown against the gate length of the device. For the batch of A2RAM cells fabricated, the

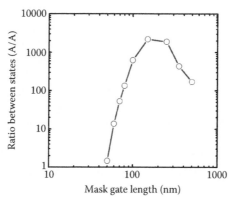

FIGURE 3.6 Current ratio between "1" and "0" states as a function of the channel length in A2RAM prototypes.

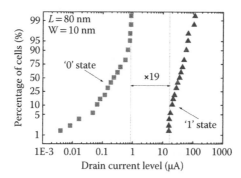

FIGURE 3.7 Statistical cumulative plots of the current levels in the "1" (triangles) and "0" (squares) 1 μs after the witting events. It is worth noting that the distribution never overlaps presenting a safety guard factor of 19.

maximum ratio occurs at $L \sim 100$ nm. This result was expected because the devices were optimized for that gate length. As the gate length decreases below 100 nm, so does the current ratio because it becomes more difficult to fully deplete the N-bridge during the "0" state (for further scaling of the device length, it is necessary to reduce the doping level).

Despite the A2RAM, devices fabricated up to now have been purely experimental demonstrators; it is interesting to show the variability of the prototyped devices. Figure 3.7 presents statistical results on the "1" and "0" current levels 10 μs after the writing events. The average current margin is about two orders of magnitude. For this particular dimensions, the maximum current in the "0" state is below 1 μA, whereas the minimum current in the "1" state is above 10 μA. The actual safety guard margin is a factor of 19.

As in any other FB-DRAM cell, in the A2RAM structure, the current margin between states does not prevail indefinitely. The parasitic BTB tunneling during the holding phase of the operation restores the hole population in the P-type body turning the "0" state into "1" state. The retention time is defined as the time required for the "0" state to recover half of the current corresponding to the stable "1" state. Results for different FB-DRAM flavors are shown in Figure 3.8. A2RAM presents very competitive retention time (average 3 ms at 85°C and 10 ms at 25°C) compared even with less scaled multigate structures.[18] The so-called bipolar Z-RAM establishes the reference, but at the expense of much larger voltages when is operated (which ultimately introduce disturbance and reliability concerns).

We would like to finish this section by considering a very important aspect that needs to be validated for any novel technology that intends to have an industrial application. The immunity to disturbance effects is one of the main challenges of any memory technology and an issue of especial relevance in the case of floating-body memories. A disturbance event can be considered a change in the bias lines of a memory matrix that may change unintentionally the state of a cell in holding state while accessing a different cell for reading or writing. In the case of the A2RAM memory, the disturbance events can affect the "0" unstable state; the "1" state cannot be turned into "0" state by the bias pattern proposed for the operation of the device (Figure 3.5). There are two cases where the "0" state may be disturbed, represented in Figures 3.9 and 3.10, depending if the change in the bias line occurs at the gate terminal or at the drain terminal.

Figure 3.9a represents a very simplistic memory matrix of A2RAM cells (three words of three bits). The little triangle at the drain of the devices represents the cell selector that controls the current flow of the cell (this device could be a diode or a transistor in an actual implementation). We are going to consider the disturbance situation of the cell ("disturbed cell," indicated with a lighting in Figure 3.9a) when we write a "1" state in the next cell sharing the drain and gate lines ("target cell," closed into the dashed line in Figure 3.9a). To write the "1" state in the target cell, it is necessary to decrease the gate bias and increase the source bias (acting as drain in this case) to trigger the BTB tunneling injection of holes. Meanwhile, the disturbed cell is holding a "0" state and may suffer from a decrease in the gate voltage that may trigger also the BTB tunneling injection of holes, turning the "0" state into "1" state. The situation previously described has been emulated in Figure 3.9b. Initially we write the "1" state to an isolated A2RAM transistor and we read the driven current. Next we write a "0" state and read it consecutively. In between each reading drain

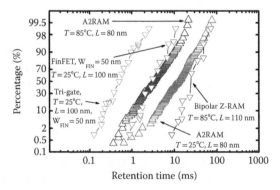

FIGURE 3.8 Statistical cumulative plots of the retention time are 25°C and 85°C compared with other floating-body-dynamic random-access memories.

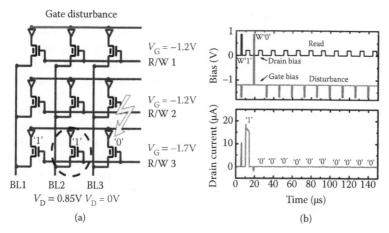

FIGURE 3.9 (a) Schematic of a simplistic A2RAM three words/three bits memory matrix to study gate disturbance phenomenon. The role of source and drain terminal is exchanged depending on the bias. (b) Experimental measurements of the current read in the "0" state during sequential reading with gate disturbance events in between them.

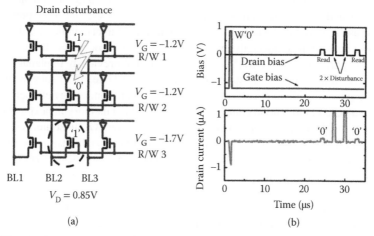

FIGURE 3.10 (a) Schematic of a simplistic A2RAM three words/three bits memory matrix to study drain disturbance phenomenon. The role of source and drain terminal is exchanged depending on the bias. (b) Experimental measurements of the current read in the "0" state before and after two drain disturbance events.

pulse, we apply a negative gate voltage pulse over the holding bias, what emulates the disturbance condition described in Figure 3.9a. As observed, the disturbance of the gate does not modify the "0" state.

The other possibility of disturbance happens when a target cell in which the "1" state must be stored (target cell, marked in dashed line in Figure 3.10a) is sharing the drain line with a cell which is holding the "0" state (disturbed cell, cell indicated with the lighting in Figure 3.10a). The increase in the source bias (acting as drain in this case) may trigger the BTB tunneling in the disturbed cell. Figure 3.10b shows the experimental conditions of this event in a single isolated device. Initially, the "0" state is written by pulsing the gate voltage. Next, the current is read by slightly increasing the drain bias, then the cell is disturbed by the applying higher voltage pulses to the drain twice and, finally the cell is read again. As observed, the current of the "0" state after the disturbance does not present any change as compared to the current of the "0" state before the disturbance.

Summarizing, we can affirm that the waveform pattern developed for the operation of this experimental A2RAM memory cell makes them also immune to disturbance events during the writing and reading processes. Only the simultaneous decrease of the gate bias and increase in the drain bias can trigger the BTB tunneling injection to turn a "0" state into a "1" state. As mentioned previously, immunity to disturbance is a paramount challenge for any solid-state storage technology and especially for floating-body based cells.

3.5 TRIDIMENSIONAL A2RAM: FINFET, TRI-GATE, AND NANOWIRE A2RAM

Several foundries (Intel among others) have recently introduced 3D Tri-Gate transistors in their 22-nm nodes, that is, three gates wrapped around the silicon channel in a 3D structure.[19] These transistors are expected to provide ultralow power benefits for use in portable devices, like smartphones and tablets, while also delivering improved performance normally expected for high-end processors. The concept of multibody partitioning 1T-DRAM cells, demonstrated previously in planar devices, can be also transferred to 3D structures (FinFET, Tri-Gate, and nanowire transistors), thus enabling memory cells with low voltage operation, energy efficiency, high performance, and fabrication compatibility as embedded memory in next technological nodes. A 3D picture of the proposed memory cell, on SOI substrate, is shown in Figure 3.11. Doping sections from source to drain and perpendicular to the BOX are shown in Figure 3.12. As observed, a conventional triple-gate field-effect transistor (FET) is modified by connecting the N^+ source and drain through an inner N-type wire (Figure 3.11). Typical total Fin width can vary from 15 to 25 nm.

The doping profiles have been generated by process simulator ATHENA, thus validating the device feasibility. The 14-nm-thick core N-bridge is 5×10^{18} cm^{-3} doped, whereas the surrounding 4-nm P-type body is maintained undoped ($N_A = 10^{14}$ cm^{-3}

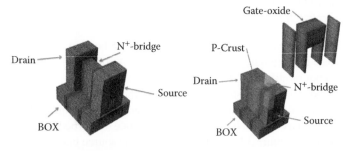

FIGURE 3.11 Schematic representation of the 3D A2RAM memory cell.

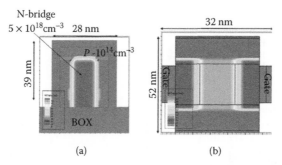

FIGURE 3.12   (a) Doping cross-section (perpendicular to the BOX, parallel to source and drain) of the Triple-gate A2RAM. The high-doped N-bridge (5×10^{18} cm^{-3}) short-circuiting source and drain is surrounded by a low-doped P-type layer (10^{14} cm^{-3}). (b) 2D-doping cross-section (parallel to the BOX) of the triple-gate A2RAM. The gate is 3 nm underlapped to improve retention characteristics.

residual boron) (Figure 3.12). The P-type external crust that surrounds the N-type core is used as the storage node while the internal core (high doped N-bridge) is used for current sense (Figure 3.12a).

The memory effect relies on the same principle as planar A2RAM cell explained in Section 3.3: when the P-type crust is charged with holes (generated by BTB tunneling and retained with a negative gate bias), the gate electric field is screened by the accumulation layer and cannot deplete the N-bridge. If V_D is increased, an electron current flows through the neutral region of the N-bridge, which behaves as a simple resistor (state "1"). The current flow through the P-body is negligible (reverse-biased junctions). When the outer P crust is empty of holes, the gate field is no longer screened and fully depletes the N-bridge: the drain current becomes extremely low ("0" state). As in planar A2RAM, differences with conventional 1T-DRAMs are twofold: (1) the drain current (defining the cell state) is due to majority carriers flowing in the volume of the bridge, (2) the coexistence of holes and electrons is ensured by the P/N junction (supercoupling suppression).

The combination of the P and N channels results in unconventional I_D–V_G characteristics. We have used numerical simulations to demonstrate the functionality as memory cell of the 3D A2RAM device; Poisson and continuity equations were solved self-consistently in 3D. First, we studied the steady state operation of the cell (continuous section of drain current curve in Figure 3.13): the combination of the parallel P and N channels results in unconventional I_D–V_G curves. In steady state, the N-bridge is always nondepleted regardless of the gate voltage. For negative V_G, the potential difference is basically absorbed by the accumulated holes in the crust P region: the more negative the gate voltage, the more holes accumulate, screening the electric field, and the drain current, which is mainly limited by the conductivity of the N$^+$-core acting as a resistor, is weakly dependent on V_G. On the contrary, for positive gate bias, the current flow comes from the parallel combination of the majority carriers (electrons) of the N-bridge (which behaves like a resistor), and the minority carriers (electrons) of the top MOSFET. As the gate bias is increased, the top inversion channel becomes dominant. Notice that this behavior is different from that of a

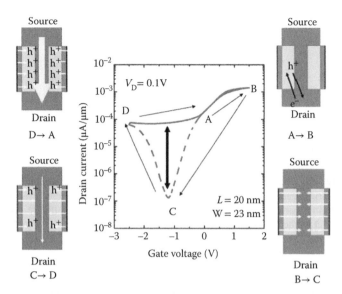

FIGURE 3.13 Hysteresis curve demonstrating the memory effect obtained by looping the gate voltage between 1.5 and −2.5 V.

depletion-mode NMOSFET, where the conduction can be effectively cut at a certain negative gate bias.

The transient behavior is analyzed in Figure 3.13:

1. From A to B, a positive voltage is applied to the gate. The upper channel becomes inverted with electrons, and the behavior of the device is similar to that of a MOSFET transistor.
2. From B to C, a negative voltage is suddenly applied to the gate. The channel becomes depleted of electrons, and as there are no sources of holes, the upper channel becomes empty of carriers. The negative electric field induced by the negative gate voltage also depletes the N-bridge and as a consequence, there is no current at all in the device.
3. If the gate voltage is decreased to even more negative values (C to D), BTB tunneling starts to appear in the source-channel and drain-channel over-lapped regions. This process injects holes into the channel that screens the negative electric field induced by the gate. As the negative gate electric field is now weaker, the N-bridge becomes partially undepleted and the drain current starts to increase. The greater the hole injection, the higher the drain current.
4. If the gate voltage is reduced to zero (D to A), BTB tunneling stops, no more holes are injected in the channel and the drain current becomes constant.

As observed in Figure 3.13, under negative gate voltages, there is a current window for the same values of gate and drain voltages, which, depending on the population of holes in the upper channel, allows the definition of two memory states:

1. "1" state: Upper channel is populated with holes that screen the negative electric field. Drain current flows through the buried N-bridge.

2. "0" state: Upper channel is in deep-depletion of carriers. The negative electric field induced by the negative gate voltage depletes the N-bridge, and no current flows between drain and source. This state is nonequilibrium state. After a long time, thermal carrier generation, junction leakage, and BTB tunneling will restore the hole population in the channel, and the "0" state will be corrupted.

Figure 3.14 shows the memory operation. The gate is biased below the threshold voltage to accommodate an accumulation of holes in the channel, which will define the "1" state current. Initially, the cell is purged by writing "0" state with a positive gate-voltage pulse. Because the pulse forward-biases the channel-source and channel-drain P–N junctions, the holes are expelled from the channel. When the gate returns to the negative holding bias, there are no carriers in the channel.

We propose two alternative mechanisms to write the "1" states (restore the hole population in the P-body): BTB tunneling by means of an over-bias pulse of the retention gate voltage, or impact ionization by applying a positive gate voltage in the gate activating the MOSFET. Nevertheless, BTB tunneling is best suited for low-power embedded applications because the writing current is typically several orders of magnitude lower than with the impact ionization mechanism (during the writing time, there is an additional contribution of current coming from the MOSFET, which is not present when using the BTB mechanism). Waveforms demonstrating the cell functionality are shown in Figure 3.14 using the BTB alternative. As observed, the "0" state virtually corresponds to zero drain current.

The potential scalability of A2RAM and its expansion to nonplanar devices gain evidence with the previous simulations. Further research and optimized prototypes may position this device in the backstage of the memory scenario waiting for a commercial application.

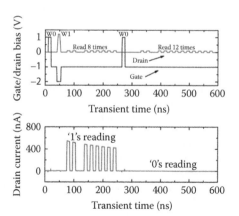

FIGURE 3.14 Full 1T dynamic random-access memory operation of a 3D-A2RAM cell. Top: Drain and gate bias pattern. The cell is initially purged (W0), then the "1" state is written by band-to-band tunneling and read eight times. Next the "0" state is written by purging again the cell and read 12 times. Bottom: Drain current reading.

3.6 CONCLUSIONS

We would like to conclude this chapter with a set of open questions that summarizes the outcome of the results presented along the previous pages.

Is a charge-based bit the best fundamental of a memory cell? If we think in terms of scalability, the common sense immediate answer will be no. The stored charge will scale down with the dimensions of the cell, therefore leading to poor sensitivity, larger variability, and easier disturbance among other issues. *Material-property*-based memories as magnetic RAMs or resistive RAMs may behave better under a very aggressive scaling scenario. This can be considered a paradoxical situation, because in the origins of the automatic treatment of stored information, the memory modules were also material based from simple punched cards to more advanced magnetics tape. *So, is the memory industry going back to a nanoscaled version of its origins?* Considering that the standard CMOS process is the milestone of the digital world, the answer, thinking in terms of embedded memory, turns more complicated. Introducing new materials with delicate thermal properties is something that the semiconductor industry will not be graceful to afford. Despite in every technology node, the continuity of standard DRAM cells has been questioned, the industry keeps pushing the limits of this 40-year-old technology. In this context, *what is the destiny of A2RAM and its floating-body counterparts?* It has been probed that A2RAM is a reality; it performs well in all the aspect studied, and considering the overall concept, it can outperform most of its floating-body counterparts. However, a change in paradigm means more than a substitution of the single DRAM cell, but a complete redesign of the memory scheme starting from the memory matrix. This is therefore something utopic as far as standard DRAM has demonstrated its scalability up to now, and probably it will continue doing so in the next node at the expense of a higher cost and complexity. A2RAM may target specific applications with need of cheap embedded storage (maybe not ultimately scaled) where it can considerably simplify the fabrication process. The last word will come from the semiconductor industry and its need from design innovation and cost reduction.

REFERENCES

1. K. Kim. Perspectives on giga-bit scaled DRAM technology generation, *Microelectronics Reliability* 40(2): 191–206, 2000.
2. F. A. Eraghi, J. Chen, R. Solomon, T. Chan, P. Ko, and C. Hu. Time dependence of fully depleted SOI MOSFET's subthreshold current, *Proceedings IEEE International SOI Conference*, Vail Vally, CO, 32–33, October 1991.
3. S. Okhonin, M. Nagoga, J. M. Sallese, and P. Fazan. A capacitor-less 1T-DRAM cell, *IEEE Electron Device Letters*, 23(2): 85–87, February 2002.
4. C. Kuo, T. J. King, and C. Hu. A capacitorless double-gate DRAM cell, *IEEE Electron Device Letters*, 23(6): 345–347, June 2002.
5. S. Okhonin, M. Nagoga, E. Carman, R. Beffa, and E. Faraoni. New generation of Z-RAM, *IEEE International Electron Devices Meeting*, Washington, DC, 925–928, 2007.
6. K. W. Song, H. Jeong, J. W. Lee, S. I. Hong, N. K. Tak, Y. T. Kim, Y. L. Choi et al. 55 nm capacitor-less 1T DRAM cell transistor with non-overlap structure, *Electron Devices Meeting in IEDM Technical Digest*, San Francisco, CA, 1–4, December 2008.

7. Z. Lu, N. Collaert, M. Aoulaiche, B. De Wachter, A. De Keersgieter, W. Schwarzenbach, O. Bonnin et al. A novel low-voltage biasing scheme for double gate FBC achieving 5s retention and 1016 endurance at 85°C, *Electron Devices Meeting (IEDM), 2010 IEEE International*, San Francisco, CA, 12.3.1–12.3.4., 2010.

8. Y. K. J. Yang, R. N. Gupta, S. Banna, F. Nemati, H. J. Cho, M. Ershov, M. Tarabbia, D. Hayes, and S. T. Robins. Optimization of nanoscale thyristor on SOI for high-performance high-density memories, *IEEE International SOI Conference*, Niagara Falls, New York, 113–114, 2006.

9. J. Wan, C. Le Royer, A. Zaslavsky, and S. Cristoloveanu. A compact capacitor-less high-speed DRAM using field effect-controlled charge regeneration, *IEEE Electron Device Letters*, 33(2): 179–181, 2012.

10. T. Hamamoto. Overview and future challenges of floating body RAM (FBRAM) technology for 32 nm technology node and beyond, *Proceedings Solid-State Device Research Conference*, Edinburgh, Scotland, 25–29, 2008.

11. J. P. Colinge. *Silicon-On-Insulator Technology: Materials to VLSI*, 3rd ed. Kluwer Academic Publishers, Dordrecht, The Netherlands, 2004.

12. M. Ertosun, P. Kapur, and K. Saraswat. A highly scalable capacitorless double gate quantum well single transistor DRAM: 1T-QW DRAM, *IEEE Electron Device Letters*, 29: 1405–1407, 2008.

13. M. Cho, C. Shin, and T. Liu. Convex channel design for improved capacitorless DRAM retention time, *International Conference on Simulation of Semiconductor Processes and Devices, SISPAD*, San Diego, CA, 1–4, 2009.

14. N. Rodriguez, F. Gamiz, and S. Cristoloveanu. A-RAM memory cell: Concept and operation, *IEEE Electron Device Letters*, 31(9): 972–974, 2010.

15. N. Rodriguez, S. Cristoloveanu, and F. Gamiz. Novel capacitorless 1T-DRAM cell for 22-nm node compatible with bulk and SOI substrates, *IEEE Transactions on Electron Devices*, 58(8): 2371–2377, August 2011.

16. N. Rodriguez, C. Navarro, F. Gamiz, F. Andrieu, O. Faynot, and S. Cristoloveanu. Experimental demonstration of capacitorless A2RAM cells on silicon-on-insulator, *IEEE Electron Device Letters*, 33(12): 1717–1719, December 2012.

17. Agilent Technologies. *Agilent B1530 User Manual*, 5th edition, August 2012.

18. C. Bassin, P. Fazan, W. Xiong, C. Cleavelin, T. Schulz, K. Schruefer, M. Gostkowski et al. Retention characteristics of zero-capacitor RAM (Z-RAM) cell based on FinFET and tri-gate devices, *Proceedings IEEE International SOI Conference*, Honolulu, HI, 203–204, October 2005.

19. S. Damaraju, V. George, S. Jahagirdar, T. Khondker, R. Milstrey, S. Sarkar, S. Scott, I, Stolero, and A. Subbiah. A 22 nm IA multi-CPU and GPU system-on-chip, *Solid-State Circuits Conference Digest of Technical Papers (ISSCC)*, 2012 IEEE International, IEEE, San Francisco, CA, 56–57, February 2012.

4 Four-State Hybrid Spintronics–Straintronics
Extremely Low-Power Information Processing with Multiferroic Nanomagnets Possessing Biaxial Anisotropy

Noel D'Souza, Jayasimha Atulasimha, and Supriyo Bandyopadhyay

CONTENTS

4.1 INTRODUCTION

There is increasing interest in implementing digital logic with single-domain nanomagnets instead of traditional transistors as the former have the potential to be extremely energy-efficient binary switches. Transistors switch by moving electrical charge into or out of their active regions. If this process is carried out nonadiabatically, then it dissipates an amount of energy equal to at least $NkT\ln(1/p)$, where N is the number of electrons (information carriers) moved into or out of the device, T the temperature, and p the *bit error probability* associated with random switching of the device (Zhirnov et al. 2003; Salahuddin and Datta 2007). On the other hand, if logic bits are encoded in two stable magnetization orientations along the easy axis of a shape-anisotropic *single-domain* nanomagnet (or the single-domain magnetostrictive layer of a multiferroic nanomagnet), then switching between these orientations can take place by dissipating only approximately $kT\ln(1/p)$ of energy, regardless of the number of spins (information carriers) in the nanomagnet (Salahuddin and Datta 2007). This results from the fact that exchange interaction between spins makes all the approximately 10^4 spins in a single-domain nanomagnet of volume approximately 10^5 nm^3 behave collectively like a giant *single* spin (Salahuddin and Datta 2007; Cowburn, et al. 1999a) (a single-information carrier) (Salahuddin and Datta 2007). Ideally, all of these spins will rotate in unison when the nanomagnet switches from one stable magnetization state to the other. This is schematically explained in Figure 4.1. As a result, for the same bit error probability p, the ratio of the minimum energy that must be dissipated to switch a nanomagnet to that dissipated to switch a nanotransistor will be approximately $1/N \ll 1$. The mutual *interaction* between spins leading to collective dynamics, which is absent in the case of charges, makes the nanomagnet switch *intrinsically* more energy-efficient than the transistor switch.*

Owing to this innate advantage, nanomagnet-based computing architectures are attracting increasing attention. In one version of nanomagnetic logic (NML) known as magnetic quantum cellular automata, Boolean logic gates are configured by placing nanomagnets in specific geometric patterns on a surface so that the dipole interactions between the nanomagnets elicit the desired logic operations on the bits encoded in their magnetization orientations (Cowburn and Welland 2000; Csaba et al. 2002). The dipole interaction also acts as an effective *wire* to concatenate successive gates and thus builds arbitrary combinational or sequential Boolean circuits. This approach builds on the Single Spin Logic paradigm, where exchange interaction between spins played the role of dipole interaction between magnets, whereas up- and down-spin polarizations encoded the two logic bits (Bandyopadhyay, Das, and Miller 1994). NML schemes can be very energy efficient if the magnets are switched in a way that dissipates very little energy in the external switching circuit.

4.2 CLOCKING NANOMAGNETIC LOGIC

Unfortunately, dipole-coupled NML schemes also have a serious drawback that limits their applications. There is no *isolation* between the input and the output ports of the magnet switch (unlike in transistors), because dipole interaction is perfectly bidirectional and does not discriminate between an *input* magnet and an *output* magnet. This hinders *unidirectional* propagation of logic bits from one stage to the

---

* There can be collective interaction between charges (electrostatic, exchange, correlation, etc.) that could lead to collective dynamics and many-body ground states similar to charge density waves, but these are not usually observed in transistors. Normally charges act independently of each other as independent degrees of freedom.

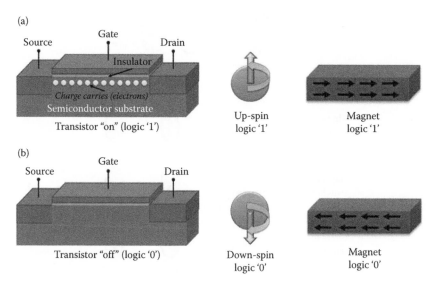

FIGURE 4.1 Transistor, single-spin, and single-domain nanomagnet encoding (a) logic state "1" and (b) logic state "0."

next—a property that is required of all Boolean logic schemes. Because unidirectionality cannot be imposed in space, one must impose it in time. That requires sequential *clocking* of the nanomagnets (much similar to bucket-brigade devices and charge-coupled device shift registers) (Schroder 1987). It is accomplished with *Bennett clocking* (Bennett 1982), which is implemented by forcibly rotating a shape-anisotropic nanomagnet's magnetization through approximately 90° from the easy to the hard axis before a bit propagates through it. The forcing agent is withdrawn just before the propagating bit reaches the magnet. At that point, the nanomagnet is temporarily at its (unstable) energy maximum, and the dipole interaction of the neighbors subsequently nudges it to the global energy minimum (correct orientation along the easy axis) and thus propagates the logic bit unidirectionally (Behin-Aein, Salahuddin, and Datta 2009; Behin-Aein et al. 2010). The forcing agent that rotates the magnetization of the chosen magnet by 90° is the Bennett clock.

Bennett clocking in NML can be carried out in two different ways (see Figure 4.2): via a global agent (e.g., a global magnetic field), which simultaneously aligns the magnetization of every nanomagnet in a logic chain along the hard axis before bit propagation through the chain (Csaba et al. 2002) (Figure 4.2a) or with a local agent (e.g., a local spin-polarized current that exerts a spin transfer torque on each nanomagnet [Bandyopadhyay and Cahay 2009; Behin-Aein, Salahuddin and Datta 2009]), which rotates every magnet's magnetization individually to align along the hard axis (Figure 4.2b). In the global scheme, the forcing agent (clock) is applied to all magnets in the chain at the same time and withdrawn before the initiation of bit propagation. In the local scheme, the forcing agent is applied to different magnets at different times (each magnet is clocked just before the bit reaching it). The advantage of the global scheme is that there is no need to individually access the magnets to deliver the force, which alleviates the lithography burden. The disadvantages

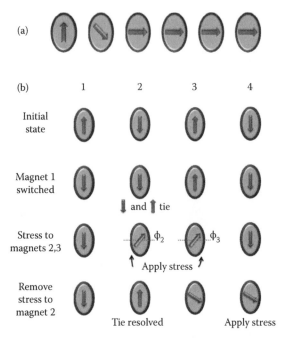

FIGURE 4.2 Bennett clocking of nanomagnetic logic. (a) Global clocking. (b) Local clocking scheme: propagating a logic bit through a chain of four dipole coupled multiferroic nanomagnets with Bennett clocking implemented with stress. First row: A chain of elliptical nanomagnets in the ground state with magnetization orientation indicated by arrows. Second row: Magnetization of the first magnet is flipped with an external agent and the second magnet finds itself in a tied state where it experiences no net dipole interaction. Third row: The second and third magnets are subjected to electrically induced stresses that rotate their magnetizations close to the hard axis. Fourth row: The second magnet is freed from stress so that its magnetization relaxes to the easy axis as a result of shape anisotropy, and it switches to the desired "up" state rather than the incorrect "down" state as the dipole interaction from the left neighbor is now stronger than that from the right neighbor so that the tie is resolved. (Adapted from Atulasimha, J. and S. Bandyopadhyay, *Appl. Phys. Lett.*, 97, 173105, 2010.)

are as follows: (1) The magnets near the end of the chain are left in their unstable energy maximum state along the hard axes for a very long time (until the traveling bit reaches them), and these magnets may spontaneously relax to one of their two energy minima (magnetization along the easy axis) before the bit reaches them. Even one magnet flipping like this will scuttle the Bennett clocking scheme, making this strategy extremely error-prone and unreliable. (2) The traveling bit must propagate through the entire chain before the global field can be applied again to reset the chain and prepare it for the next bit. This makes the computing architecture *nonpipelined* and hence impractically slow (Bandyopadhyay and Cahay 2009). Attempts to alleviate the first problem by using magnets with biaxial anisotropy that have shallow energy minima at the hard axes have been reported in the literature (Carlton et al. 2008), but have been shown to be ineffective at room temperature (Spedalieri et al. 2011).

4.3 CLOCKING NANOMAGNETIC LOGIC WITH MULTIFERROIC NANOMAGNETS

Local clocking of magnets, that is, rotating the magnetization from the easy to the hard axis, can be accomplished through a variety of means such as generating a local magnetic field directed along the hard axis (Behin-Aein, Salahuddin, and Datta 2009), applying a spin transfer torque by passing a spin-polarized current through the magnet with the spin polarization in the direction of the hard axis (Behin-Aein et al. 2010), or by inducing domain wall motion with a spin-polarized current (Yamanouchi et al. 2004; Fukami et al. 2009). Recently, an extremely energy-efficient strategy for local Bennett clocking of NML was proposed (Atulasimha and Bandyopadhyay 2010), where electrically generated mechanical strain rotates the magnetization of a magnetostrictive layer. It can be implemented by applying a small voltage to a multiferroic nanomagnet consisting of two elastically coupled piezoelectric and magnetostrictive layers (Eerenstein, Mathur, and Scott 2006) as shown in Figure 4.3. The applied voltage generates strain in the piezoelectric (PZT; lead zirconate titanate) layer, which is transferred almost entirely to the magnetostrictive layer by elastic coupling if the latter layer is much thinner than the former (Atulasimha and Bandyopadhyay 2010). This strain/stress can cause the magnetization of the magnetostrictive layer to rotate by a large angle and has been demonstrated in recent experiments (Brintlinger et al. 2010; Hockel et al. 2012; Dusch et al. 2013), although not in single-domain nanomagnets. Voltage-controlled resistive switching of the magnetization vector by approximately 90° was also theoretically demonstrated to be feasible in ferromagnetic multilayers and spin valves mechanically coupled to a ferroelectric substrate, with one of the ferromagnetic layers possessing a small degree of cubic magnetocrystalline anisotropy (Pertsev and Kohlstedt 2010).

These rotations are sufficiently large to fulfill the requirements of Bennett clocking in logic chains (Atulasimha and Bandyopadhyay 2010; Salehi Fashami, Atulasimha, and Bandyopadhyay 2012). This particular scheme is known as "hybrid spintronics and straintronics" as the application of mechanical strain, generated by a tiny voltage, makes the magnetization rotate. Normally, strain can rotate the magnetization of an isolated magnet by up to approximately 90°, because it moves the energy minimum of the magnet from the easy to the hard axis. However, if the strain is *withdrawn at the right juncture*, as soon as the 90° rotation has been completed, the magnetization will continue to rotate and the magnetization will end up rotating by 180° (Roy, Bandyopadhyay, and Atulasimha 2012). This will result in a complete bit flip. Rotation by approximately 90° is sufficient for Bennett clocking in logic

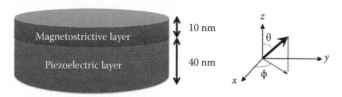

FIGURE 4.3 A bilayer multiferroic nanomagnet composed of a magnetostrictive layer and a piezoelectric layer.

chains, but for use in memory, the 180° rotation is required. This issue, however, will not be discussed further here as this chapter is focused on four-state devices and not the traditional two-state memory elements.

In the studies conducted by Atulasimha and Bandyopadhyay (2010), Bennett clocking of two-state logic chains was considered where the logic switches in the chain are ellipsoidal multiferroic nanomagnets of major axis = 105 nm and minor axis = 95 nm. The PZT layer of the multiferroic was 40 nm thick and the magneto-strictive layer (Terfenol-D) was 6 nm thick. Bennett clocking was implemented by applying tiny rectangular voltage pulses of approximately 15 mV amplitude across the PZT layer of the multiferroic that generated strain and rotated the magnetization of the magnetostrictive layer by approximately 90°. Transient simulations showed that for approximately 1-ns pulse period, the total energy dissipated in the clocking circuitry and in a nanomagnet is approximately 200 kT per rotation at room temperature (Roy, Bandyopadhyay, and Atulasimha 2012; Roy, Bandyopadhyay, and Atulasimha 2013). Thermal noise was ignored in the simulations, but it does not have much effect in the case of Bennett clocking. Dipole coupling between such Bennett-clocked nanomagnetic switches was also shown to propagate information unidirectionally along a chain while dissipating a few 100 kT per bit at approximately 1 GHz clock rate (Salehi Fashami et al. 2011). Finally, a dipole-coupled NML NAND gate with fan-in and fan-out was designed with multiferroic nanomagnets and shown to dissipate a total energy of only approximately 1000 kT per NAND operation (Salehi Fashami, Atulasimha, and Bandyopadhyay 2012). This makes multiferroic nanomagnets one of the most energy-efficient digital switches, and the hybrid spintronic/straintronic scheme one of the minimally dissipative memory and logic paradigms extant.

4.4   FOUR-STATE NANOMAGNETS IMPLEMENTED WITH BIAXIAL ANISOTROPY

Although two-state devices may suffice for Boolean computing, four-state logic devices (Figure 4.4) have features that can be exploited for non-Boolean computing applications. These include *associative* memory that is useful for image reconstruction and pattern recognition and four-state nanomagnetic implementations of neurons for neuromorphic computing. It is also possible to realize a four-state universal Boolean logic gate (D'Souza, Atulasimha, and Bandyopadhyay 2011a) for increased logic density and propagate logic bits down a chain of four-state switches using a somewhat more complex clocking sequence. Elementary image processing functionality implemented with four-state nanomagnets has been theoretically demonstrated and described in Section 4.6.4 (D'Souza, Atulasimha, and Bandyopadhyay 2012).

Section 4.5 discusses the use of magnetocrystalline and geometric (or shape) anisotropy to achieve biaxial anisotropy in planar nanomagnets. Section 4.6 discusses some potential applications that include memory, Boolean logic, complex information processing, and neuromorphic computing. Section 4.7 summarizes the current state of the art as well as the important technological barriers and issues that must be addressed before such four-state nanomagnets implemented with biaxial anisotropy can usher in new technology.

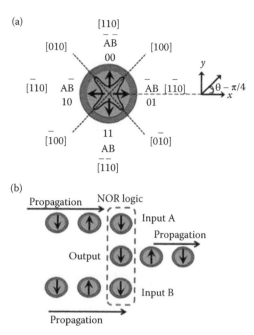

FIGURE 4.4 (a) Four-state multiferroic nanomagnets. (b) Scheme illustrating logic and propagation of four-state NOR logic along a chain of multiferroic nanomagnets.

4.5 ACHIEVING BIAXIAL ANISOTROPY IN NANOMAGNETS

4.5.1 MAGNETOCRYSTALLINE ANISOTROPY

A four-state memory element can be implemented with a magnetostrictive layer (e.g., single-crystal nickel [Ni]), which would exhibit biaxial magnetocrystalline anisotropy in the (001) plane. Epitaxial films of single-crystal (001) Ni can be grown on a suitable substrate using molecular beam epitaxy (MBE) (Naik et al. 1993; Chow 1991). Although there are no reports of growing single-crystal Ni on a piezoelectric substrate, there are no obvious technological barriers to prevent this. The crystal structure for Ni (face-centered cubic) is shown in Figure 4.5a. If the thickness of the (001) Ni layer is considerably smaller than the lateral dimensions, it would be energetically costly for the magnetization vector to rotate out of plane and, therefore, the magnetization always lies in the (001) plane. As a result, the *easy* axes of single-crystal Ni in the ground/unstressed state are [110], [1$\bar{1}$0], [$\bar{1}\bar{1}$0], and [$\bar{1}$10], in Miller notation. Thus, there are four possible (degenerate) energy minima in which four 2-bit combinations can be encoded (00, 01, 11, 10), as shown in the saddle-shaped curve of Figure 4.5b. The resulting energy minima occur along the ±45° and ±135° directions. Therefore, a rotation of +45° is introduced ($\theta \rightarrow \theta - \pi/4$), where θ is the angle made by the magnetization direction with the +x axis. This is equivalent to rotating the Cartesian coordinate axes by an angle of 45° about the axis normal to the nanomagnet's plane. In the new coordinate system, the energy minima occur along the x and y axes (0°, ±90°, ±180°). The bit assignments ($\overline{AB}$, $\overline{A}B$, AB, $A\overline{B}$) are also shown.

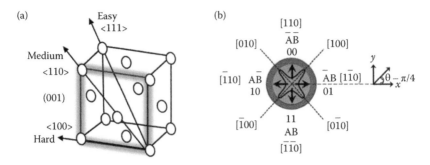

FIGURE 4.5 (a) Crystal structure of nickel. (b) Biaxial anisotropy implemented with magnetocrystalline anisotropy.

4.5.2 Shape Anisotropy

In addition to the magnetocrystalline anisotropy, the magnetic behavior of a magnetic material is also affected by the shape of the nanomagnetic element. Consider a polycrystalline specimen with no preferred grain orientation and, therefore, no net crystal anisotropy. A spherical element of this specimen would be magnetized to the same extent in any direction based on the applied magnetic field direction. However, elliptically shaped elements, for instance, would result in a major (long) and a minor (short) axis. As the demagnetizing field along the minor axis is stronger than along the major axis, the applied field would have to be stronger along the minor axis to produce the same field in the specimen (Cullity and Graham 2009). Therefore, it is easier for the magnetization to align itself along the major (easy) axis, resulting in two stable orientations for the magnetization vector, parallel and antiparallel to the easy axis.

By manipulating the shape of the nanomagnet, the magnetic properties of the element can be engineered, with different shapes giving rise to different anisotropic behaviors. For instance, Cowburn et al. (1999b) experimentally demonstrated that supermalloy ($Ni_{80}Fe_{14}Mo_5$) nanomagnets with triangular, square, and pentagonal geometries (corresponding to rotational symmetries of order three, four, and five, respectively) exhibit anisotropy with 6-fold, 4-fold, and 10-fold symmetries, respectively. The anisotropies of these nanomagnets are measured using the modulated field magneto-optical anisotropy technique (Cowburn et al. 1997) and are shown in Figure 4.6a through c.

Biaxial anisotropy in magnetic thin films has been previously shown in single-crystal films (Boyd 1960), coupled films (Wang 1968), double-layer films (Siegle 1965), as well as in a four-pointed star-shaped film (Lee 1968) (Figure 4.7a), with the latter showing biaxial anisotropy caused by shape effects. Recently, studies have been conducted on the effects of configurational anisotropy on concave nanomagnets, with small variations in parameters such as the thickness and radius of curvature/concavity giving rise to large but predictable variations in the direction and strength of the easy axes of magnetization (Figure 4.7b) (Lambson et al. 2013).

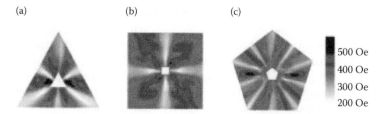

FIGURE 4.6 Polar plots demonstrating experimentally measured anisotropy field inside the 5-nm-thick nanomagnets of triangular (a), square (b), and pentagonal (c) symmetry. High field values correspond to easy anisotropy axes and low values to hard anisotropy axes. (Reprinted with permission of IOP Publishing, R. P. Cowburn et al., *Europhysics Letters (EPL)*, 48, 221–7, © 1999 http://iopscience.iop.org/0295-5075/48/2/221)

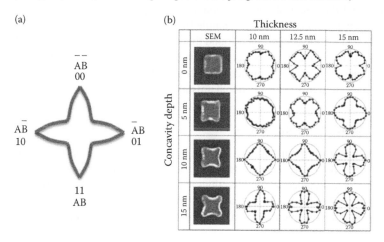

FIGURE 4.7 (a) Four-pointed star-shaped structure demonstrating biaxial anisotropy (From Lee, F., *IEEE Trans. Magn.*, 4, 502–506, 1968). (b) Scanning electron micrographs and anisotropy profiles of nanomagnet squares with concavities of varying depths and film thicknesses. The edge length of the squares is 100 nm. (With kind permission from Springer Science+Business Media: [*Applied Physics A*, Concave nanomagnets: investigation of anisotropy properties and applications to nanomagnetic logic, 111, 2013, 413–21, B. Lambson, Z. Gu, M. Monroe, S. Dhuey, A. Scholl, and J. Bokor © Springer. All rights reserved.])

4.6 APPLICATIONS OF NANOMAGNETS WITH BIAXIAL ANISOTROPY

4.6.1 MEMORY

Besides applications in logic, neuromorphic computing and image recovery, nanomagnetic elements possessing biaxial anisotropy (spatially separated from a uniaxially anisotropic, fixed ferromagnetic layer by an ultrathin nonmagnetic spacer) have also been used to exhibit voltage-controlled resistive switching by approximately 90° for giant magnetoresistance (GMR) (Pertsev and Kohlstedt 2010) and magnetic tunnel junction (MTJ) (Pertsev and Kohlstedt 2009) applications. For data storage purposes, the free ferromagnetic layer in the GMR scheme should possess low forward and backward switching fields when compared with the dielectric breakdown and ferroelectric coercive fields of the piezoelectric substrate. Cubic ferromagnetic materials such as Ni and $Co_{40}Fe_{60}$ are preferred for these applications owing to their high magnetostriction and weak magnetocrystalline anisotropy. This hybrid GMR scheme, shown in Figure 4.8, shows an electric-write nonvolatile magnetic memory cell with nondestructive readout.

Another technique used to modify the anisotropy of nanomagnets, similar to shape anisotropy, involves the introduction of multiple easy axes by introducing small modifications to the uniform magnetization of nanomagnets of a specific symmetric shape (Cowburn et al. 1999b). For instance, experiments were performed by Lambson et al. (2013) to study triangular-, square-, and pentagonal-shaped nanomagnets and the effect of configurational anisotropy on their magnetic properties. By modifying parameters such as sample thickness and concavity of an indentation introduced along the edges, it was found that the magnitude and direction of the easy axes can be individually adjusted (Figure 4.7b). NML devices that require ultrahigh energy efficiency and performance reliability could gainfully exploit the anisotropy control present in this configurational anisotropy scheme.

4.6.2 UNIVERSAL LOGIC GATES

As mentioned in Section 4.5.1, by introducing magnetocrystalline anisotropy in the magnetostrictive layer, the multiferroic nanomagnet can be made to possess four possible stable magnetization directions ("up," "right," "down," "left") in which four 2-bit combinations can be encoded ("00," "01," "11," "10"). Logic operations are carried out by implementing a *clock* that flips the nanomagnet's magnetization

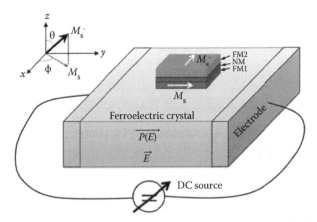

FIGURE 4.8 Ferromagnetic (FM) and nonmagnetic (NM) multilayer scheme (FM1/NM/FM2) fabricated on thick ferroelectric substrates for the electric control of giant magnetoresistance. (N. Pertsev, and H. Kohlstedt, Resistive switching via the converse magnetoelectric effect in ferromagnetic multilayers on ferroelectric substrates, *Nanotechnology* 21(47), 475202, 2010 © IOP Publishing. Reproduced with permission from IOP Publishing. All rights reserved.)

orientation in response to one or more inputs to generate the desired output. By manipulating the dipole interactions between neighboring nanomagnets, various types of logic gates can be implemented. The four-state scheme with multiferroic nanomagnets possessing biaxial magnetocrystalline anisotropy can be exploited to realize four-state NOR logic (D'Souza, Atulasimha, and Bandyopadhyay 2011a) (NOR represents a universal logic gate in digital circuits as it, similar to NAND logic, can be used to create all other logic gates). This is accomplished by applying a clock cycle consisting of a sequence of stresses to the output nanomagnet with an input nanomagnet on each side of it. Therefore, the final state of the output magnet is determined by its dipole interactions with the input magnets, whereas the stress cycle and an applied DC bias magnetic field provide the conditions necessary for NOR logic.

To understand how a four-state NOR gate is realized, consider a three-nanomagnet array (linear along the x axis) as shown in Figure 4.9a through d. A small DC bias magnetic field (pointing "up") is also applied, the significance of which will be explained later in this section. The input nanomagnets (AB, CD) are placed on either side of the output nanomagnet (EF). In this scenario, when two nanomagnets are

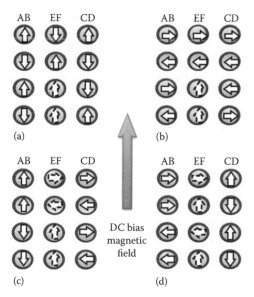

FIGURE 4.9 (a) The input combinations have magnetization directions perpendicular to the magnet array axis, resulting in the output direction having two possible orientations "up" or "down". (b) The magnetization directions of the inputs are parallel to the magnet axis. Consequently, the output orientations are either "left" or "right" or "up" (tie condition). (c) The left input magnet, AB, is either "up" or "down" whereas the right input, CD, is either "left" or "right." The output is, therefore, either "left" or "right" for nontie cases and "up" (determined by bias field) when a tie condition arises. (d) AB is either "left" or "right" whereas CD is "up" or "down." Similar to (c), the outputs are either "left," "right," or "up" (tie condition). (N. D'Souza, J. Atulasimha, and S. Bandyopadhyay, *J. Phys. D: Appl. Phys.*, 44(26), 265001, 2011 © IOP Publishing. Reproduced with permission from IOP Publishing. All rights reserved.)

placed next to each other, two types of dipole interaction arise. In the first case, when the magnetizations of both nanomagnets are perpendicular to the array axis (i.e., along the y axis), the dipole interaction favors an antiparallel ordering (or anti-ferromagnetic coupling) of adjacent magnetizations. The second type occurs when the magnetizations are parallel to the array axis (i.e., along the x axis). In this case, dipole interaction favors a parallel ordering (or ferromagnetic coupling) of adjacent magnetizations. These dipole interactions are exploited, along with a small global DC magnetic field to resolve tie situations, to realize a four-state NOR universal logic gate as explained in the following paragraph.

All combinations for the inputs, encoding bits AB and CD, along with the output EF required for NOR logic, are shown in Figure 4.9a through d. Four different scenarios are investigated, with each row representing a particular input combination. The input magnets are assumed to have fixed magnetizations whereas the output bit EF is subjected to a stress cycle (described in detail later in the this section) that causes its magnetization to rotate into its final, preferred orientation. In the first scenario (Figure 4.9a), the magnetization direction of the input magnets is perpendicular to the axis of the nanomagnet array. When both input magnetizations are oriented "up" (or "down") (first two rows in Figure 4.9a), the resulting dipole coupling favors a "down" (or "up") orientation of the output nanomagnet EF. When one input magnetization points "up" and the other points "down" (last two rows in Figure 4.9a), the output nanomagnet is in a *tied* or frustrated state. This is resolved by the small global DC magnetic field that forces the output magnetization to point "up." The second scenario, shown in Figure 4.9b, represents the case when the input magnetizations are parallel to the nanomagnet array axis. When both input magnetizations are oriented "left" or "right" (first two rows in Figure 4.9b), the dipole coupling favors a "left" or "right" output nanomagnet orientation, respectively. However, a tied state arises when one input magnetization points "left" and the other points "right" (last two rows in Figure 4.9b). In this case, the output nanomagnet ends up pointing neither "left" nor "right" but orients itself upward because of the global bias magnetic field. The third (Figure 4.9c) and fourth (Figure 4.9d) scenarios involve mixed inputs, where one input magnetization points perpendicular ("up/down") and the other parallel ("left/right") to the axis of the array, thereby favoring a "down/up" or "left/right" orientation of the output magnetization. If one of the inputs is "down," the bias field adds to the dipole field ensuring that the output is "up," whereas if one of the inputs is "up," the bias field counters the dipole field, causing the output to point either "left" or "right" depending on the second input. The importance of the magnitude of the bias field is explained later. A Karnaugh map (K-map) is used to simplify the logical expression of the output (EF), based on the binary representations of the input (AB, CD) and output (EF) configurations. The K-map output table is shown in Figure 4.10. Simplification yields the expressions $E = \bar{B} \cdot \bar{D}$ (equivalently, $E = \overline{B+D}$) and $F = \bar{A} \cdot \bar{C}$ (or $F = \overline{A+C}$), which represents NOR logic. Here, $\bar{A}$ represents the logical inverse (NOT) of A, "+" symbolizes the logical OR operation, and "·" the logical AND operation.

In this illustrative example, the dimensions of the nanomagnet were chosen to meet the following requirements: (1) the magnetocrystalline energy barrier of the nano-magnets was sufficiently high (~0.55 eV or ~22 kT at room temperature) to reduce

AB CD	00	01	11	10
00	11	01	00	10
01	01	01	00	00
11	00	00	00	00
10	10	00	00	10

E

F

EF

AB CD	00	01	11	10
00	1	0	0	1
01	0	0	0	0
11	0	0	0	0
10	1	0	0	1

AB CD	00	01	11	10
00	1	1	0	0
01	1	1	0	0
11	0	0	0	0
10	0	0	0	0

FIGURE 4.10 A Karnaugh-map representation of the input (AB, CD) combinations is illustrated, with the output *EF* indicated in the dotted rectangle, which is then separated into individual E and F sub-K-maps in order to determine their logical expressions and $F = \overline{A}\,\overline{C}$. (N. D'Souza, J. Atulasimha, and S. Bandyopadhyay, *J. Phys. D: Appl. Phys.*, 44(26), 265001, 2011 © IOP Publishing. Reproduced with permission from IOP Publishing. All rights reserved.)

the bit error probability because of spontaneous magnetization flipping, (2) the stress anisotropy energy (~1.5 eV) generated in the magnetostrictive Ni because of a strain of 500×10^{-6} transferred from the PZT layer could rotate magnetization by overcoming the magnetocrystalline anisotropy, (3) the dipole interaction energy (~0.2 eV) was kept lower than the shape anisotropy energy to prevent the magnetization from switching spontaneously without the application of the electric-field induced stress for clocking.

In studying the energetics of this scheme, detailed simulations were performed to confirm that the magnetization of the output magnet, independent of its initial orientation, always represents the NOR function of the inputs when the PZT layer is clocked with a voltage across its thickness to generate the proper strain cycle. The voltage strains the PZT layer via the d_{31} coupling. Here, the uniaxial tension or compression is always ensured by mechanically restraining expansion or contraction in the direction perpendicular to the $+45°$ direction. The same stress cycle could also be achieved by applying the electric field along the $+45°$ direction, which generates stress in that direction via the d_{33} coupling. For purposes of simulation, the total energy of the output nanomagnet when its magnetization vector subtends an angle θ_2 with the positive x axis is given by the following equation (Cullity and Graham 2009):

$$E_{\text{total}}(\theta_2) = \underbrace{\frac{\mu_0}{4\pi R^3}[M_s^2\Omega^2][-2\cos\theta_2(\cos\theta_3 + \cos\theta_1) + \sin\theta_2(\sin\theta_3 + \sin\theta_1)]}_{E_{\text{dipole}}}$$

$$+\underbrace{\frac{K_1\Omega}{4}\cos^2(2\theta_2)}_{E_{\text{magnetocrystalline}}} - \underbrace{\frac{3}{2}\lambda_{100}\sigma\Omega\cos^2(\theta_2 - \pi/4)}_{E_{\text{stress anisotropy}}} - \underbrace{\frac{\mu_0}{4\pi}[M_s\Omega]H_{\text{applied}}\sin\theta_2}_{E_{\text{magnetic}}}$$

(4.1)

where the first term represents the dipole interaction energy between the output magnet and its neighbors subtending angles θ_1 and θ_3 with the positive x axis, the second term represents the magnetocrystalline anisotropy energy with K_1 as the first-order magnetocrystalline anisotropy constant, the third term is the stress anisotropy energy because of stress applied along the [100] direction (45° with the x axis) with λ_{100} being the magnetostrictive constant in the direction of stress, and the final term is the interaction energy because of the DC bias field $H_{applied}$ pointing in the "up," or [110], direction. Here, μ_0 is the permeability of free space, M_s the saturation magnetization, Ω the nanomagnet volume, and R the distance between the centers of two adjacent nanomagnets. The stress σ is positive for tension and negative for compression.

As explained in Section 4.5.1, considering single-crystal Ni ($K_1< 0$) as the magnetostrictive layer, the magnetic easy axes are along <111> and hard axes are along <100>. Also, as it is assumed that the two-dimensional geometry of the nanomagnet precludes out-of-plane excursion of the magnetization vector because of a large magnetostatic energy penalty, the magnetization is confined to the (001) plane whereas [110], [1$\overline{1}$0], [$\overline{1}$$\overline{1}$0], and [$\overline{1}$10] are the unstressed/ground states, which respectively correspond to the 90°, 0°, −90°, and 180° orientations as shown in Figure 4.5b. When the input bit changes its state, the dipole and DC magnetic field interaction energies are not large enough to deflect the magnetization away from these minima. However, upon applying a stress, the magnetizations are pushed out of these minima. As the magnetostrictive coefficient λ_{100} of Ni is negative, a tensile stress along [100] rotates the magnetization to either the −45° or the +135° state (depending on which is closer to its initial state), whereas a compressive stress along [100] direction rotates the magnetization to either the −135° or +45° states. When stress is removed, the dipole interactions and applied DC bias field determine which of the two adjacent easy directions the output magnetization settles into. This realizes the NOR function. Note that to rotate the magnetization through 180°, one needs both a tensile and compressive stress cycle, with each half cycle producing a +90° rotation. However, applying this tension and compression cycle need not always cause a 180° rotation. The final amount of rotation is determined uniquely by the states of the two input magnets (left and right neighbors). This is highlighted for one particular scenario and is described next.

As mentioned earlier, the applied DC bias magnetic field plays an important role in determining NOR logic. The magnitude of this field (~1000 A·m^{-1}, applied along the $+y$ axis) is significant for all input combinations, particularly for the cases in Figure 4.9c and d, which have mixed inputs. This is because the dipole field (~1500 A·m^{-1}) acting along the magnet's $\pm x$ axis ($\theta = 0°$ or 180° due to ferromagnetic coupling along the magnet axis) has twice the magnitude of the dipole field (~750 A·m^{-1}) acting along the magnet's $\pm y$ axis ($\theta = 90°, 270°$; antiferromagnetic coupling perpendicular to the magnet axis). As the dipole energy term in $[-2\cos\theta_2(\cos\theta_3 + \cos\theta_1) + (\sin\theta_3 + \sin\theta_1)\sin\theta_2]$ suggests, the *cosine* term contributes twice as much to the dipole interaction energy as the *sine* term. As a result, the output nanomagnet would inherently favor ferromagnetic coupling (along the $\pm x$ axis) over anti-ferromagnetic coupling (along the $\pm y$ axis) by a factor of 2:1. The magnitude of the bias field (~1000 A·m^{-1}) is chosen to compensate for this inequality. If $H_{applied}$<< 1000 A·m^{-1} (for instance ~250 A·m^{-1}), the parallel component would dominate (final two rows of Figure 4.9c) the perpendicular component, and force a rotation along the x axis instead of the desired "up" state, as required by the NOR logic scheme. However, if the DC bias field is too large (>1500 A·m^{-1}), it would force a magnetization rotation to the "up" direction irrespective of the states of the inputs, thereby invalidating the NOR logic scheme.

Equation 4.1 is used to calculate the total potential energy of the output nanomagnet EF as a function of θ_2 when subjected to the stress cycle: Tension (T) → Relaxation (R) → Compression (C) → Relaxation. This sequence is carried out in increments/decrements of 0.1 MPa up to the maximum stress amplitude of 100 MPa. One particular case is described here, in which the inputs, AB and CD, are "right" ("01") and "down" ("11"), respectively. The initial orientation of the output nanomagnet *EF* is taken to be "down" ("11"), although any other starting direction would also result in the correct final state for NOR logic. The energy profiles of *EF* are shown in Figure 4.11 when the stress cycle (T→R→C→R) is applied to it. The "*" markers represent the magnetization orientation of the nanomagnet at the corresponding stress value, whereas the dotted lines represent the energy profile (circle indicates the

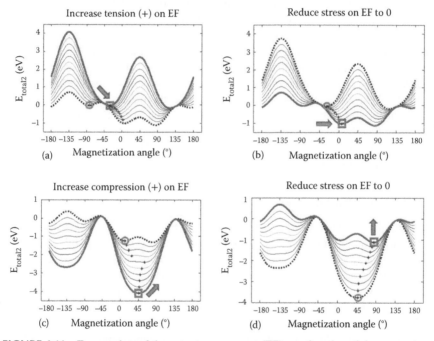

FIGURE 4.11 Energy plots of the output nanomagnet (EF) as a function of the magnetization angle. The initial conditions used are: AB = "01," CD = "'11," and EF = "11." (a) With no stress applied to magnet EF initially, the magnetization direction begins at −90°. The first stage of the stress cycle (tension at +45°) is then initiated, which causes the magnetization to rotate away from the stress axis to the closest energy minima (−45°). (b) Upon relaxation of stress on EF (stage 2), the closest energy minimum is at 0° and therefore, the magnetization rotates into that position. (c) Stage 3 of the stress cycle involves the application of a compressive stress (negative, +45°) on EF. The energy minima are located along the stress axis; therefore, the magnetization rotates and settles at +45°. (d) Finally, the stress on EF is relaxed, which causes the magnetization to rotate to the closest energy minimum (+90°). The arrows indicate the approximate magnetization orientation at the end of a particular stage. The "*" markers represent the magnetization orientation at a particular stress; the circle and square depict the initial and final magnetization states of EF at the beginning and at the end of the stress cycle, respectively.

magnetization orientation) of EF at the onset of a particular stress cycle whereas the thick, solid line represents the energy landscape (square highlights the magnetization orientation) of EF at the end of that particular stress cycle. The thin solid lines represent the energy profiles of EF at intermediate stress values in the cycle.

In Figure 4.11a, when tension (+100 MPa, in increments of 0.1 MPa) is applied to EF along the [100] direction ($\theta_2 = +45°$), the magnetization rotates from its initial "down" state ($\theta_2 = -90°$) and settles at approximately −40° (the dipole coupling and the bias magnetic field result in the slight shift away from −45°). This is because Ni has negative magnetostriction and, thus, a tensile stress (positive) applied along the +45° direction tends to create energy minima along a direction perpendicular to it (+135° and −45°). As −45° is closer to the initial orientation (−90°), the magnetization rotates toward that energy minimum instead of that at +135°.

The tensile stress is gradually relaxed to zero (in 0.1 MPa decrements) in the next stage (Figure 4.11b), causing a magnetization rotation from approximately −40° to approximately 0°. In the subsequent stage, a compressive stress (negative) of −100 MPa is gradually applied to EF (Figure 4.11c) along the [100] direction (+45°), creating an energy minima along the stress axis (+45° and −135°). As +45° is closest to the previous orientation (~0°), the magnetization rotates toward that axis. In the final stage of the stress cycle (Figure 4.11d), the stress is gradually relaxed to zero, causing the magnetization to settle to the desired orientation along the +90° direction, thereby demonstrating successful NOR operation for this case. All other cases have been comprehensively studied to verify NOR logic functionality (see http://stacks.iop.org/JPhysD/44/265001/mmedia).

4.6.3 Logic Propagation

Although four-state NOR logic was discussed in Section 4.6.2, a novel scheme for propagation of this multistate logic is discussed here (D'Souza, Atulasimha and Bandyopadhyay 2011b). Its feasibility is shown by modeling the rotation of each nanomagnet's magnetization subjected to a cycle of tensile and compressive stresses generated by positive and negative electrostatic potentials applied across the PZT layer of each multiferroic nanomagnet.

When the magnetizations of two adjacent nanomagnets are parallel to the x axis, the ordering will be ferromagnetic (F), but when the magnetizations are perpendicular to the x axis, the ordering will be antiferromagnetic (AF) because of dipole interaction. As a result, if the magnetization orientation of the first bit in a linear array (along the x axis) is switched from its initial state to one of the three other stable states, three possible arrangements result. As each nanomagnet has four possible magnetization orientations, there are 12 distinct configurations that may arise when the first bit is switched, as shown in Figure 4.12. The remaining four configurations are trivial cases representing the situation when the input bit is not flipped, and hence are not included in this discussion.

Consider the AF arrangement of Figure 4.12a. The magnetization orientation of the first nanomagnet acts as the input bit to the line. In this configuration, the input magnetization can be switched from its initial "up" state to the "down," "right," or "left" state. The *final state* column represents the corresponding nanomagnet states

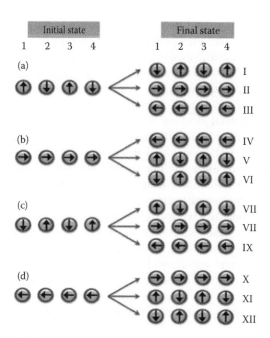

FIGURE 4.12   The 12 distinct scenarios encountered during logic propagation. The "Initial State" column represents the "ground" state of a four-magnet array or "wire" with magnet 1 taken as the input. The "Final State" column corresponds to the expected state of the wire when the input magnet is switched from its initial state to any of the three other possible states. (a) Antiferromagnetic arrangement with input = "up," (b) ferromagnetic array with input = "right," (c) antiferromagnetic arrangement with input = "down," and (d) ferromagnetic configuration with input = "left." (© 2012 IEEE. Reprinted, with permission, IEEE, N. D'Souza, J. Atulasimha, and S. Bandyopadhyay, An energy-efficient Bennett clocking scheme for 4-state multiferroic logic, *IEEE Trans. Nanotech.*, 11(2), 418–425, 2012.)

based on whether F-coupling (along the nanomagnet array axis) or AF-coupling (perpendicular to this axis) occurs. Therefore, when the input bit is flipped from "up" to "down," the change is propagated along the array, and with the appropriate clocking scheme, the input magnetization direction is replicated in every odd-numbered nanomagnet from the left as a result of AF ordering. If the input is switched to either "left" or "right," F-coupling will ensure that all the nanomagnets assume the "left" or "right" orientation, respectively. Similar switching mechanisms apply to the other three configurations in Figure 4.12b through d. Here, only numerical results corresponding to row VII in the arrangements of Figure 4.12c are presented. These pertain to the AF arrangement with the input magnetization oriented "down." All other cases have been exhaustively examined to confirm successful operation, and are discussed in detail in the simulations conducted by D'Souza et al. (2011c).

Figure 4.12c shows that in an AF-coupled array, the first bit is replicated in every odd-numbered nanomagnet (and has therefore propagated through the line) if the array *can reach ground state* once the first bit is flipped. This scenario can occur only if the array does not get stuck in a metastable state, which prevents it from reaching the ground state. It can be shown that dipole interaction itself cannot guarantee that the array will reach the ground state. That makes it necessary to use a multiphase clock, which will drive the system out of any potential metastable state. The dipole interaction energy is also usually not sufficient to overcome the magnetocrystalline anisotropy energy and rotate a nanomagnet out of its current orientation to a different orientation to propagate the bit. This is another reason why a clock is necessary so as to supply the energy needed to overcome the magnetocrystalline anisotropy energy barrier. In Bennett clocking schemes, the clocking agent (local magnetic field, STT, strain, etc.) will rotate the magnetization into an unstable state, usually an energy maximum. The dipole interaction with its neighbors will push it into the desired stable state and ensure unidirectional propagation of a logic bit. The effect of thermal fluctuations, which will induce switching errors at room temperature, is beyond the scope of the analysis performed here, but it is never so harmful that the basic scheme is invalidated. The clock cycle and stress sequences required to propagate a logic bit along a nanomagnet array are discussed in the following paragraph.

Consider the nanomagnet array of Figure 4.13 consisting of four nanomagnets in their respective ground states (row I). The magnetization of nanomagnet 1 (column 1, leftmost nanomagnet) is the input bit. It is flipped from its initial "up" to the "down" state at time $t = 0$. Subsequently at time $t = 0+$ (row II), nanomagnet 2 experiences equal and opposite dipole interactions from its two nearest neighbors

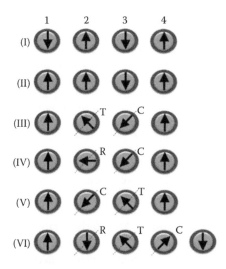

FIGURE 4.13 The clock cycle and stress sequences involved are demonstrated on the antiferromagnetic case when the input is switched from its initial "down" (row I) to the "up" state, which results in a tie-condition (row II). To counteract this, a four-stage "clock" scheme is applied to magnets 2 and 3 (rows III–VI) consisting of tension (T), compression (C) and relaxation (R). The stress sequence applied to magnet 2 is T → R → C → R, whereas magnet 3 undergoes a C → C → T → T stress cycle. At the end of a single clock cycle (row VI), the magnetization of magnet 2 is rotated from its initial "up" state (rows I and II) to a "down" state. The exact clock cycle is then repeated on the next set of magnets in the array (3 and 4). (© 2012 IEEE. Reprinted, with permission, IEEE, N. D'Souza, J. Atulasimha, and S. Bandyopadhyay, An energy-efficient Bennett clocking scheme for 4-state multiferroic logic, *Nanotechnology, IEEE Trans.*, 11(2), 418–425, 2012.)

(nanomagnets 1 and 3) that are magnetized in opposite directions. The net dipole interaction experienced by nanomagnet 2 is, therefore, zero. Because the nanomagnet would not flip its magnetization in response to the change in the first nanomagnet's bit, propagation of the input bit along the array is prevented and it is stuck in a *metastable state*, unable to reach the ground state.

To break this deadlock and allow the logic bit to propagate beyond nanomagnet 2, the following clock cycle is applied: T → R → C → R (to nanomagnet 2) and C → C → T → T (to nanomagnet 3). Nanomagnets 1 and 4 are assumed to be stiff (a good estimation if the magnetocrystalline anisotropy energy is significantly larger than the dipole interaction energy), whereas nanomagnets 2 and 3 rotate when stressed. Also, note that the stress is applied along the [100] direction (+45° to the x axis).

Stage 1 (Tension [T]/Compression [C] [Row III]): After the input nanomagnet is switched, nanomagnet 2 is subject to a tensile stress (in steps of 0.1 MPa up to a maximum value of +100 MPa). As Ni has negative magnetostriction, a tensile stress raises the energy along the axis of applied stress whereas lowering the energy along the axis perpendicular to this direction; a compressive stress does the exact opposite. Therefore, the tension that is applied to nanomagnet 2 along the [100] direction will result in energy minima (easy axis) along the −45° and +135° (−225°) directions whereas raising the energy barrier in the +45°and −135° (+225°) directions. Because the initial state of nanomagnet 2 is along the 90° direction and the energy barrier is raised along the −135° (+225°) direction, the only possible magnetization rotation that can take place is from +90° to +135°. Energy profiles showing the raising and lowering of energy levels of nanomagnets 2 and 3 are shown in the simulation results in the work of D'Souza et al. (2011c). A compressive stress is simultaneously applied to nanomagnet 3 (gradually increased to a maximum value of −100 MPa) along the [100] axis. This causes its magnetization to rotate from the initial −90° state to the −135° state (row III). In all cases, stresses are simultaneously applied on nanomagnets 2 and 3.

Stage 2 (Relaxation [R]/Compression [C] [Row IV]): In this next step, the tensile stress on nanomagnet 2 is gradually reduced to zero while keeping the compressive stress on nanomagnet 3 fixed. The magnetization of nanomagnet 3 remains oriented in the −135° direction, but the magnetization of nanomagnet 2 rotates from +135° to +180°. These rotations take place so as to lower the energy of a nanomagnet to the minimum energy state.

Stage 3 (Compression [C]/Tension [T] [Row V]): A compressive stress (up to −100 MPa) is now applied on nanomagnet 2. The compressive stress on nanomagnet 3 is simultaneously relaxed to zero and is immediately followed by the application of a tensile stress (up to +100 MPa, on nanomagnet 3) while still keeping the compressive stress on nanomagnet 2 on. Nanomagnet 2 rotates to the preferred lowest energy state along −135°. The relaxation of stress on nanomagnet 3 pushes its magnetization to +180° (ferromagnetic coupling is preferred over AF coupling as the former has a stronger dipole interaction), whereas the subsequent tensile stress results in rotation of the magnetization to +135°.

Stage 4 (Relaxation [R]/Tension [T] [Row VI]): Finally, the compressive stress on nanomagnet 2 is relaxed while keeping the tensile stress on nanomagnet 3 fixed. The magnetization of nanomagnet 2 then rotates to its final desired state of −90° ("down"). Thus, the clocking sequence successfully flips the magnetization of nanomagnet 2 in response to the flipping of the input nanomagnet 1 and allows the logic bit to propagate past nanomagnet 2. The same sequence of stresses is then applied to the next set of nanomagnets (3 and 4, with 2 and 5 now assumed to be stiff, etc.), which results in nanomagnet 3 eventually settling in the "up" orientation (+90°), mirroring the state of the input bit. By continuing this cycle, the input bit can be propagated down the entire chain, resulting in successful unidirectional logic propagation.

4.6.4 IMAGE RECONSTRUCTION AND PATTERN RECOGNITION

Besides doubling the logic density of conventional two-state logic devices, the four-state multiferroic scheme with biaxial magnetocrystalline anisotropy can be exploited for complex tasks such as image recovery and recognition, associative memory, and neuromorphic computing. In this section, image recovery, after a target image has been corrupted by noise, is discussed. It is implemented through the use of planar nanomagnetic elements (single-crystal magnetic layer and not multiferroic element). In addition, an image recognition scheme is also described (D'Souza, Atulasimha, and Bandyopadhyay 2012). In all these schemes, it is the physics of magnetization dynamics that elicits signal processing activity. Thus, these are *all-hardware* implementations with no involvement of any software. Consequently, these schemes are several orders of magnitude faster than traditional approaches that are hamstrung by the execution of software instruction sets. The disadvantage is, of course, that processors built on this principle are invariably application-specific integrated circuits (ASICs) with limited scope for reconfiguration. However, that is a small price to pay for unprecedented speed and energy saving.

Consider a single-crystal, single-domain nanomagnet (Cowburn et al. 1999a) in the shape of a circular disk having no in-plane shape anisotropy, but possessing biaxial magnetocrystalline anisotropy (Cullity and Graham 2009). This can be fabricated by growing an epitaxial (001) film using MBE (Naik et al. 1993; Chow 1991). The nanomagnet would then have four stable magnetization directions (D'Souza, Atulasimha, and Bandyopadhyay 2011a) ("up," "right," "down," "left") as shown in Figure 4.14a. An array of such nanomagnets can be used to store a black–gray–white (or any three-color) image by storing each pixel's shade in one of the four stable states of a nanomagnet. Thus, a two-dimensional image can be stored in a two-dimensional array of nanomagnets. If this stored image is corrupted by moderate noise that deflects the magnetizations of the nanomagnets from their stable orientations by small angles, the ensuing magnetization dynamics automatically recovers the stored image by bringing the magnetizations back to their original orientations (D'Souza, Atulasimha, and Bandyopadhyay 2012). Therefore, there is an inherent built-in error correction. When integrated with MTJs, the same array can be used to recognize images by comparing them with stored images pixel by pixel (D'Souza, Atulasimha, and Bandyopadhyay 2012). As the image recovery is performed in parallel and is completely hardware based, these tasks can be executed at ultrahigh speeds.

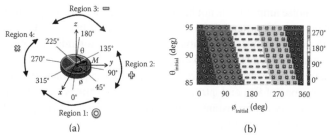

FIGURE 4.14 (a) The spherical coordinate system used to define the magnetization vector and the shading scheme used to designate quadrants. (b) Plot of the magnetization vector's final state for all allowed values of the initial state ($\theta_{initial}$, $\varphi_{initial}$). (© 2012 IEEE. Reprinted, with permission, IEEE, N. D'Souza, J. Atulasimha, and S. Bandyopadhyay, 1. D'Souza, N., Atulasimha, J. & Bandyopadhyay, S., An ultrafast image recovery and recognition system implemented with nanomagnets possessing biaxial magnetocrystalline anisotropy. *Nanotechnology, IEEE Trans.*, 11(5), 896–901, 2012.)

The four stable states of a magnet encode the colors black, gray, white, and gray. Although two different shades of gray could be encoded in the four-state scheme, the three-shade encoding is used for simplicity and symmetry. Noise can corrupt the stored image by perturbing the magnetization vector and deflecting it away from its initial stable state. The Landau–Lifshitz–Gilbert (LLG) equation is used to study the resulting magnetization dynamics to determine the final state of every perturbed magnet (or pixel). For this purpose, it is assumed that the magnets are spaced far enough (~1 μm) so that dipole interaction between them can be ignored. By simulating the magnetization dynamics of each four-state magnet, image recovery is demonstrated in a 512 × 512 pixel black–white–gray image in only approximately 2 ns. By contrast, conventional complementary metal–oxide–semiconductor (CMOS)–based image filters would have taken several microseconds to perform the same task (Kajiwara et al. 2010).

To show image recovery, an isolated circular nanomagnet of Ni with diameter 100 nm and thickness 10 nm is considered. As described in Sections 4.6.1–4.6.4, when considering a thin film of Ni in the (001) plane, its easy axis of magnetization lies along the <110> direction, with the hard axis along the <100> direction (D'Souza, Atulasimha, and Bandyopadhyay 2011a; D'Souza, Atulasimha and Bandyopadhyay 2011b). The coordinate system used to study the magnetization dynamics of this scheme is shown in Figure 4.14a. The easy axes of single-crystal Ni in the ground/unstressed state are [110], [1$\bar{1}$0], [$\bar{1}\bar{1}$0], and [$\bar{1}$10]), in Miller notation. Therefore, there are four possible (degenerate) energy minima in which four 2-bit combinations can be encoded (00, 01, 11, 10). The coordinate axis system is rotated by 45° in the magnet's plane (i.e., $\varphi \to \varphi - \pi/4$), where φ is the angle subtended by the in-plane component of the magnetization vector with the $+x$ axis. This rotational transformation ensures that the energy minima are along the $\pm x$-axes ($\varphi = 0°$ or $\pm 180°$) and $\pm y$-axes ($\varphi = \pm 90°$). The angle subtended by the magnetization vector with the z axis is θ. When $\theta = 90°$, the magnetization vector is in the plane of the magnet.

If noise deflects the magnetization vector of a magnet away from its initial stable state to a new state, the latter evolves with time in accordance with the LLG equation.

As long as the noise amplitude is not large enough to deflect the magnetization vector closer to another stable orientation or cause large out-of-plane excursion, it always returns to the initial stable orientation. The timescale for this operation is on the order of a few nanoseconds (for realistic parameters).

The dynamics of the system is governed by the LLG equation that determines the temporal evolutions of $\theta(t)$ and $\varphi(t)$ and therefore determines the orientation of the magnetization vector at any instant of time. The total magnetic energy of the single-domain nanomagnet is given by the equation:

$$
\begin{aligned}
E_{\text{total}}(t) &= E_{\text{magnetocrystalline-anisotropy}}(t) + E_{\text{shape-anisotropy}}(t) \\
&= \frac{K_1}{4}\sin^2 2\theta(t) + \left[\frac{K_1}{4}\sin^4\theta(t) + \frac{K_2}{16}\sin^2\theta(t)\sin^2 2\theta(t)\right] \\
&\quad \cos^2 2\phi(t) + \frac{\mu_0}{2}M_s^2\left[N_z\cos^2\theta(t) + N_x\sin^2\theta(t)\right]
\end{aligned}
\tag{4.2}
$$

where K_1 and K_2 are the first- and second-order magnetocrystalline anisotropy constants, respectively. The permeability of free space is represented by μ_0, M_s is saturation magnetization and N_x, N_y, and N_z the components of the demagnetization factor along the x, y, and z axis, respectively. As a circular nanomagnet is considered, the values of N_x and N_y are identical. These demagnetizing factor components are determined from the oblate spheroid estimation (Cullity and Graham 2009). The magnetization M is assumed to have a constant magnitude with its motion described by a unit direction vector $n_m = M/|M| = \hat{e}_r$ at any instant of time, where $\hat{e}_r$ is the unit vector in the radial direction of the spherical coordinate system and makes an angle θ with the z axis.

The torque experienced by the magnetization per unit volume with the potential energy given in Equation 4.2 can be written as follows:

$$
T_E(t) = -n_m \times \nabla E_{\text{total}}\big(\theta(t),\phi(t)\big) = -\hat{e}_r \times \nabla E_{\text{total}}(\theta(t),\phi(t))
\tag{4.3}
$$

To study the dynamics of the magnetization, M, the following LLG equation is solved.

$$
\frac{dn_m(t)}{dt} \quad \alpha\left(n_m(t) \times \frac{dn_m(t)}{dt}\right) = -\frac{2}{hM_s\Omega}\,^{B}\,T_E(t)
\tag{4.4}
$$

where α is the Gilbert damping coefficient (0.045 for Ni), μ_B the Bohr magneton constant, and Ω the volume of the nanomagnet. Equations 4.2 through 4.4 are solved, beginning with the initial values of θ and φ, to find their values $[\theta(t), \varphi(t)]$ at any instant of time t.

The states $\varphi = 45°$, $135°$, $225°$, and $315°$ are the highest energy states when $\theta = 90°$. If the magnetization is driven to any of these states by noise (unlikely as these are the highest energy states), it has equal probability of decaying to either one of its two neighboring minimum energy (stable) states. Therefore, a region of $1°$ is excluded around these critical points.

In the simplest case, when the magnetization vector lies in-plane ($\theta = 90°$, no out-of-plane excursion), the four regions are depicted by different shades, as shown in Figure 4.14a. The plot of Figure 4.14b shows the magnetization vector's final, settled state for all values of $\varphi_{initial}$ ($0°$ to $360°$) and with $\theta_{initial}$ between $85°$ and $95°$. It can be seen that for $\theta_{initial} = 90°$, the final states for the four regions $316° <\varphi_{initial}< 44°$, $46° <\varphi_{initial}< 134°$, $136° <\varphi_{initial}<224°$, and $226° <\varphi_{initial}< 314°$ are $0°$, $90°$, $180°$, and $270°$, respectively. However, when $\theta_{initial} = 89°$, the regions shift by $+5°$: ($320° <\varphi_{initial}< 50°$), ($41° <\varphi_{initial}< 131°$), ($131° <\varphi_{initial}< 221°$), and ($221° <\varphi_{initial}< 311°$). Similarly, when $\theta_{initial} = 91°$, a shift of $-5°$ is seen. This shift occurs because the out-of-plane excursion of the magnetization vector generates an additional precessional torque acting on the magnetization vector owing to the coupled $\theta - \varphi$ dynamics. As long as the out-of-plane excursion $\Delta\theta_{initial}$ ($\theta_{initial} - 90°$) is small, the shift is linearly proportional to $\Delta\theta_{initial}$. For $\Delta\theta_{initial} > 5°$, this shift increases nonlinearly, as shown in Figure 4.15, where the dynamics of the magnetization vector are examined when it is initially out of plane for various values of $\theta_{initial}$ ($85°$, $81°$, $77°$, $72°$, $10°$) and $\varphi_{initial} = 50°$. The plot of Figure 4.15a shows this nonlinear behavior of the magnetization vector's final orientation whereas Figure 4.15b through f show the trajectories of the magnetization vectors for each scenario. For the case with $\theta_{initial} = 85°$ (Figure 4.15b), the tip of the magnetization vector initially lies at $\varphi = 50°$ and precesses around and finally settles at $\varphi = 0°$ ($x = 1$, $y = 0$, $z = 0$) and not at the desired position ($\varphi = 90°$). This can be attributed to an additional counterclockwise torque that the magnetization vector experiences, resulting in it straying into the ambit of another stable state and settling at $\varphi = 0°$. If the magnetization vector is initially out of plane by $-9°$ ($\theta_{initial} = 81°$), the final settled state of the magnetization vector is at the incorrect state of $\varphi = -90°$ ($x = 0$, $y = -1$, $z = 0$), as shown in Figure 4.15c. In Figure 4.15d, the initial out-of-plane direction of the magnetization vector is $\theta_{initial} = 77°$. This results in an even greater torque than that experienced by the magnetization in the earlier two cases, resulting in the magnetization settling at $\varphi = 180°$ ($x = -1$, $y = 0$, $z = 0$). For the case with $\theta_{initial} = 72°$ (Figure 4.15e), the magnetization rotates in a counterclockwise direction from $\varphi_{initial} = 50°$ to finally precess and settle at $\varphi = 90°$. Although this is the desired final state, the magnetization vector precesses in the incorrect direction, with the torque causing the magnetization to rotate past the other three undesired states before fortuitously settling into the correct state. Figure 4.15f shows the erratic magnetization vector precession for the case when it is initially out of plane by a large degree ($\theta_{initial} = 10°$). When the magnetization initially lifts out of plane in the other direction ($\theta_{initial} > 90°$), the magnetization experiences an increased torque in the clockwise direction, making it more prone to switching to an incorrect state in that direction, based on the extent of the initial out-of-plane excursion.

Figure 4.16a shows the color scheme used to encode pixel shades in a black/gray/white image. The magnetization's orientation encodes three different shades as follows: $\varphi = 0°$ (black), $\varphi = 90°$ (gray), $\varphi = 270°$ (gray), and $\varphi = 180°$ (white). The shades can also be assigned numerical values: black = 0, gray = 0.5, and white = 1. A 512×512 pixel image (Figure 4.16b) is then encoded according to this scheme and the numerical value of each pixel is randomly varied to simulate the effect of noise (Figure 4.16d). The out-of-plane excursion is limited to $\pm1°$, as this allows for a reasonable amount of azimuthal deflection. Therefore, the range of in-plane deflection that can be corrected by this scheme is then restricted to $\pm40°$. The ensuing

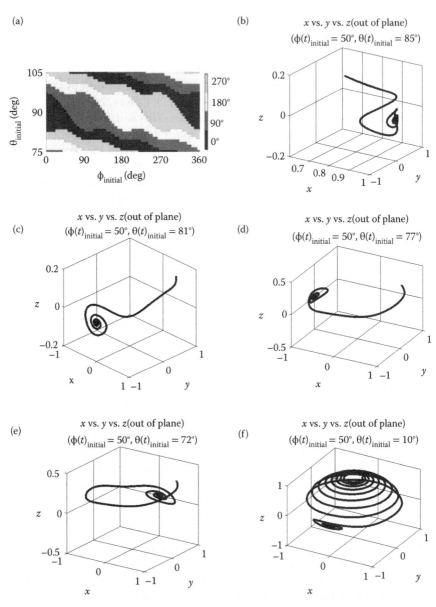

FIGURE 4.15 (a) Plot illustrating a nonlinear shift in the final settled state (region) of the magnetization vector for $\Delta\theta_{initial} > 5°$. Dynamics (trajectory) of the magnetization vector with $\varphi_{initial} = 50°$ and $\theta_{initial} =$ (b) 85°, (c) 81°, (d) 77°, (e) 72°, and (f) 10°.

magnetization dynamics for the image (focusing on the eye for clarity) is shown in Figure 4.16d through f, with the final image recovered in approximately 2 ns.

This scheme can be used to perform image recognition as well, as shown in Figure 4.17. Reading and writing of data to each pixel is carried out by incorporating synthetic antiferromagnets (SAFs) to create an MTJ–SAF scheme in which each Ni

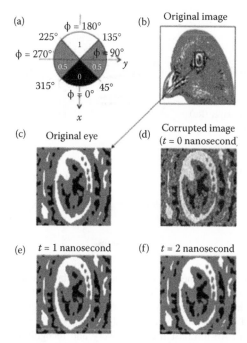

FIGURE 4.16 (a) Pixel's shade is given a value of 0 (black), 0.5 (gray), or 1 (white). (b) Original 512 × 512 pixel image. Each pixel's shade is encoded into the magnetization state of the corresponding magnet. (c) Eye of the bird is blown up for clarity. (d) Noise corrupts the image at time $t = 0$ by deflecting the magnetization vectors from their initial states. The out-of-plane excursion is restricted to 1°. (e) Partially recovered image at $t = 1$ ns. (f) Fully recovered image at $t = 2$ ns. (© 2012 IEEE. Reprinted, with permission, IEEE, N. D'Souza, J. Atulasimha, and S. Bandyopadhyay, 1. D'Souza, N., Atulasimha, J. & Bandyopadhyay, S., An ultrafast image recovery and recognition system implemented with nanomagnets possessing biaxial magneto-crystalline anisotropy. *Nanotechnology, IEEE Trans.*, 11(5), 896–901, 2012.).

nanomagnet is sandwiched between two SAF layers—the bottom SAF layer is used for the read operation whereas the pixel-write operation occurs through the top SAF layer. When writing to a pixel, an appropriate spin-polarized current is amplified and passed through the upper SAF layer depending on the incoming input electrical signal or "color" ("white," "gray," or "black") from a charge-coupled device (CCD). The spin-polarized current, in turn, imparts the corresponding amount of spin-angular momentum to the magnetization in the Ni layer based on the color (white, gray, or black), causing it to rotate to an appropriate state (based on the color scheme of Figure 4.16a).

4.6.5 NEUROMORPHIC COMPUTING

Although Section 4.6.4 discussed a specific image reconstruction and pattern recognition attribute of four-state magnets with biaxial anisotropy, the ultimate aim is to develop a more generalized computing architecture. The key is to exploiting certain functionalities of four-state multiferroic architecture, such as minimum

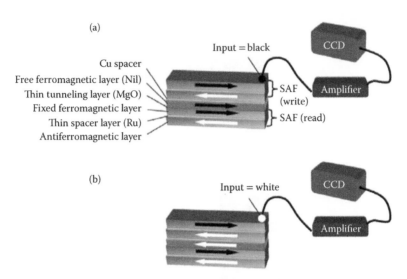

FIGURE 4.17 MTJ-SAF scheme used to read (lower SAF layer) and write (upper SAF layer) images to each pixel (Ni layer). (a) CCD sends a "black" input to be written to the pixel. (b) CCD sends a "white" input to be written. (© 2012 IEEE. Reprinted, with permission, IEEE, N. D'Souza, J. Atulasimha, and S. Bandyopadhyay, 1. D'Souza, N., Atulasimha, J. & Bandyopadhyay, S., An ultrafast image recovery and recognition system implemented with nanomagnets possessing biaxial magnetocrystalline anisotropy. *Nanotechnology, IEEE Trans.*, 11(5), 896–901, 2012.)

threshold stress/dipole energy for flipping (neuron firing); dipole coupling acting as *synapse* between neighboring nanomagnets (neurons) to realize these neuromorphic architectures.

Neuromorphic hardware implemented with transistors and other charge-based devices such as memristors (Snider 2011) has been stymied by the difficulty of implementing synapses with low power consumption. Memristors are extremely energy hungry and their switching characteristics do not show the sharp threshold behavior required of neurons. On the other hand, the "transfer function" of a multiferroic nanomagnet exhibits strong threshold behavior. Here, the input function is the stress or the voltage applied to the nanomagnet, whereas the output function is the magnetization. When the stress is high enough that the stress anisotropy energy overcomes the shape anisotropy energy, the magnetization switches *abruptly*. Figure 4.18a shows the magnetization orientations of the second and third magnets in the four-magnet chain of Figure 4.2b when the third magnet is unstressed and the second magnet is gradually stressed. Figure 4.18b shows the equivalence between a four-state multiferroic logic element and their dipole coupling and a *neuron* and *synapse*, respectively.

The threshold behavior can be effectively used for neurons. The voltage of the PZT layer could be the sum of signals from source neurons and the magnet can be the sink neuron that fires (magnetization changes suddenly) when the voltage on the PZT layer exceeds a threshold. The advantage of this implementation is that the voltage required for firing is extremely small (few mV), which makes these neurons

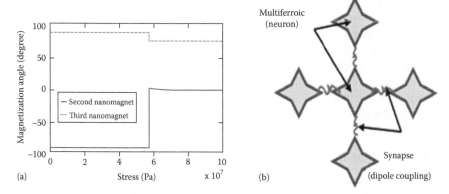

FIGURE 4.18 (a) Magnetization angles (angle subtended by the magnetization vector with the hard axis) of the second and third magnets in the chain of Figure 4.2b as a function of stress when the third magnet is unstressed and the second is gradually stressed. (b) Dipole coupled multiferroic nanomagnets as elements of a neuromorphic network. (Based on Snider, G., *Nanotechnology*, 22(1), 015201, 2011.)

comparable to those in the human cortex. Further, this threshold will change when the dipole coupling is different as the magnitude and direction of the effective dipole fields because of the neighbors can help switching or make it harder (need more stress and consequently voltage) to switch. This is how the *dipole coupling* will act as synapses between the neurons.

4.7   SUMMARY AND CONCLUSIONS

In conclusion, a brief overview of extremely energy-efficient multiferroic nanomagnetic switches, comprising piezoelectric and magnetostrictive layers, was provided. By incorporating magnetocrystalline anisotropy in the magnetostrictive layer or with the use of geometry to create appropriate shape anisotropy, these multiferroic elements can be made to possess biaxial anisotropy and, thereby, exhibit four stable magnetization orientations that can store and encode four bits of information in one element. This scheme can be exploited to create a universal four-state NOR logic gate and a "logic wire" to propagate this information. Through the use of materials such as Terfenol-D (high magnetomechanical coupling), these operations can be made ultraenergy efficient with the energy dissipated in each clocking cycle made as low as approximately 100 kT/bit. However, one key issue is that magnetization dynamics at room temperature tends to be error-prone, and dipole-coupled magnetic logic architectures may not be sufficiently reliable because dipole strengths are usually too weak to ensure reliable switching of the magnets in the presence of thermal noise (Spedalieri et al. 2011; Carlton et al. 2012; Salehi Fashami et al., unpublished). This realization has led some groups to investigate alternate non-Boolean logic schemes that are more forgiving of errors. These include neuromorphic architectures (Snider 2011) and Bayesian computation (S. Khasnavis et al., unpublished data), which are intrinsically more error tolerant than Boolean logic. Thus, the greater potential for application of four-state nanomagnets with biaxial anisotropy in the future is in non-Boolean computing applications.

These include their use as neuromorphic elements as described in Section 4.3, in Bayesian computation (S. Khasnavis et al., unpublished data), associative memory (Roychowdhury et al. 1996), and ASICs such as for ultrafast image recovery and recognition chips as described in Section 4.3. In such applications, the ultralow energy dissipation in nanomagnets and their ability to implement certain functionalities (such as a neuron) could make them triumph over CMOS technology, which is excessively dissipative. Ultimately, this may play a key role in perpetuating Moore's law downscaling of computing devices beyond the year 2020.

ACKNOWLEDGEMENTS

All authors acknowledge the support of the National Science Foundation (NSF) under the NEB2020 Grant ECCS-1124714 and SHF grant CCF-1216614 as well as the Semiconductor Research Company (SRC) under NRI task 2203.001. J. Atulasimha acknowledges support from NSF CAREER grant CCF-1253370.

REFERENCES

Alam, M. T., M. Siddiq, G. H. Bernstein, M. T. Niemier, W. Porod, and X. S. Hu. 2010. On-chip clocking for nanomagnet logic devices. *Nanotechnology, IEEE Transactions On*, 9(3): 348–351.

Atulasimha, J. and S. Bandyopadhyay. 2010. Bennett clocking of nanomagnetic logic using multiferroic single-domain nanomagnets. *Applied Physics Letters* 97: 173105.

Bandyopadhyay, S. and M. Cahay. 2009. Electron spin for classical information processing: A brief survey of spin-based logic devices, gates and circuits. *Nanotechnology* 20(41): 412001.

Bandyopadhyay, S., B. Das, and A. E. Miller. 1994. Supercomputing with spin-polarized single electrons in a quantum coupled architecture. *Nanotechnology* 5(2): 113–133.

Bandyopadhyay, S. and V. Roychowdhury. 1996. Computational paradigms in nanoelectronics: Quantum coupled single electron logic and neuromorphic networks. *Japanese Journal of Applied Physics* 35(Part 1, no. 6A): 3350–3362.

Behin-Aein, B., D. Datta, S. Salahuddin, and S. Datta. 2010. Proposal for an all-spin logic device with built-in memory. *Nature Nanotechnology* 5(4): 266–70.

Behin-Aein, B., S. Salahuddin, and S. Datta. 2009. Switching energy of ferromagnetic logic bits. *Nanotechnology, IEEE Transactions on* 8(4): 505–514.

Bennett, C. H. 1982. The thermodynamics of computation—a review. *International Journal of Theoretical Physics* 21(12): 905–940.

Boyd, E. L. 1960. Magnetic anisotropy in single-crystal thin films. *IBM Journal of Research and Development* 4(2): 116–129.

Brintlinger, T., S.-H. Lim, K. H. Baloch, P. Alexander, Y. Qi, J. Barry et al. 2010. In situ observation of reversible nanomagnetic switching induced by electric fields. *Nano Letters* 10(4): 1219–1223.

Carlton, D., N. Emley, E. Tuchfeld, and J. Bokor. 2008. Simulation studies of nanomagnetbased logic architecture. *Nano Letters* 8(12): 4173–4178.

Carlton, D., B. Lambson, A. Scholl, A. Young, P. Ashby, S. Dhuey et al. 2012. Investigation of defects and errors in nanomagnetic logic circuits. *Nanotechnology, IEEE Transactions on* 11(4): 760–762.

Chow, P. P. 1991. Molecular beam epitaxy. In *Thin Film Processes II*, edited by J. L. Vossen and W. Kern, pp. 133–176. Boston: Academic Press.

Cowburn, R. P., A. Ercole, S. J. Gray, and J. A. C. Bland. 1997. A new technique for measuring magnetic anisotropies in thin and ultrathin films by magneto-optics. *Journal of Applied Physics* 81(10): 6879.

Cowburn, R. P., D. K. Koltsov, A. O. Adeyeye, M. E. Welland, and D. M. Tricker. 1999a. Single-domain circular nanomagnets. *Physical Review Letters* 83(5): 1042–1045.

Cowburn, R. P., D. K. Koltsov, A. O. Adeyeye, and M. E. Welland. 1999b. Designing nano-structured magnetic materials by symmetry. *Europhysics Letters (EPL)* 48(2): 221–227.

Cowburn, R. P. and M. E. Welland. 2000. Room temperature magnetic quantum cellular automata. *Science* 287(5457): 1466–1468.

Csaba, G., A. Imre, G. H. Bernstein, W. Porod, and V. Metlushko. 2002. Nanocomputing by field-coupled nanomagnets. *Nanotechnology, IEEE Transactions On* 1(4): 209–213.

Cullity, B. D. and C. D. Graham. 2009. *Introduction to Magnetic Materials*. Wiley.

D'Souza, N., J. Atulasimha, and S. Bandyopadhyay. 2011a. Four-state nanomagnetic logic using multiferroics. *Journal of Physics D: Applied Physics* 44(26): 265001.

D'Souza, N., J. Atulasimha and S. Bandyopadhyay. 2011b. An energy-efficient Bennett clock-ing scheme for 4-state multiferroic logic. *Nanotechnology, IEEE Transactions On* 11(2): 418–425.

D'Souza, N., J. Atulasimha, and S. Bandyopadhyay. 2011c. arXiv Preprint arXiv:1105.1818v1.

D'Souza, N., J. Atulasimha, and S. Bandyopadhyay. 2012. An ultrafast image recovery and recognition system implemented with nanomagnets possessing biaxial magnetocrystal-line anisotropy. *Nanotechnology, IEEE Transactions On* 11(5): 896–901.

Dusch, Y., N. Tiercelin, A. Klimov, S. Giordano, V. Preobrazhensky, and P. Pernod. 2013. Stress-mediated magnetoelectric memory effect with uni-axial TbCo$_2$/FeCo multilayer on 011-cut PMN-PT ferroelectric relaxor. *Journal of Applied Physics* 113(17): 17C719.

Eerenstein, W., N. D. Mathur, and J. F. Scott. 2006. Multiferroic and magnetoelectric materi-als. *Nature* 442 (7104): 759–65

Fukami, S., T. Suzuki, K. Nagahara, N. Ohshima, Y. Ozaki, S. Saito, et al. 2009. Low-current perpendicular domain wall motion cell for scalable high-speed MRAM. VLSI Technology, 2009 Symposium on: 230–231.

Hockel, J. L., A. Bur, T. Wu, K. P. Wetzlar, and G. P. Carman. 2012. Electric field induced magnetization rotation in patterned Ni ring/Pb(Mg$_1$/3Nb$_{2/3}$)O$_3$]$_{(1-0.32)}$-[PbTiO$_3$]$_{0.32}$ het-erostructures. *Applied Physics Letters* 100(2): 022401.

Kajiwara, Y., K. Harii, S. Takahashi, J. Ohe, K. Uchida, M. Mizuguchi et al. 2010. Transmission of electrical signals by spin-wave interconversion in magnetic insulator. *Nature* 464(7286): 262–266.

Lambson, B., Z. Gu, M. Monroe, S. Dhuey, A. Scholl, and J. Bokor. 2013. Concave nano-magnets: Investigation of anisotropy properties and applications to nanomagnetic logic. *Applied Physics A* 111(2): 413–421.

Lee, F. 1968. Shape-induced biaxial anisotropy in thin magnetic films. *Magnetics, IEEE Transactions on* 4(3): 502–506.

Naik, R., C. Kota, J. S. Payson, and G. L. Dunifer. 1993. Ferromagnetic-resonance studies of epitaxial Ni, Co, and Fe films grown on Cu(100)/Si(100). *Physical Review B, Condensed Matter* 48(2): 1008–1013.

Pertsev, N. and H. Kohlstedt. 2009. Magnetic tunnel junction on a ferroelectric substrate. *Applied Physics Letters* 95(16): 163503.

Pertsev, N. and H. Kohlstedt. 2010. Resistive switching via the converse magnetoelectric effect in ferromagnetic multilayers on ferroelectric substrates. *Nanotechnology* 21(47): 475202.

Roy, K., S. Bandyopadhyay, and J. Atulasimha. 2012. Energy dissipation and switching delay in stress-induced switching of multiferroic nanomagnets in the presence of thermal fluc-tuations. *Journal of Applied Physics* 112(2): 023914.

Roy, K., S. Bandyopadhyay, and J. Atulasimha. 2013. Binary switching in a symmetric potential landscape, *Scientific Reports*, 3, 3038.

Roychowdhury, V. P., D. B. Janes, S. Bandyopadhyay, and X. Wang. 1996. Collective computational activity in self-assembled arrays of quantum dots: A novel neuromorphic architecture for nanoelectronics. *Electron Devices, IEEE Transactions On* 43(10): 1688–1699.

Salahuddin, S. and S. Datta. 2007. Interacting systems for self-correcting low power switching. *Applied Physics Letters* 90(9): 093503.

Salehi Fashami, M., J. Atulasimha, and S. Bandyopadhyay. 2012. Magnetization dynamics, throughput and energy dissipation in a universal multiferroic nanomagnetic logic gate with fan-in and fan-out. *Nanotechnology* 23(10): 105201.

Salehi Fashami, M., K. Munira, S. Bandyopadhyay, A. W. Ghosh, and J. Atulasimha. 2013. Switching of dipole-coupled multiferroic nanomagnets in the presence of thermal noise: Reliability of nanomagnetic logic. *Nanotechnology, IEEE Transactions on* 12(6): 1206–1212.

Salehi Fashami, M., K. Roy, J. Atulasimha, and S. Bandyopadhyay. 2011. Magnetization dynamics, Bennett clocking and associated energy dissipation in multiferroic Logic. *Nanotechnology* 22(15): 155201.

Schroder, D. K. 1987. *Advanced MOS Devices*. Addison-Wesley, Reading, MA, USA.

Siegle, W. T. 1965. Exchange coupling of uniaxial magnetic thin films. *Journal of Applied Physics* 36(3): 1116.

Snider, G. 2011. Instar and outstar learning with memristive nanodevices. *Nanotechnology* 22(1): 015201.

Spedalieri, F. M., A. P. Jacob, D. E. Nikonov, and V. P. Roychowdhury. 2011. Performance of magnetic quantum cellular automata and limitations due to thermal noise. *Nanotechnology, IEEE Transactions on* 10(3): 537–546.

Wang, C. P. 1968. A coupled magnetic film device for associative memories. *Journal of Applied Physics* 39(2): 1220.

Yamanouchi, M., D. Chiba, F. Matsukura, and H. Ohno. 2004. Current-induced domain-wall switching in a ferromagnetic semiconductor structure. *Nature* 428(6982): 539–42.

Zheng, H., J. Wang, S. E. Lofland, Z. Ma, L. Mohaddes-Ardabili, T. Zhao, et al. 2004. Multiferroic $BaTiO_3$-$CoFe_2O_4$ nanostructures. *Science* 303(5658): 661–663.

Zhirnov, V. V., R. K. Cavin, J. A. Hutchby, and G. I. Bourianoff. 2003. Limits to binary logic switch scaling—a gedanken model. *Proceedings of the IEEE* 91(11): 1934–1939.

5 Improvement and Applications of Large-Area Flexible Electronics with Organic Transistors

Koichi Ishida, Hiroshi Fuketa, Tsuyoshi Sekitani,
Makoto Takamiya, Hiroshi Toshiyoshi,
Takao Someya, and Takayasu Sakurai

CONTENTS

5.1 INTRODUCTION

With the advantage of the circuit integration on a plastic thin film with a low-temperature, low-cost, printable or printing process, organic transistor is widely recognized as a promising candidate for realizing large-area flexible electronics such as displays [1–3], power/data transmission sheets [4,5], and sensor arrays [6–8]. The technology of organic transistors has been significantly improving.

First of all, the improvement of process maturity and yield realizes highly integrated circuits such as an 8-bit microprocessor and analog-to-digital converters (ADCs) [9–11]. For the low-voltage operation, a self-assembled monolayer (SAM) technology lowered the operation voltage toward 1 V [12]. The fastest n-type metal–oxide–semiconductor (nMOS) transistors with C_{10}-DNTT [13] achieved the signal propagation delay of 420 ns per stage in the ring oscillator at supply voltage of 3 V [14]. For high-frequency applications, some radio-frequency identifications (RFIDs) have been proposed [15,16]. Those are implemented with only high-voltage p-type metal–oxide–semiconductor (pMOS) transistors because the mobility of nMOS transistors is much slower than that of pMOS. To address this issue, a hybrid organic (pMOS)/solution-processed metal–oxide (nMOS) RFID tag was recently proposed [17]. However, the operating speed of organic transistors is still not sufficient for the operation at hundreds megahertz order even if high voltages of 20 V or higher are supplied to the circuit. Therefore, the system-level integration of organic transistors and silicon complementary metal–oxide–semiconductor (CMOS) circuits is one of the practical solutions [18].

This chapter briefly surveys recent progress of organic transistor technologies. In addition, examples of large-area flexible electronic applications, such as an electromagnetic interference (EMI) measurement sheet [18], a user customizable logic paper (UCLP) [19], a pedometer with energy harvesters [20], and 100 V AC Energy Meter [21] are introduced.

5.2 RECENT PROGRESS OF ORGANIC TRANSISTORS

Organic pMOS transistors have been getting better yield and time-dependent reliability, and therefore highly integrated organic pMOS circuits were presented such as an 8-bit microprocessor and ADCs. In terms of low-voltage operation, SAM realized 1–3 V operation enabling direct silicon-organic circuit interface in [18]. In terms of printed electronics, some printed CMOS circuits were presented [21–24]. This section introduces resent progress of technologies of organic transistors and their circuit blocks briefly.

5.2.1 8-Bit Microprocessor on Plastic Foil

The improvement of yield of organic pMOS transistors enables the integration of thousands of transistors, and an 8-bit microprocessor on a plastic foil for the first time is presented [9]. As there are no previous organic processors, the comparison with Intel 4004 processor is provided in the last resort [9]. It is, however, interesting that the first organic processor is with shorter gate length and lower operation voltage than those of Intel 4004 processor. The core of processor, an 8-bit arithmetic and logical unit (ALU), consists of NAND gates and inverters in 5 μm organic pMOS process. The processor operates with 40 Hz clock at V_{DD} of 10 V. Number of transistors (area) of the processor foil and the running averager instruction code foil are 3381 (1.96×1.72 cm^2) and 612 (0.72×0.64 cm^2), respectively. Although the performance should be improved for practical applications, it shows good technical feasibility of the organic microprocessor.

5.2.2 Fully Integrated $\Delta\Sigma$ Analog-to-Digital Converter

A continuous time $\Delta\Sigma$ ADC with 5 μm organic pMOS transistors was presented [10]. It is a fully integrated organic ADC for the first time. The V_{TH} mismatch of input differential pair transistor in an operational amplifier is a critical issue for an analog circuit, because its offset voltage directly degrades the performance. In silicon circuits, the common centroid layout will improve the matching properties. On the contrary, it is not sufficient for organic circuits. To compensate V_{TH} mismatch of input differential pair transistors in the amplifier, the circuitry employs back-gate biasing. By using the mismatch compensation scheme, the gain of the operational amplifier (opamp) is also improved. Measured SNR and bandwidth of ADC are 26.5 dB and 15.6 Hz, respectively. Clock frequency is 500 Hz at V_{DD} of 15 V. The circuit consists of 129 transistors in 13×20 mm^2 area. Now the performance of the ADC is mainly limited by the opamp gain of 23 dB. On the other hand, a 40 dB gain organic amplifier was recently reported by another research group [22]. This fact is attracting attention for further improvement of the organic $\Delta\Sigma$ ADC.

5.2.3 Low-Voltage Organic Complementary Metal–Oxide–Semiconductor Circuit

A low-voltage organic CMOS circuit using SAM-dielectric gate oxide was presented [12]. The gate oxide consists of two layers, a 1.8- to 3.8-nm-thick aluminum oxide layer and 2.1-nm-thick SAM. The nanometer-order gate oxide thickness realizes 1.5–3 V operation, which is around 10 times lower than that of conventional organic transistors. Semiconductor materials for pMOS transistors and nMOS transistors are pentacene and hexadecafluorocopperphthalocyanine (F16CuPc), respectively. This improvement enables a direct silicon-organic circuit interface in [18]. The fastest low-voltage nMOS transistors with SAM and C_{10}-DNTT [13] achieved the signal propagation delay of 420 ns per stage in the ring oscillator at supply voltage of 3 V [14].

5.2.4 Fully Printed Organic Circuit

There are several approaches to realize fully printed organic circuits. A pMOS ring oscillator is proposed that aims fully printed circuits not only devices but also interconnection [23]. On the other hand, a complete printed organic CMOS technology on a plastic foil was proposed [24]. The devices are fabricated by 5 μm line/space resolution screen-printing process and annealing at 100°C. Oscillation frequency of a seven-stage ring oscillator that consists of $W/L = 1000/20$ μm CMOS transistors varies from 70 Hz at 40 V to 16 Hz at 20 V. A 4-bit ADC implemented with fully printed organic CMOS [25] was presented [11]. Although they use a sputter and laser ablation processes for the 30-nm-thick gold metal layer forming, CMOS transistors themselves are printed, and therefore the first printed ADC on a flexible substrate is claimed. The ADC consists of more than 100 CMOS transistors and an integrated resistive DAC in 24.5 cm^2 area. Measured bandwidth and SNDR are 2 Hz and 19.6 dB SNDR, respectively.

5.3 LARGE-AREA FLEXIBLE ELECTRONICS APPLICATIONS

This section introduces some examples of large-area flexible organic electronics such as an EMI measurement sheet, a UCLP, an insole pedometer with piezoelectric energy harvester, and a 100 V AC energy meter.

The EMI measurement sheet demonstrates a direct silicon-organic circuit interface for the first time. A 0.18 μm CMOS silicon large-scale integration (LSI) chip directly drives a 2 V organic CMOS circuitry through a stretchable interconnection. USLP consists of the 2 V organic CMOS circuits and ink-jet printed interconnections on a piece of paper and provides a new field customizability to users. The insole pedometer also employs the 2 V organic transistors driven by small DC power generated by piezo-electric films. These three applications have a similarity of utilizing the advantages of 2 V organic transistors. In contrast, the 100 V AC energy meter employs 20–100 V organic transistor. The drawback that the device requires high-voltage power supply is now turned to its advantage because the energy meter should handle high voltages. In this section, details on the above-mentioned applications are described.

5.3.1 A STRETCHABLE ELECTROMAGNETIC INTERFERENCE MEASUREMENT SHEET

EMI that degrades the dependability of electronic devices is becoming a serious issue. As modern electronic devices have 3D structures and the packaging is dense, it is difficult to analyze the EMI generation points in electronic devices such as mobile phones and large flat-panel displays. EMI largely depends on the circuit board layout. Localizing either an EMI source or a critical wiring is difficult by simulation. EMI measurement is, therefore, important for the development of electronic systems. However, there is no method of measuring EMI on the surface of 3D structures. In the conventional method, a pencil-like magnetic-field probe with X–Y scanning equipment and spectrum analyzers are used for the EMI measurement [26]. In the method, the surface of the electronic device should be scanned repeatedly with the probe. However, the scanning equipment can only move in a flat plane. In another conventional method, a measurement system with an integrated array of magnetic-field loop antennas is used [27]. Although the method captures the distribution of a magnetic field, it is not applicable to 3D structures as the antenna array is implemented on a flat and rigid printed circuit board (PCB). To solve this problem, an EMI measurement sheet [18] was proposed, which enables the measurement of EMI distribution on the surface of 3D structures by wrapping the devices with the measurement sheet. Once EMI noise is roughly localized with the measurement sheet, one can easily scan EMI noise using the probe method for precise localization or better quantification of the electromagnetic field. The sheet can measure not only a magnetic field but also an electric field suitably by simply changing its antenna connection [18].

Figure 5.1 shows a prototype of the stretchable EMI measurement sheet. Each PCB includes 2 × 2 antenna coils and a silicon EMI measurement LSI. The sheet consists of 4 × 4 PCBs, and therefore 8 × 8 antennas are located in 12 × 12 cm² area. The antennas and LSIs are controlled using 2 V organic CMOS decoder and selector. Each module is electrically connected with a stretchable interconnect made of carbon nanotubes (CNTs) [28]. The overall system is sealed with a rubber sheet made of silicone elastomer. The sheet is, therefore, flexible and stretchable. Figure 5.2 shows the block diagram of the EMI measurement sheet. Three-bit address and select signals are

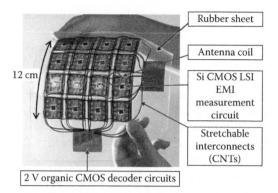

FIGURE 5.1 Stretchable electromagnetic interference measurement sheet. (K. Ishida et al., Stretchable EMI measurement sheet with 8 × 8 coil array, 2 V organic CMOS decoder, and 0.18 μm silicon CMOS LSIs for electric and magnetic field detection, *IEEE Journal of Solid-State Circuits*, 45(1), 249–259, © 2010 IEEE.)

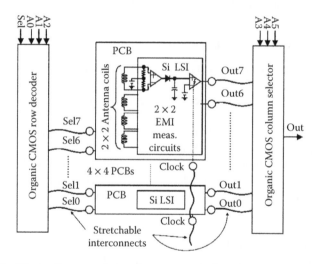

FIGURE 5.2 Block diagram of the electromagnetic interference measurement sheet. (K. Ishida et al., Stretchable EMI measurement sheet with 8 × 8 coil array, 2 V organic CMOS decoder, and 0.18 μm silicon CMOS LSIs for electric and magnetic field detection, *IEEE Journal of Solid-State Circuits*, 45(1), 249–259, © 2010 IEEE.)

applied to the organic CMOS row decoder. The decoder selects one row of the arrayed EMI measurement circuit. Each select signal is connected to a PCB with rubber-like stretchable interconnects. Each PCB has a 2 × 2 antenna array to pick up EMI, an LSI, and six stretchable interconnects. The four antennas share the LSI to measure EMI. The outputs of silicon LSIs are connected to the organic CMOS column selector with stretchable interconnects. In the conventional integrated array of magnetic-field antennas on a solid board, the processing function of the measured results is not integrated in the board. In contrast, the EMI measurement sheet has the LSIs distributed near the antennas for EMI measurement. It is the first demonstration of a distributed in situ EMI measurement, which has a potential to improve the measurement speed and accuracy. The diameter of the loop antenna is 9.8 mm. The antennas are made on a rigid PCB, because the antennas on a flexible film provide unstable antenna characteristics depending on the mechanical bending. When the scale of the array is large, a low-cost and large-area decoder is required to reduce the number of interconnects, and therefore organic transistors are suitable for the decoder. The performance of the sheet will improve as the operational speed of organic devices improves. In particular, a faster row decoder and column selector will realize high-speed scanning of the array.

The system-level integration of silicon CMOS technology, 2 V organic CMOS technology [12], and stretchable interconnects including CNTs makes a stretchable EMI measurement sheet possible, and the proposed LSI demonstrates EMI noise measurement up to 1 GHz using a rectifier and comparator operating at only 100 kHz. By changing the connection of the antenna to the LSI, the electric and magnetic fields are successfully measured separately. The minimum detectable magnetic-field noise power is −70 dBm and the maximum detectable noise frequency is 1 GHz. The minimum detectable electric-field noise power is −60 dBm and the maximum detectable noise frequency is 700 MHz.

5.3.2   USER CUSTOMIZABLE LOGIC PAPER

UCLP is proposed for both prototyping of larger-area electronics and educational applications [19]. In particular, learners can study and experience the operation of integrated circuits by fabricating custom integrated circuits, using at-home ink-jet printers to print conducting interconnects on paper that contains prefabricated arrays of organic transistors. The feasibility of UCLP is demonstrated with the newly proposed sea-of transmission-gates (SOTG) of organic CMOS transistors, providing field customizability through the use of the printable electronics technology. UCLP is applicable to a wide range of products of printable electronics including flexible displays and electronic paper, as well as for educational purposes. This technology provides a new means to add programmability for integrated circuits used in large-area electronics. Figure 5.3 shows a prototype of UCLP [19]. In UCLP, paper that contains an array of vias and an organic SOTG film are stacked.

Figure 5.4 shows the cross-sectional view of UCLP. The 2 V organic CMOS transistors [12] are fabricated on a polyimide film, and they are covered with a protective layer of parylene. Connection pads to the paper are formed with gold on top of this protective layer. The interconnects are ink-jet printed with a silver nanoparticle ink [29] onto the paper by users. In this UCLP, a 3-mm pad- and via-pitch rule is adopted, and each via extension is 300 μm. The line/space design rule (L/S = 200 μm) of the printed interconnect is determined by both printing resolution and sheet resistance. Typical printers realize 100-μm resolution with the silver nanoparticle ink. However, the minimum line width should be determined by the sheet resistance of the ink. The sheet resistance depends on the room temperature and relative humidity during printing. A sheet resistance of 0.14 Ω/square can be achieved by family use printers at a high room temperature of 30°C and relative humidity of 80%. However, we assume a sheet resistance of 0.2 Ω/square by considering a practical room temperature of 17°C and relative humidity of 35%. We adopt 200 μm wide interconnects in this study, and thus resistance between two pads can be estimated around 2.2 Ω. The line space of

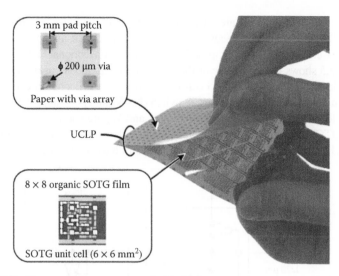

FIGURE 5.3   Prototype of a user customizable logic paper.

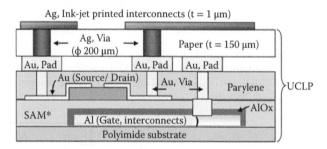

FIGURE 5.4   Cross-sectional view of a user customizable logic paper. (K. Ishida et al., User customizable logic paper (UCLP) with sea-of transmission-gates (SOTG) of 2-V organic CMOS and ink-jet printed interconnects, *IEEE Journal of Solid-State Circuits*, 46(1), 285–292, © 2011 IEEE.)

200 µm is determined by crosstalk. The crosstalk of the proposed printed interconnects is caused by a resistive coupling rather than a capacitive coupling. The sheet resistance of the precoated nanoconductive base is typically 9×10^9 Ω/square, but satellite ink drops by family use printers lower the actual sheet resistance to around 10% of this value. Thus, the isolation of 20-mm-long interconnects with a space of 200 µm can be estimated at around 9 MΩ. The minimum via-hole diameter should be larger than 200 µm to achieve an acceptable via resistance of 2.7 Ω. Finally, to implement practical circuits, the number of interconnects between the pads should be around five. As a result, a 3-mm pad- and via-pitch rule is determined.

The large pad and via pitch present a design challenge for UCLP to increase integration density. Gate array (G/A) architectures have been widely used in silicon technologies. The G/A architecture includes two pMOS transistors, two nMOS transistors, and nine via holes in a logic cell. In silicon technologies, a narrow via spacing rule is available, and therefore, the number of vias is not critical in the cell area. On the other hand, as a 3-mm pad- and via-pitch are adopted in UCLP, pads and vias now dominate the cell area. The number of vias becomes a critical issue in UCLP, and to solve the problem, an area-efficient SOTG is proposed instead of the conventional G/A approach.

Figure 5.5 shows the schematic of the SOTG unit cell. SOTG uses a type of pass transistor logic, and a single SOTG cell has six transistors. Each unit cell has a pair of complementary transmission gates, n-switch (NSW) and p-switch (PSW), and four terminals for ink-jet printed interconnects. V_{DD} and V_{SS} are common to every SOTG cell. In SOTG, the output (OUT) is connected to either PSW or NSW depending on the input (IN). Figure 5.6 shows several examples of basic logic gate implementation

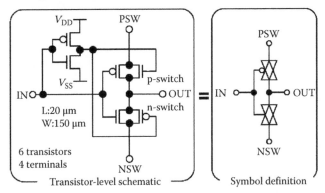

FIGURE 5.5 Schematic and symbol definition of sea-of transmission-gates unit cell. (K. Ishida et al., User customizable logic paper (UCLP) with sea-of transmission-gates (SOTG) of 2-V organic CMOS and ink-jet printed interconnects, *IEEE Journal of Solid-State Circuits*, 46(1), 285–292, © 2011 IEEE.)

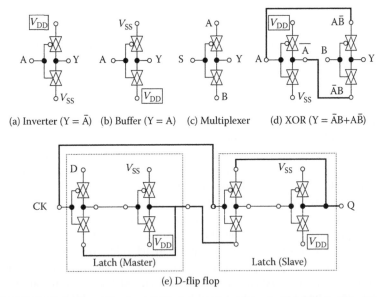

FIGURE 5.6 Examples of logic gate implementation with sea-of transmission-gates cells. (a) Inverter, (b) buffer, (c) two-input multiplexer, (d) two-input XOR, and (e) D-flip flop.

with the proposed SOTG cells. An inverter requires one logic cell that consists of six transistors, as shown in Figure 5.6a. In terms of saving on rewiring cost or area, the SOTG architecture makes it easy to customize the UCLP. In fact, either a buffer or a two-input multiplexer can be implemented with one logic cell, as shown in Figure 5.6b and c. Thanks to the complementary transmission gates, not only the XOR as shown in Figure 5.6d but also any two-input logical operation can be realized with two logic cells. In addition, a D-flip flop can be implemented with only four logic cells as shown in Figure 5.6e. A total of 64 logic cells are implemented in 73×73 mm^2 area.

5.3.3 Insole Pedometer with Piezoelectric Energy Harvester

Energy harvesting is an enabling technology for realizing an ambient power supply for wireless sensor nodes and mobile devices. By using flexible photovoltaic cells and piezoelectric films, we can readily harvest ambient energy if flexible energy harvesters can be realized. Conventional silicon circuits, however, are not best suited to realizing flexible large-area energy harvesters because they are not mechanically conformable to uneven surfaces such as shoes. To address this challenge, an organic insole pedometer with a piezoelectric energy harvester is proposed as the first step toward ambient energy harvesting using organic flexible electronics [20].

The main challenge in the design of organic circuits for piezoelectric energy harvesting is the robust operation of pMOS-only circuits at a low-supply voltage. In energy-harvesting applications, the harvested power is small and the rectified voltage is low (e.g., 2 V). In organic circuit design, pMOS-only circuits are often used, because the mobility of pMOS transistors is much higher than that of nMOS transistors in our process. The operation of pMOS-only circuits is not robust, and the noise margin is small because of their rationed-logic nature. A pseudo-CMOS inverter that consists of four pMOS transistors [30] has a high gain, but it requires a negative voltage bias. In energy-harvesting applications, however, a single power supply is typical. Therefore, in this work, to increase the noise margin of pMOS-only logic circuits, a negative voltage is generated by a charge pump and is applied as the bias of pseudo-CMOS inverters. We use a 2 V pMOS process with SAM technology [12] and dinaphtho[2,3-b:2',3'-f]thieno[3,2-b]thiophene (DNTT) [31] for pMOS transistors.

Figure 5.7 shows the photograph of the prototype insole pedometer that includes the piezoelectric energy harvester and the 2 V organic pMOS rectifier and counter. A polyvinylidene difluoride (PVDF) sheet is used as the piezoelectric energy harvester. Twenty-one rolls of PVDF film are embedded in the insole. Each time the insole is pressed by the foot during walking, the harvested energy is rectified by the organic rectifier, and the number of the steps is counted by the organic counter. Figure 5.8 shows a block diagram of the proposed organic insole pedometer. It consists of four circuit blocks. The all-pMOS full-wave rectifier supplies a voltage V_{DD} of approximately 2 V to all circuit blocks. In the clock generator, the output of the PVDF harvester is half-wave-rectified and a Schmitt trigger inverter converts the half-wave-rectified signal into a clock signal. The generated clock signal is sent to both the pMOS negative voltage generator and the 14-bit pseudo-CMOS step counter with gate-boosted pMOS switches. The step counter records the number of steps up to 16,383 steps at the maximum frequency of 4.4 Hz using the harvested power. The

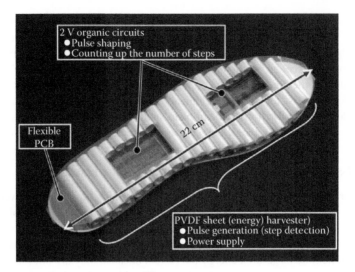

FIGURE 5.7   Photograph of the prototype insole pedometer.

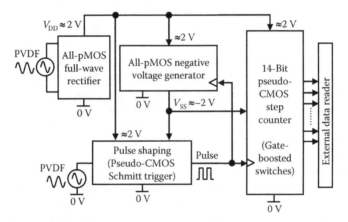

FIGURE 5.8   Block diagram of the insole pedometer. (K. Ishida et al., Insole pedometer with piezoelectric energy harvester and 2 V organic circuits, *IEEE Journal of Solid-State Circuits*, 48(1), 255–264, © 2013 IEEE.)

negative voltage generator supplies a voltage V_{SS} (e.g., −2 V) to the counter and can provide 12 µW power with 65% efficiency. The organic insole pedometer consists of 462 transistors, and its size is 22 × 7 cm².

5.3.4   100 V AC ENERGY METER

A smart meter is essential for realizing the power grid. To reduce the energy loss in the power grid, an extremely fine-grain power monitoring system is desirable, and it will require an enormous number of low-cost power meters. Existing power meters, however, do not meet the cost and size requirements. On the other hand, organic devices on flexible films have a great potential to realize low-cost power meters. In this context, a 100 V AC power meter based on a system-on-a-film (SoF) concept is presented in Ishida et al. [21].

Figure 5.9 shows the photograph of the proposed 100 V AC power meter on a flexible film, including (1) analog circuits composed of a 20 V organic CMOS opamp for AC current sensing, (2) logic circuits composed of a 20 V organic CMOS frequency divider for integrating the measured current, (3) AC-to-DC power converter composed of a 100 V organic pMOS rectifier to generate 20 V DC power for the power meter, (4) an organic light-emitting diode (OLED) [32] bar indicator, and (5) an AC connector inserted between the power plug and the AC outlet are fully integrated on a 200 × 200 mm² flexible film. The entire sheet can be folded, and the total size of the proposed AC power meter can be shrunk to 70 × 70 mm². Figure 5.10 shows the block diagram of the proposed 100 V AC power meter. The measured 100 V 50 Hz AC load current i_L is first converted into the sense voltage (v) by means of the sense resistor (R). The converted sense voltage v is then amplified by the amplifier and rectified into V_{SENSE}, which is compared with the triangular waveform (V_{TRI}) by the comparator. The output of the comparator enables or disables the 10-bit counter. Five most significant bits in the counter are connected to the OLED bar indicator.

To get the accumulated results, the maximum integration time of the power meter is designed to be 43 minutes. The 0.05 Hz clock for the input of the triangular waveform generator and the 0.4 Hz clock for the counter are generated by a 10-bit frequency divider, for which the clock is generated by a half-wave rectifier from 100 V 50 Hz AC signal. The required DC power for the power meter is provided by converting the 100 V 50 Hz AC power into 20 V DC power by the full-wave rectifier implemented using 100 V organic pMOS. The current consumption of the system, mainly consumed by the five-digit OLED bar indicator, is around 2 mA. As the

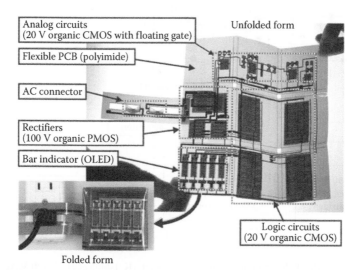

FIGURE 5.9 Prototype of the organic 100 V AC energy meter on a flexible film. (K. Ishida et al., A 100 V AC energy meter integrating 20 V organic CMOS digital and analog circuits with a floating gate for process variation compensation and a 100 V organic PMOS rectifier, *IEEE Journal of Solid-State Circuits*, 47(1), 301–309, © 2012 IEEE.)

driving capability of organic nMOS is weaker than that of pMOS by an order, an all-pMOS full-wave rectifier topology is chosen. The gate length and width of each transistor are 20 μm and 100 mm, respectively, which can supply up to 2 W DC power, the highest power level ever reported.

In our 20 V CMOS, DNTT-based pMOS transistor has eight times higher carrier mobility than that of Naphthalenetetracarboxylic diimide-based nMOS. In addition, our CMOS inverter gain was only 3.2 at 20 V and this leads to functional errors in the large-scale logic circuits. To solve the problem, we designed the frequency divider with high-gain pseudo-CMOS inverters [30]. The pseudo-CMOS inverter uses only pMOS. The gain of 148, static noise margin of 6.7 V, and 156-Hz oscillation frequency of a three-stage ring oscillator can be achieved at 20 V supply voltage. In the divider, nMOSs are still used only for transmission gates, in which high gain is not required. Thanks to the high gain of pseudo-CMOS, the divider successfully operates at 50 Hz and 20 V. In the frequency divider, the dynamic slave latch, which consists of only an inverter and the parasitic capacitance as the charge keeper, is used to reduce the number of transistors.

A major challenge in organic analog circuit design is to compensate for large process variations. The offset voltage in the differential pair of the amplifier due to the device mismatch should be reduced to lower than the sense voltage generated by the sense resistor R in Figure 5.10. The AC energy meter consists of 609 transistors and the total area excluding AC connector is 200×200 mm^2 (unfolded form) or 70×70 mm^2 (using form).

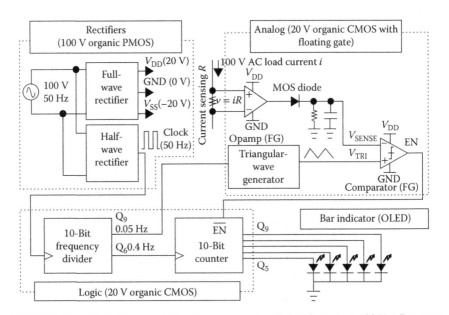

FIGURE 5.10 Block diagram of the AC energy meter. (K. Ishida et al., A 100 V AC energy meter integrating 20 V organic CMOS digital and analog circuits with a floating gate for process variation compensation and a 100 V organic PMOS rectifier, *IEEE Journal of Solid-State Circuits*, 47(1), 301–309, © 2012 IEEE.)

5.4 CONCLUSIONS

In this chapter, resent progress of organic transistors and their circuit blocks are briefly surveyed. The improvement of the organic technologies enables the implementation of highly integrated circuit blocks such as microprocessors and ADCs on a flexible film. Self-assembled monolayer technology drastically lowers the operation voltage of organic transistors down to 1 V. Fully printed technology will reduce the total cost of the large-area electronics in the near future.

Some feasibility studies of the organic flexible applications are also introduced. The EMI measurement sheet, the UCLP, and the insole pedometer utilize the advantage of 2 V organic transistors. In particular, the direct silicon-organic circuit interface will bring more possibilities of new application. In contrast, the 100 V AC energy meter employs 20–100 V organic transistor. The drawback, that the device requires high-voltage power supply, turns to an advantage because the energy meter should handle high voltages. As previously mentioned, flexible electronics with organic transistors will open up new application fields.

REFERENCES

1. P. Andersson, D. Nilsson, P.-O. Svensson, M. Chen, A. Malmström, T. Remonen, T. Kugler, and M. Berggren. Active matrix displays based on all-organic electrochemical smart pixels printed on paper, *Advanced Materials*, 14(20), 1460–1464, 2002.
2. A. Sugimoto, H. Ochi, S. Fujimura, A. Yoshida, T. Miyadera, and M. Tsuchida. Flexible OLED displays using plastic substrates, *IEEE Journal of Selected Topics in Quantum Electronics*, 10(1), 107–114, 2004.
3. Y. Fujisaki, Y. Nakajima, T. Takei, H. Fukagawa, T. Yamamoto, and H. Fujikake. Flexible active-matrix organic light-emitting diode display using air-stable organic semiconductor of dinaphtho[2, 3-b: 2′, 3′-f]thieno[3, 2-b]-thiophene, *IEEE Transactions on Electron Devices*, 59(12), 3442–3449, 2012.
4. M. Takamiya, T. Sekitani, Y. Miyamoto, Y. Noguchi, H. Kawaguchi, T. Someya, and T. Sakurai. Design Solutions for Multi-Object Wireless Power Transmission Sheet Based on Plastic Switches, IEEE International Solid-State Circuits Conference, San Francisco, CA, 362–363, 2007.
5. L. Liu, M. Takamiya, T. Sekitani, Y. Noguchi, S. Nakano, K. Zaitsu, T. Kuroda, T. Someya, and T. Sakurai. A 107pJ/b 100kb/s 0.18um Capacitive-Coupling Transceiver for Printable Communication Sheet, IEEE International Solid-State Circuits Conference, San Francisco, CA, 292–293, 2008.
6. H. Kawaguchi, T. Someya, T. Sekitani, and T. Sakurai. Cut-and-paste customization of organic fet integrated circuit and its application to electronic artificial skin, *IEEE Journal of Solid-State Circuits*, 40(1), 177–185, Jan. 2005.
7. H. Marien, M. S. J. Steyaert, E. van Veenendaal, and P. Heremans. Analog building blocks for organic smart sensor systems in organic thin-film transistor technology on flexible plastic foil, *IEEE Journal of Solid-State Circuits*, 47(7), 1712–1720, 2012.
8. H. Fuketa, K. Yoshioka, Y. Shinozuka, K. Ishida, T. Yokota, N. Matsuhisa, Y. Inoue et al. 1 μm-Thickness 64-Channel Surface Electromyogram Measurement Sheet with 2 V Organic Transistors for Prosthetic Hand Control, IEEE International Solid-State Circuits Conference, San Francisco, CA, 104–105, 2013.
9. K. Myny, E. van Veenendaal, G. H. Gelinck, J. Genoe, W. Dehaene, and P. Heremans. An 8-bit, 40-instructions-per-second organic microprocessor on plastic foil, *IEEE Journal of Solid-State Circuits*, 47(1), 284–291, 2012.

10. H. Marien, M. Steyaert, E. van Veenendaal, and P. Heremans. A fully integrated ΔΣ ADC in organic thin-film transistor technology on flexible plastic foil, *IEEE Journal of Solid-State Circuits*, 46(1), 276–284, 2011.

11. S. Abdinia, M. Benwadih, R. Coppard, S. Jacob, G. Maiellaro, G. Palmisano, M. Rizzo et al. A 4b ADC Manufactured in a Fully-Printed Organic Complementary Technology Including Resistors, IEEE International Solid-State Circuits Conference (ISSCC), 106–107, San Francisco, CA, 2013.

12. H. Klauk, U. Zschieschang, J. Pflaum, and M. Halik. Ultralow-power organic complementary circuits, *Nature*, 445, 745–748, 2007.

13. M. J. Kang, I. Doi, H. Mori, E. Miyazaki, K. Takimiya, M. Ikeda, and H. Kuwabara. Alkylated Dinaphtho[2,3-b:2′,3′-f]Thieno[3,2-b]Thiophenes (Cn-DNTTs): Organic semiconductors for high-performance thin-film transistors, *Advanced Materials*, 23(10), 1222–1225, 2011.

14. U. Zschieschang, R. Hofmockel, R. Rödel, U. Kraft, M. J. Kang, K. Takimiya, T. Zaki et al. Megahertz operation of flexible low-voltage organic thin-film transistors, *Elsevier Organic Electronics*, 14(6), 1516–1520, 2013.

15. E. Cantatore, T. C. T. Geuns, G. H. Gelinck, E. van Veenendaal, A. F. A. Gruijthuijsen, L. Schrijnemakers, S. Drews, and D. M. de Leeuw. A 13.56-MHZ RFID system based on organic transponders, *IEEE Journal of Solid-State Circuits*, 42(1), 2007.

16. K. Myny, M. J. Beenhakkers, N. A. J. M. van Aerle, G. H. Gelinck, J. Genoe, W. Dehaene, and P. Heremans. A 128b Organic RFID Transponder Chip, Including Manchester Encoding and Aloha Anti-Collision Protocol, Operating with a Data Rate of 1529b/s, IEEE International Solid-State Circuits Conference, San Francisco, CA, 206–207, 2009.

17. K. Myny, M. Rockele, A. Chasin, D. Pham, J. Steiger, S. Botnaras, D. Weber et al. Bidirectional Communication in an HF Hybrid Organic/Solution-Processed Metal-Oxide RFID Tag, IEEE International Solid-State Circuits Conference, San Francisco, CA, 312–314, 2012.

18. K. Ishida, N. Masunaga, Z. Zhou, T. Yasufuku, T. Sekitani, U. Zschieschang, H. Klauk, M. Takamiya, T. Someya, and T. Sakurai. Stretchable EMI measurement sheet with 8 × 8 coil array, 2 V organic CMOS decoder, and 0.18 μm silicon CMOS LSIs for electric and magnetic field detection, *IEEE Journal of Solid-State Circuits*, 45(1), 249–259, 2010.

19. K. Ishida, N. Masunaga, R. Takahashi, T. Sekitani, S. Shino, U. Zschieschang, H. Klauk, M. Takamiya, T. Someya, and T. Sakurai. User customizable logic paper (UCLP) with sea-of transmission-gates (SOTG) of 2-V organic CMOS and ink-jet printed interconnects, *IEEE Journal of Solid-State Circuits*, 46(1), 285–292, 2011.

20. K. Ishida, T.-C. Huang, K. Honda, Y. Shinozuka, H. Fuketa, T. Yokota, U. Zschieschang et al. Insole pedometer with piezoelectric energy harvester and 2 V organic circuits, *IEEE Journal of Solid-State Circuits*, 48(1), 255–264, 2013.

21. K. Ishida, Tsung-Ching Huang, K. Honda, T. Sekitani, H. Nakajima, H. Maeda, M. Takamiya, T. Someya, T. Sakurai. A 100 V AC energy meter integrating 20 V organic CMOS digital and analog circuits with a floating gate for process variation compensation and a 100 V organic PMOS rectifier, *IEEE Journal of Solid-State Circuits*, 47(1), 301–309, 2012.

22. G. Maiellaro, E. Ragonese, A. Castorina, S. Jacob, M. Benwadih, R. Coppard, E. Cantatore, and G. Palmisano. High-Gain Operational Transconductance Amplifiers in a Printed Complementary Organic TFT Technology on Flexible Foil, *IEEE Transactions on Circuits and Systems I: Regular Papers*, 60(12), 3117–3125, Dec. 2013.

23. A. C. Hübler, F. Doetz, H. Kempa, H. E. Katz, M. Bartzsch, N. Brandt, I. Hennig et al. Ring oscillator fabricated completely by means of mass-printing technologies, *Elsevier Organic Electronics*, 8(5), 480–486, 2007.

24. A. Daami, C. Bory, M. Benwadih, S. Jacob, R. Gwoziecki, I. Chartier, R. Coppard et al. Fully Printed Organic CMOS Technology on Plastic Substrates for Digital and Analog Applications, IEEE International Solid-State Circuits Conference, San Francisco, CA, 328–330, 2011.

25. S. Jacob, M. Benwadih, J. Bablet, I. Chartier, R. Gwoziecki, S. Abdinia, E. Cantatore et al. High Performance Printed N and P-type OTFTs for CMOS Applications on Plastic Substrate, European Solid-State Device Research Conference (ESSDERC), Bordeaux, France, 173–176, 2012.

26. N. Masuda, N. Tamaki, T. Kuriyama, J.-C. Bu, M. Yamaguchi, and K, Arai. High frequency magnetic near field measurement on LSI chip using planar multi-layer shielded loop coil, 2003 IEEE International Symposium on Electromagnetic Compatibility, Istanbul, Turkey, 80–85, May 2003.

27. B. Archambeault. Predicting EMI emission levels using EMSCAN, 1993 IEEE International Symposium on Electromagnetic Compatibility, Dallas, TX, 48–50, Aug. 1993.

28. T. Sekitani, Y. Noguchi, K. Hata, T. Fukushima, T. Aida, and T. Someya. A rubberlike stretchable active matrix using elastic conductors, *Science*, 321, 1468–1472, 2008.

29. S. Shino. Conductive Member, and Its Developing Method, Japan Patent JP2008-004375, 2008.

30. T.-C. Huang, K. Fukuda, C.-M. Lo, Y.-H. Yeh, T. Sekitani, T. Someya, and K.-T. Cheng. Pseudo-CMOS: A design style for low-cost and robust flexible electronics, *IEEE Transactions Electron Devices*, 58(1), 141–150, 2011.

31. T. Yamamoto and K. Takimiya. Facile synthesis of highly pi-extended heteroarenes, dinaphtho[2,3-b:2′,3′-f]chalcogenopheno[3,2-b]chalcogenophenes, and their application to field-effect transistors, *Journal of American Chemical Society*, 129, 8, 2224–2225, 2007.

32. H. Nakajima, S. Morito, H. Nakajima, T. Takeda, M. Kadowaki, K. Kuba, S. Hanada, and D. Aoki. Flexible OLEDs poster with gravure printing method, *Society for Information Display 2005 Digest*, VI, 1196–1199, 2005.

6 Soft-Error Mitigation Approaches for High-Performance Processor Memories

Lawrence T. Clark

CONTENTS

6.1   INTRODUCTION

Radiation-hardened microprocessors for use in aerospace or other high-radiation environments [1] have historically lagged behind their commercial counterparts in performance. The RAD750 (BAE Systems Inc., Arlington, VA), released in 2001 on a 250-nm rad-hard process, can reach 133 MHz [2]. Recent updates of this device to a 150-nm process have improved on this, but only marginally [3]. This device, built on a radiation-hardened process, lags in part due to the difficulty in keeping such processes up to date, for relatively low-volume devices [4]. The SPARC AT697 (Atmel Corp., San Jose, CA) introduced in 2003 has an operating frequency of 66 MHz, uses triple modular redundancy (TMR) for logic, and error detection and correction (EDAC) and parity protection for memory, soft-error protection [5,6]. More recent radiation hardened by design (RHBD) processors have reached 125 MHz [7]. In contrast, unhardened embedded microprocessors contemporary to these designs achieve dramatically better performance on similar generation processes. For instance, the XScale microprocessor, fabricated on a 180-nm process, operates at clock frequencies over 733 MHz [8]. Ninety-nanometer versions of the XScale microprocessors achieved 1.2 GHz [9] with the cache performance being even higher [10]. More modern designs, such as those in 32-nm cell phone system on chip (SOC) devices, are multicore, out-of-order microprocessors, running at over 1.5 GHz [11]. As portable devices have become predominant, power dissipation has become the overriding concern in microprocessor design. The most effective means to achieving low power is clock gating, which limits circuit active power dissipation by disabling the clocks to sequential circuits such as memories. In caches and other memories, this means that the operation of clocking and timing circuits must also be protected from radiation-induced failures, including erroneously triggered operations.

6.1.1   STATIC RANDOM-ACCESS MEMORY/CACHE CIRCUITS

Modern microprocessors and SOC ICs have numerous embedded memories ranging in size and access time requirements. Examples include small fast register files (RFs) comprising the data RF, the write buffers, translation look-aside buffers (TLBs), and various queues. These RF memories, with capacities ranging from 1 to 3 kB, often have access times under 500 ps. Single-cycle Level 1 (L1) caches are possible with very aggressive circuit design [12]. Because of the tight timing constraints and the need for fine write (byte write) granularity, EDAC has been infrequently used in commercial microprocessor RFs or L1 caches, due to the deleterious impact it has on the access time. Larger arrays benefit from EDAC with essentially no performance cost with appropriate pipelining, for example, the L2 caches in Ricci et al [9].

Caches are small, fast SRAM memories that allow any address to be mapped to them, enabling the cache to simulate the full memory, but with greatly improved access time. Cache memory is actually composed of two logically distinct (and usually physically distinct) memory arrays—the tag and data arrays as shown in Figure 6.1. Each entry in the tag array holds a portion (the tag) of the memory address, so that the location of the data held in the cache can be found by comparing the tag to the memory address requested. Each tag entry has a corresponding block or *line* of associated data that is a copy of the memory residing in the corresponding address location. The block or line size generally ranges from four to sixteen words of memory. To access the cache, the tag memory is accessed and if the value stored in memory matches the address, a cache hit is signaled and the associated cache data array value is returned. Otherwise, a cache *miss* is asserted. In this case, the needed data is fetched from the main memory or the next hierarchical cache level.

L1 caches nearest the core must return data in one, or more usually, two clock cycles. To accomplish this, the tag and data are read out simultaneously in what is known as a *late-way-select* architecture. The tag comparison (indicated by the question marks in Figure 6.1) determines if one of the tags match the requested memory address. The logical OR of these comparisons indicates a *hit*, that is, the program referenced memory location does reside in the cache. The match signal selects the data from the correct way to be read out, via the way multiplexer. The number of ways determines the associativity. With an associativity of one (direct-mapped) only one location in the tag and data arrays can hold an address. Greater associativity reduces conflicts for these locations. Power and latency increase with associativity, making this a key trade-off. L1 caches typically have associativities of one to four, although high associativity caches have been used, particularly in low power designs [8].

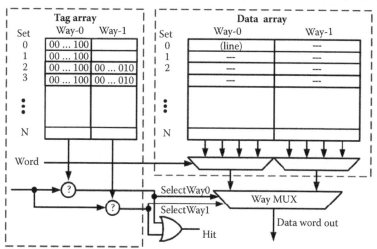

FIGURE 6.1 Architectural view of a two-way set associative cache memory. The data array organization can be different, with separate arrays for each word. This moves the word multiplexing between arrays.

6.1.2 SOFT-ERROR MECHANISMS

Single event effect (SEE) is the general term describing device failures due to impinging radiation that deposits charge in the circuit upsetting the circuit state. Depending on the incident particle, devices, and circuit types, there can be different failure modes from the collected charge. These range from benign to catastrophic, but the focus of this chapter is on those that cause soft errors, that is, circuit state upsets that do not permanently damage the circuit. The primary collection mechanisms are drift, charge funneling (a drift-based mechanism), and diffusion (Figure 6.2). The drift components are very fast, whereas diffusion is slower. Charge deposited in the space-charge or depletion region (dashed outline in the inset of Figure 6.2a) is rapidly collected toward the P–N junction by drift. N-type diffusions collect electrons and thus drive the node low, whereas P-type diffusions collect holes and drive the node high. An important prompt collection mechanism is charge *funneling* whereby the ion track distorts the electric field lines, and allows collection by drift beyond the nominal depletion regions [13]. Diffusion is the least important collection mechanism, because it is slower and diffuses the charge across more collecting nodes. Being slower, it is less likely to overwhelm the driving circuit. Nonetheless, it is clearly responsible for some upset mechanisms and significantly contributes to multiple node collection (MNC) and multiple bit upset (MBU) [14].

The actual waveform is highly dependent on the amount of charge deposited as well as the driving or restoring circuit, which must remove the deposited charge. In the case of a latch, the deposited charge may be sufficient to upset the bistable circuit state, a charge amount termed the Q_{crit} of the latch, causing a single event upset (SEU) [15]. In this case, the restoring circuit is turned off. For combinational circuits, the

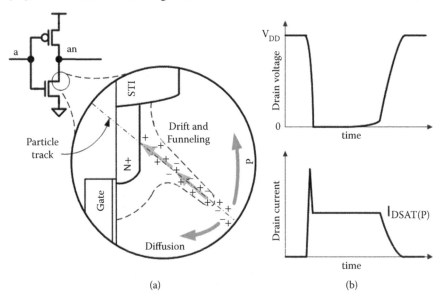

(a)   (b)

FIGURE 6.2 (a) Soft-error mechanisms and (b) their impact on circuit behavior. A particle impinging on the N-type metal–oxide–semiconductor drain produces a negative voltage glitch (single event transient) whose recovery time is a function of the restoring current provided by the P-type metal–oxide–semiconductor transistor.

effect is termed a single event transient (SET) that can inadvertently assert signals or be captured as machine state when sampled by a latch [16,17]. During an SET, as the collecting node is driven to the supply rail, the diode is no longer at a favorable voltage to collect charge until sufficient charge has been removed by the driver. Some charge may recombine, but due to long lifetimes in modern Si substrates, it tends to linger in a manner analogous to a diode's diffusion capacitance [18]. Consequently, the SET may be prolonged. In Figure 6.2, the P-type metal–oxide–semiconductor (PMOS) transistor provides its maximum current, $I_{DSAT(P)}$, until sufficient deposited charge has been removed so that the N-type metal–oxide–semiconductor drain diode current is less than $I_{DSAT(P)}$. At this time, the transient of the circuit output node at or near V_{SS}, transitions back to V_{DD}. The resulting voltage pulse (the SET) may cause erroneous circuit operation, particularly in dynamic circuits.

6.1.2.1 Charge Collection Physics

An ionizing particle can generate charge through direct ionization, whereby the particle loses energy through columbic interaction with the Si lattice. The energy loss rate is related to the material stopping power dE/dx or linear energy transfer (LET) of the material that it is passing through. The LET is given by

$$LET = \left(\frac{1}{\rho}\right)\frac{dE}{dx} \tag{6.1}$$

where ρ is the material density (2.42 g/cm³ for silicon), E is the energy, and x is the distance traveled through silicon. LET is thus the appropriate measure of the particle charge generation capability. The charge produced is

$$Q = L_{track}\left(\frac{\rho LET}{E_{ehp}}\right)q \tag{6.2}$$

where L_{track} is the particle track length and E_{ehp} is the energy required to create an electron–hole pair. Thus, in silicon, one LET generates

$$Q = 1\mu m\left(\frac{10^{-4}\mu m/cm(2.42g/cm^3)1(MeV\cdot cm^2/g)}{3.6eV(10^{-6}MeV/eV)}\right)1.602(10^{-7})pC \tag{6.3}$$

in 1 μm of track length, which is approximately 10 fC of charge per micron of track length per MeV·cm²/mg of LET. Note that a 32-nm technology generation transistor has less than 1 fC of total capacitance per micron of gate width. Charge collection is at diodes, so electrons are collected by N-type diffusions and holes by P-type diffusions. In a bulk complementary metal–oxide–semiconductor (CMOS) process, the latter are in N-type wells of limited depth, reducing the charge due to collected holes.

Angled strikes may allow more charge to be collected at a circuit node (Figure 6.2). The effective LET, LET_{eff}, is given by

$$LET_{eff} = \frac{LET}{\cos\theta} \tag{6.4}$$

where θ is the angle at which the particle strikes the die with $0°$ being orthogonal to the surface. Originally, LET_{eff} explained greater charge collection at a single node when the particle incidence deviated from normal. Because modern junctions can be very small, the charge track may cross under many sensitive nodes, leading to more separate nodes collecting charge, for example, MBUs. In these cases, LET_{eff} may not be appropriate.

In spacecraft, commonly occurring heavy ions can have LETs up to about 50 Mev·cm²/mg. In Earth orbit, protons are a much greater threat. Until recently, protons have caused soft errors primarily through indirect mechanisms. If the proton undergoes an inelastic collision with a Si nucleus, alpha or gamma particles may be emitted, or a spallation reaction where the nucleus is fragmented may occur [19]. Protons may produce enough charge through direct ionization to be of importance in highly scaled, for example, sub-65-nm devices, where the Q_{crit} can be as low as 0.3 fC. For example, highly scaled SRAMs have been shown to be susceptible to upset due to direct ionization by low-energy protons [20].

In terrestrial environments, the primary causes of soft errors are alpha particles and neutrons. Efforts to mitigate soft errors in terrestrial systems have transitioned over time from those caused by alpha particles to those caused by high-energy neutrons, as materials used in packaging and fabrication have been reduced in alpha emissivity to fluxes below 0.001 α/cm²/hr. Most terrestrial neutrons pass through the IC without interacting with the circuits, but a small fraction of them interact. When the neutron energy exceeds the necessary threshold energy, nuclear reactions may occur, producing Al and Mg ions that have significant LET. The neutron energy threshold for $^{28}Si + n \rightarrow p + ^{28}Al$ is roughly 2.7 MeV, and for $^{28}Si + n \rightarrow \alpha + ^{25}Mg$ is around 4 MeV. Beyond 15 MeV, nuclear reactions tend to create more charged ions like $^{28}Si + n \rightarrow ^{24}Mg + \alpha + n$. The charged aluminum (Al) and magnesium (Mg) ions generally travel short distances but produce a large charge deposition. Process improvements to mitigate neutron-induced SEEs include removing borophosphosilicate glass (BPSG) as an intermetal dielectric. BPSG that has a high ^{10}B concentration and ^{10}B has a high neutron-capture cross-section. When absorbing a neutron, ^{10}B splits into a 1.5 MeV alpha particle and a lithium nucleus.

6.1.2.2 Circuit Cross-Section Measurements

ICs are generally ground tested in neutron, proton, or ion beams to determine their susceptibility to SEE. The probability of hitting a target is measured by its apparent target size or cross-section in units of area, given by

$$\sigma = \frac{Errors}{Fluence} \tag{6.5}$$

where fluence is measured in particles/cm². The primary goal of SEE mitigation is to limit the errors and hence the apparent cross-section. For heavy ions, the cross-section is typically plotted versus LET using a Weibull fit. The onset of errors is defined as the threshold LET and the saturation cross-section occurs where essentially every particle above that energy is likely to cause an upset. Note, however, that half of the time the type of collection may drive the node to the same logic state that it is already in, for example, N-type diffusions driving a node at V_{SS} low have no impact.

6.1.2.3 Static Random-Access Memory Single Event Effect

Many different latch designs have been proposed to mitigate SEU. The most common approaches are based on the dual interlocked cell that uses circuit redundancy—storage nodes are duplicated, and if only one node is upset, the latch self-corrects [21]. However, as noted previously, ionizing particle deposited charge may be collected by multiple junctions. As fabrication process scaling pushes circuit nodes closer together, spacing the redundant nodes sufficiently far apart, so that multiple redundant circuit nodes are not upset by one impinging particle, increasingly makes using these techniques in SRAM cells impractical. In these MNC cases, the charge is collected by multiple SRAM cells—two or more physically adjacent bits are upset at the same time creating an MBU [22]. Heavy ions and neutrons have been shown to readily produce MBUs [23,24]. Figure 6.3 shows measured SRAM array upsets due to broad beam ion irradiation, with a large proportion of the upsets comprising MBUs. The likelihood that a strike generates an MBU is dependent on the stored data pattern and the ion incident angle, with grazing angles obviously creating the most upsets as the deposited charge is through or near more collecting nodes. Consequently, recent SRAM hardening work has increasingly focused on using EDAC rather than hardening individual cells. It is thus important that the SRAM design interleave storage cells belonging to the same word or parity group, so that MNC does not cause an MBU in the same word. SRAM designers have generally used at least four-cell interleaving to accomplish this and 65-nm data show this continues to be effective [24].

McDonald et al. observed decades ago that SETs in SRAM decoders and control logic could cause upsets that are nonrandom and thus not amenable to mitigation using EDAC [25]. This issue has been observed in SRAMs hardened by process as well, where resistor-hardened SRAM cells exhibited no static errors, but did exhibit dynamic (operating) errors at relatively low LET [26]. These errors can cause the wrong word line (WL) to be asserted, causing a silent data corruption (SDC) error as the parity or EDAC bits read out may match the data. Figure 6.3 also clearly shows one

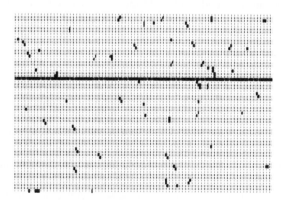

FIGURE 6.3 Measured multibit upsets in a 90-nm static random-access memory. Inset shows example of specific failure patterns. The diagram is taken over time. However, each set of errors has an associated time stamp. Each multiple bit upset is due to a different particle. The upset row is due to a single event transient–induced word line mis-assertion, which wrote the bit line state to the inadvertently selected row.

row completely upset. This is due to an SET-induced WL assertion, which wrote the bit line (BL) values into the cells selected by that WL. Mavis et al. reported local WL mis-assertions in an SRAM with hierarchical WLs [27]. Other SET-induced errors can include writes to multiple cells or writes from read-out cells to others, in the case where a WL glitches on after the BLs are fully driven but before BL precharging occurs. Simply increasing driver size and node capacitance has been suggested to mitigate SET [28]. However, experimental measurements of the sizes required, and analysis of larger (merged) SRAM array sizes shows that this may be largely impractical [29].

6.1.3 MITIGATING SINGLE EVENT UPSET IN THE CACHE HIERARCHY AND ITS IMPACT ON LATENCY

Memories make extensive use of dynamic circuits. Dynamic circuits are more prone to upset—each dynamic circuit is a (half) latch, and as described later can thus *catch* a voltage transient. Consequently, cache SEE mitigation is difficult, as well as critical to the overall microprocessor hardness. In a cache, SEUs may alter the processor architectural state and SETs can cause inadvertent operations, either of which can produce an unrecoverable error or worse, SDC. Moreover, the hardening must be accomplished without adding substantial circuit delay or power. Generally, cache access time limits the microprocessor frequency and caches have accounted for as much as 43% of total embedded microprocessor power dissipation [30]. Although clock gating is the most effective means of limiting power dissipation, some hardening techniques can make clock gating difficult or impossible [31], and the clocks themselves may be inadvertently asserted by SETs [32,33]. Therefore, memory hardening must comprehend the possibility of clock mis-assertions, as well as other types of circuit errors.

L1 caches, that is, those closest to the core itself, most often use the late-way-select architecture shown in Figure 6.4, where multiple tag and data locations are accessed in parallel. The late-way-select scheme also hides the TLB delay, which occurs during the tag set decode and read times. This series TLB/cache tag compare allows the cache tags to use physical, rather than virtual addresses. Note that the decoders receive a portion of the address that is not mapped by the TLB, that is, the cache is virtually indexed.

For SEU protection, L1 caches often use only parity protection, which requires less latency to calculate and facilitates handling byte writes (by having a parity bit per byte). Using an error correcting code (ECC) for cache data, EDAC requires much larger granularity to be efficient; it is usually applied to an entire line. Thus, for EDAC, the existing data must first be read, the write data inserted, and the ECC is recalculated before the actual write. This requires multiple clock cycles and requires additional write buffering for L1 caches. In L2 caches, this is usually the case. In addition, the original data must have its ECC checked and any errors must be corrected before inserting the write data. Memory (and cache) data comprises processor architectural state—a corruption can catastrophically affect program operation. In a cache, which is composed of many subarrays in both the tag and the data storage blocks, as apparent in Figure 6.4, many other potential SET-induced errors can occur, including mis-assertion of the comparators or on the way select circuits, in addition to the SRAM-specific errors in decoders and or sense amplifiers and write circuits.

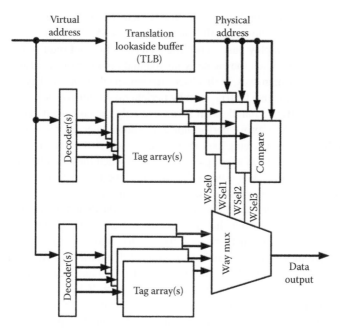

FIGURE 6.4 Four-way set associative L1 cache memory array and circuit organization. Separate tag and data arrays can be accessed simultaneously (there may be more than four, and one array can contain more than one way's data or tag). The late way select design allows time for the virtual to physical address translation by the translation look-aside buffer before comparing the tag values. The arrays are indexed using address bits within the page, which are not mapped by the translation look-aside buffer, that is, are the same for both virtual and physical addresses.

L2 caches typically access the tag first, and then only the array or multiple arrays that actually hold the data, to save power. The greater latency is tolerable, because the L2 cache is accessed only on an L1 cache miss. Moreover, significant latency is unavoidable—large L2 caches require pipeline stages just to allow for signal delay through on-chip wiring, for example, the L2 cache in Ricci et al. [9] has an eight core clock-cycle latency. A write-through cache simultaneously schedules a write to the main memory (or next level cache) when writing. A write-back (WB) cache only writes data to the main memory when data is deallocated. The latter minimizes bus utilization while a write-through cache can be protected from soft errors by parity checking rather than EDAC, because there is a redundant copy of the data in the next level of the memory hierarchy. In the event of a parity error, the L1 cache line can be invalidated and the good copy will be reloaded from the (presumably ECC protected) main memory when next requested.

6.2 RADIATION HARDENING BY DESIGN LEVEL 1 CACHE DESIGN

6.2.1 CACHE ORGANIZATION

This section presents a cache designed to provide hardness by detecting SEE-induced errors. The scheme allows the write-through cache to be invalidated when an error is detected, avoiding potentially otherwise undetectable errors propagating to the processor architectural state. The 16 KB cache is four-way set associative, and uses write-through and no-write-allocate policies [34,35]. Write-through operation allows the cache to employ parity checking. Thus, it does not require EDAC in the timing critical path. When the cache is invalidated on an error, it naturally reloads the data on subsequent misses. In this cache design, in addition to parity errors due to SEU, invalidation can also be triggered by peripheral circuit errors caused by SETs. Alternatively, a software handler can invalidate only the corrupted entries, but this does not significantly improve performance, as shown later. Stores are not gated by the tag hit. Stores perform a tag lookup in the cycle where the write is scheduled. On a hit, the data array write is subsequently executed when there is a free cache cycle. A one entry write buffer delays holds the data until the free cycle occurs.

The cache supports four basic operations: lookup, read, write, and global cache invalidation. Global cache invalidation simply clears the dual redundant valid bits in the tag array. This occurs when the cache is reset, or can be used to rapidly invalidate the contents when a cache SEE-induced error is detected. The dual redundant valid bits are in separate data arrays, so that one control signal mis-assertion cannot upset both. Single cycle writes of 8 bits (store byte) up to 128 bits (cache line fill) are supported. The cache is organized as 1024 cache lines, with 16 bytes (four 32-bit words) in each cache line. It is virtually indexed and physically tagged. Single-cycle fills maximize the memory to cache bandwidth, which at one line per clock is up to 16 GB/s at a 1 GHz clock rate.

The tag array and the two data arrays have 176,128 total bits of storage. Key tag comparator and hit circuits are dual redundant. The SRAM cell has the usual six-transistor structure, but is 3.6 times the size of a cell with smaller two-edge transistors drawn on the same logic design rules—this was done to provide total ionizing dose (TID) hardening (TID is beyond the scope of this chapter). The cell is wide and short to increase critical node spacing, reducing the probability of an MBU affecting the same parity group.

6.2.2 CIRCUIT DESIGN AND OPERATION

6.2.2.1 Circuit Design

Where possible, the hardened L1 cache tag and data arrays use identical circuits and layout cells, that is, the tag array circuits are mostly identical to those in the data array. The data array is split, with the most significant 18 bits of four words on the left and the least significant 18 bits of the same words on the right [36]. In each half data array, there are four words, with four banks for each word. A subbank contains thirty-two 72-bit rows as shown in Figures 6.5 and 6.6. Two subbanks share sense and write circuits, which are between them. There are two bytes, two WLs, and two ways in a data array row. The bit line swing is full V_{DD} to 0 V, so no sense amplifiers are required—NAND gates are used. This lack of a requirement for replica-timed sense amplifier enable signals and no BL multiplexer selects reduces the number of signals that may cause erroneous memory array operation due to SETs. The major data array circuits include the WL decoder, SRAM cell arrays, precharging circuits for the local subarray BLs, and global bit lines (GBLs) which act as the high speed

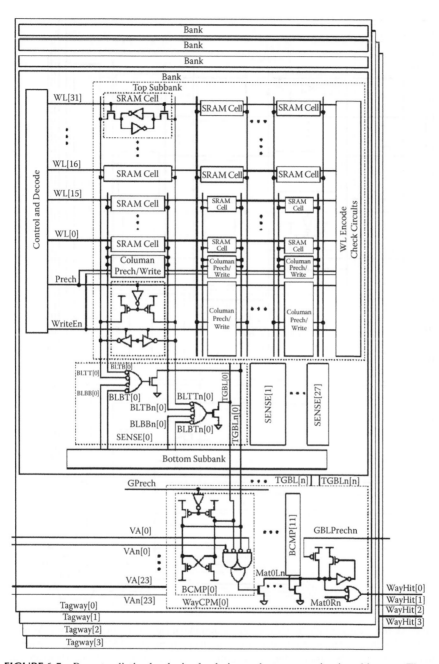

FIGURE 6.5   Recent radiation hardening by design cache tag array circuit architecture. There are 16 SRAM cells on each bit line to speed up the way selection signal timing. The 16-cell bit line groups are stacked with the top feeding through the bottom SRAM cells. The tag banks drive dynamic differential domino global bit lines, that are compared to the differential virtual addresses from the TLB. Differential signals ensure that there is no circuit race between the TLB and tag array.

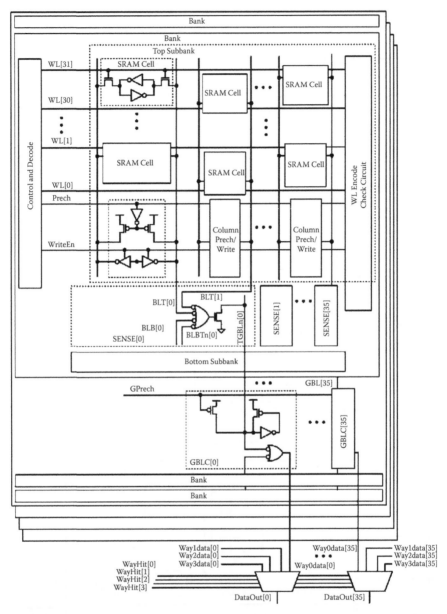

FIGURE 6.6 Recent radiation hardening by design data array circuit architecture. Ways 0, 1, 2, and 3 are interleaved. In the data array sub-bank, there are 32 SRAM cells on each bit line—the tag must be faster as the timing critical path is through the selects. There are two data arrays, each providing 36 bits (LSB array shown).

output bus and multiplexer, the write control, and way multiplexer, as shown in Figures 6.5 and 6.6.

Since the tag critical timing path through the comparators is limiting, making this path faster is beneficial. To this end, the tag array has 16 cells per BL, by using coincident BLs in a 32-cell column, whereas the data arrays have 32 cells per BL (Figure 6.5). This does, however, increase the peripheral circuit height, as each BL pair requires its own precharge and write circuits, as shown. The tag uses differential reads (Figure 6.5), whereas the data array reads are single ended (Figure 6.6). Both shorter BLs and differential readout reduce the tag read delay, improving the cache access time. The former nearly halves the delay from WL assertion to sense amplifier output. The latter eliminates one inversion delay and allows the one falling edge to the comparators to begin the compare in what would otherwise be a timing race.

6.2.2.2 Performance

Simulated cache circuit operation and critical path timing are shown in Figure 6.7. The timing critical path includes all the circuits that perform the address lookup in the cache. The access begins with the subbank clock assertion (top waveform). The BL and GBL precharge signals are deasserted to allow the BL readout to proceed. The WL decoder then asserts one WL high in banks that are selected (implicitly) by a gated clock assertion. The WL decoder resides after the clock, providing maximum time for address generation in the previous pipeline stage. The tag WL is asserted coincident with the data WL (not shown). One side of the high skew NAND4 BL sensing gates in the tag path drives the GBL pull down transistor on, driving the GBL low, as its complement GBLn stays high. Three of the four dynamic comparators miss and thus discharge, finishing the way selection. The GBLs of the top four banks and bottom four banks associated with each way pair are combined by a

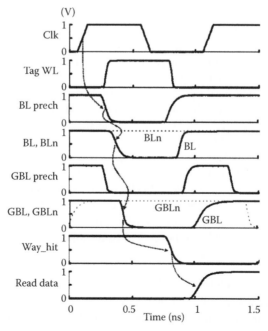

FIGURE 6.7 Simulated recent radiation hardening by design cache critical timing path simulation from clock to read data out. The data array timing is not shown as the critical path is through the way selection signals (buffered versions of WayHit[x]) to the way multiplexers.

NAND2 gate (see Figure 6.6). The static way multiplexer selects the data from the way that hit in the tag array for output. A static way multiplexer is required because all the dynamic comparators precharge to the hit condition. The final output (data out) from the cache way select multiplexer is 830 ps after the rising edge of clock that instituted the operation.

6.2.2.3 Power Dissipation

Beside radiation hardness, low power is another primary goal in the cache design. The design is divided into 32 data and 32 tag subbank halves receiving individually gated clocks. The gated clocks dramatically reduce the overall cache activity factor—a subbank is activated only when needed, with a maximum of one tag and four data subbanks active for a fill operation, and eight active for a load operation. Clocks are hardened as described in Chellapa et al. [33]. Additional clocks drive the central tag comparator circuits and data array output latches. The WL decoders minimize the address line capacitance and dissipate negligible power in unselected banks. Finally, the cache layout minimizes wire lengths, improving both propagation delays and power dissipation due to signaling. Physically, the tag array resides between the data arrays minimizing the hit signal propagation distance to the data array way multiplexers, and reducing the load (or instruction fetch) power and delay [34]. Redundant tag comparators drive the left and right array way selects, respectively, minimizing the load and delay of each, while allowing checks that determine if one mismatches the other.

6.2.3 Hardening Features

The cache arrays are SEU protected by parity and the bit column interleaving, which for each parity group is maximized by the physical organization. As illustrated in Figure 6.8a, the data arrays use byte parity, which matches the required write granularity. By interleaving bits belonging to two bytes, two WLs, and two ways, the bits belonging to the same parity group (same byte, WL, and way) are interleaved by eight, that is, separated by a distance equal to the width of seven SRAM cells. The tag arrays contain 28 bits per row, using nibble-sized parity groups to maximize the distance, a three SRAM cell separation as shown in Figure 6.8b. The parity is checked on any read values from the data array, but only the hit value in the tag. Any SET-discharged BL or GBL is detected by a parity mismatch. This requires that the critical node spacing of the SRAM cells is carried through all of the cache the column circuits. Alternate groups use even and odd parity. This ensures that a bank that fails to read due to a failed clock assertion will cause a parity error, as all BLs at logic one (V_{DD}) are not a valid output code. Mismatching valid bits constitute one of many detected error conditions.

6.2.3.1 Cache Error Detection Circuits

As mentioned, because an SET can cause an incorrect read or write, the cache uses extensive error detection circuits that monitor potentially erroneous operations due to an SET. These in turn are used to trigger an invalidation operation on the assumption that the architectural state may have been corrupted. The detectors create no new critical timing paths, so the approach taken here allows circuit delays and clock

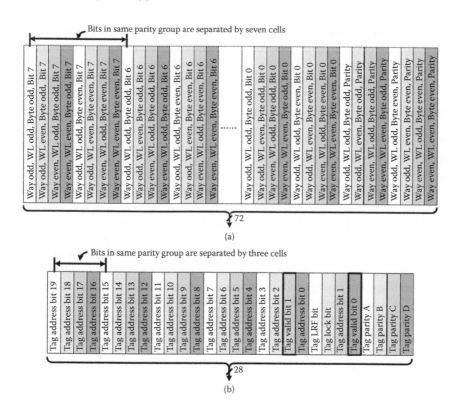

FIGURE 6.8 Physical bit interleaving for (a) data arrays and (b) tag arrays. Bits of the same color are in the same parity group. Spacing is greater in the data arrays, at seven cells between bits in the same parity group. In the tag arrays, the dual redundant valid bits are not included in the parity checks, but a mismatch is indicative of an error.

rates that rival unhardened designs. Because SETs are transient, that is, they can dissipate before a sampling clock edge, static checking circuits are inappropriate. The checking circuits are *one's catching*, with a set–reset type of operation. The basic dynamic error checker (EC) circuit is based on a domino gate, which as mentioned, with a keeper comprises a half latch—the half latch catches values in one direction. The EC output is subsequently latched locally through a classical dynamic to static conversion, or combined with other error flags, for example, from other memory banks (subarrays), and transmitted to the error control logic where it is latched. An SEE-induced error in the ECs may cause a false error. However, because they compose a very small portion of the cache area, and even in a satellite system, ionizing radiation strikes are infrequent compared to the operating frequency, the impact of both actual and false errors on the overall cache performance is negligible.

6.2.3.2 Single Event Upset Error Checking in the Periphery Circuits
Incorrect operations due to a SET that may upset the cache architectural state are protected against by a variety of error checking circuits to detect such events. While BLs are dynamic, and thus prone to SET discharge, the interleaving guarantees that

these will result in parity errors so long as the stored data is correct. The focus is on catching control signal, rather than BL errors, although some ECs rely on verifying the correct BL response. Many of the ECs are redundant—a production design may not require all of them, but the research design presented here used the philosophy of *better safe than sorry*. This section describes some of the WL assertion and BL precharge ECs with illustrations that are exemplary. The other circuits are described.

6.2.3.2.1   Word Line Assertion

If there is a WL error during a write, a write can be missed or data can be written to the wrong row as shown in Figure 6.3. Depending on the array organization, this can also cause the parity or ECC bits to be correct despite the data being incorrect, that is, correct data in the wrong location. To verify correct WL assertion, a dynamic WL encoder is used in each subarray (Figures 6.5 and 6.6) to regenerate the address based on the WL(s) actually asserted (Figure 6.9). The encoded address is subsequently compared with the input (cache set) address. A simple encoding comparison would catch a single WL mis-assertion but cannot detect all multiple WL mis-assertions. For example, an erroneous assertion of WL1 = 000001, would be masked by a correct assertion of WL3 = 000011 if both occurred. To ensure that all such cases are detected, dual complementary encoders are used. This approach will detect any combination of WL mis-assertions. Referring to Figure 6.9, the WL encoder regenerates the address as output WL_encode[n:0] and complement address WL_encodeN[n:0]. The B address is then inverted and, along with the A address, compared against the original set address. Physically, the encoders are interleaved because a blank in one corresponds to a bit in the other—there are no empty cells in the layout as evident by each BL pull down in the schematic driving the left (actual) or right (complement) versions. The layout structure is very similar to the regular SRAM columns (the pull downs are just the access devices). Because the positive and inverted polarity versions are not complementary, that is, they may not match, individual rather than cross-coupled keepers are used.

Because the cache outputs are dynamically multiplexed, that is, only the accessed bank should attempt to discharge the GBLs, incorrectly asserted WLs in unselected subbanks can also cause errors. This is addressed by the WL NOR checker, also shown in Figure 6.9. The right side of Figure 6.9 is the NOR WL EC that detects WL assertions during the precharge phase. This is the exemplary EC circuit. The high fan-in domino gate is triggered by a timing window, where the keeper half latch catches the error. The timing window duration is the time that PrechgN is asserted. The output is captured for subsequent processing and repair. The EC is explicitly reset to remove the error as shown.

6.2.3.2.2   Bit Line Precharge

Failure to precharge the SRAM BLs will cause the next read to be logically ORed with the previous read data. The BL precharge EC circuit monitors the BL and BLn of each column, from a top or bottom subbank (Figure 6.10). If a precharge assertion is missed due to a SET, one of BL or BLn will remain low during precharge. This condition is detected similarly, with a chopped clock (delayed assertion, but synchronous deassertion) to time the detection at the end of the precharge phase of operation. Figure 6.11 shows

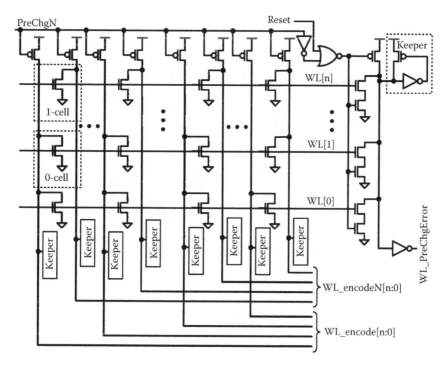

FIGURE 6.9 Word line encoder (left) and NOR checker (right). The encoders are interleaved, that is, layout positions of the transistors match the positions shown here.

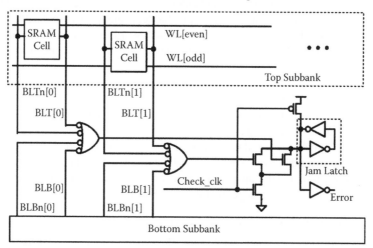

FIGURE 6.10 Local and global bit line precharge suppression detection circuit.

simulation of a suppressed BLn precharge being detected. The condition of the precharge failing to deassert in time will cause a timing problem, that is, the read will be slow. If the precharge fails to deassert completely, the read will be *snuffed* and the read-out data will be all 1's on the BLs. A similar checker is used to detect this case as well.

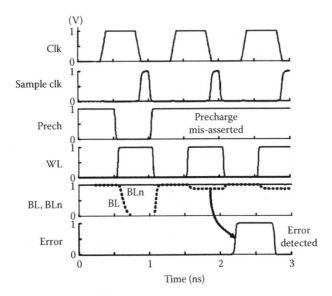

FIGURE 6.11   Simulated bit line precharge mis-assertion and error detection. Prech fails to de-assert in the second clock cycle beginning at 1.5 ns. This in turn, causes the bit line to fail to discharge. The checker detects that neither bit line or BLn discharged on the read and asserts an error.

The write driver is sufficiently strong so that it can override the BL precharge so writes occur correctly if both are simultaneously asserted. However, if the read data is incorrect due to a late or mis-asserted precharge, a single parity polarity will not signal an error in all cases. Consequently, in each word, two bytes use even parity and two bytes use odd parity (one in each side) so that if a read is delayed or fails completely, a parity error occurs.

6.2.3.2.3   Global Bit Line Precharge

The GBL output bus is dynamic as mentioned. Because the subbank read is dynamic, the GBL is a D2 domino bus that lags the local BLs in both read and precharge timing. The GBL precharge can thus be affected by a SET on its precharge control or contention due to a subbank driving its output incorrectly. Two checkers detect errors on the GBLs: one is for the unintended precharge assertions and the other detects failures of the GBLs to precharge, that is, suppression errors.

6.2.3.2.4   Write Bit Line and Enable Assertion

An inadvertently asserted write enable may affect an array that is leaving an accessed state or a reading array by incorrectly writing to it. A WREN and a NOR EC are used to detect these cases. These circuits are similar to that used for the WLs. Because the number of WREN signals is small, encoding is not required. The WREN NOR checker detects errors if any WREN is asserted during the precharge phase. A replica column checks that the BLs were asserted long enough to flip one SRAM cell as well. The cell is reset in each precharge phase.

6.2.3.2.5   Bit Line Read

A number of SET errors can cause incorrect or incomplete BL or BLn signal development. For instance, there may be an unintended short pulse in the BL precharge control signal, or delayed WL development, for example, from a delayed WL enable clock. Any such SET-induced control signal assertion or timing error can cause a delayed, incomplete or incorrect read. These conditions are detected by the read ECs, which reside in each subbank. This EC uses a replica column with a single active SRAM cell that has one side tied to V_{DD} internally, ensuring a read of 0V on the other (single-ended readout) BL. The error condition is signaled if both BL and BLn are high during the check window.

6.2.3.2.6   Tag Compare and Hit Determination

The tag comparators are dynamic so an SET can discharge the comparator domino node. By making these circuits, as well as the hit generation signals dual redundant, they are compared to ensure that any SEE-induced discharge is caught. A multiway hit would result in two ways being selected simultaneously—the merged data from a multiple way selection can have correct parity. Consequently, multiway hits also constitute an error condition. The overhead of the dual redundancy is minimized by using one comparator driver pair for each of the two data memory arrays. Thus, they are each nearly half the size of the one that would be required otherwise.

6.2.3.2.7   Cache Bank Clocks

Cache clock assertions are verified by redundant enables that are checked against actual clock assertions in the clock spine [33].

6.2.3.3   Error Checker Area Impact

The checking circuits comprise 15% of the cache area. This checking-based scheme also allows liberal use of dynamic circuits, which are protected by the checkers or parity. Such circuits are important in high performance cache design, as they allow higher performance in almost every stage, particularly the output busses and the tag comparisons.

6.2.4   MEASURED SPEED AND POWER

The RHBD cache design has been fabricated on both the standard and low standby power versions of a standard bulk CMOS 90-nm foundry process. The test chip is composed of a cache, programmable test engine, unhardened phase lock loop (PLL), and RHBD clock spine [34]. All test engine circuits, including I/O, are implemented as TMR circuits that self-correct by voting, with the exception of the foundry provided PLL. The PLL is shielded during heavy ion tests. The self-correcting TMR circuits have demonstrated exceptional hardness and thus do not contribute to the measured SEE-induced error rate.

The test chips verify the efficacy of the error checking approaches at speeds up to 200 MHz at $V_{DD} = 1.32$ V, limited by the test engine circuitry. A test that works around these timing limitations was written to experimentally determine the cache speed. This test demonstrates that the cache data arrays can write and read at 1.02 GHz at 1.3 V. At $V_{DD} = 1.45$ V, the measured cache power dissipation is 22.8, 25.1, and 26.4 mW at 85.7, 92.3, and 100 MHz, respectively. The standby power on this die is 71 μW. The 1 GHz peak cache power is calculated to be 266.1 mW, by linear extrapolation. At nominal V_{DD} = 1.2 V, the peak cache power dissipation is projected to be 182 mW at 1 GHz at 100% activity factor. In actual operation, stalls and other architectural inefficiencies will make the typical power lower. For a data cache, because loads and stores are about one-third of operations, typical cache power dissipation will be below 0.1 mW/MHz.

6.2.5 SINGLE EVENT EFFECT HARDNESS

6.2.5.1 Single Event Effect Testing

The test die has been exposed to the 15 MeV/u ion beam at Texas A&M University in air at room temperature, with V_{DD} = 1.45 V as well as at the Lawrence Berkeley NL cyclotron with 10 MeV/u ions in vacuum at V_{DD} = 1.2 V. Our previous broad beam SRAM tests on this fabrication process (e.g., used for the SRAM data in Figure 6.3) had shown that high V_{DD} exacerbates multibit upsets. N, Ne, Cu, Ar, Kr, and Au ions were used in the beam tests. Fluences from 5×10^5 to 2×10^7 particles/cm^2 were used, adjusted so that high LET MBUs would not overwhelm the 8 k entry field-programmable gate array error recording buffers. The maximum angle was 79° (0° being normal incidence) and the beam was incident on the die front that is, metallization side. The device package shielded part of the cache during those tests, as evident in upset bit maps, so in most tests the maximum angle used was 65°. During each beam test, the programmable test engine exercises the cache with lookups (processor load instructions) and (optionally) store operations at 100% activity factor. When a checking circuit detects an error, the test is interrupted, the reported error is recorded for analysis, and the tag and data array contents are dumped and analyzed to determine the impact of the error on the cache contents. The test is then resumed.

6.2.5.2 Tag Array Results

6.2.5.2.1 Memory Array Soft Errors

The tag array measured cross-sections versus LET are shown in Figure 6.12. Least squares Weibull fits to the incorrect hit/miss response, tag parity error, and valid bit upset are also shown. The bit cell upset errors exhibit a rapid rise and saturation with a uniformly low threshold LET that follows the bit cell cross-section, indicating that most are SEU. Note, however, that a local or global BL discharge will also register as a parity error as mentioned previously. The tag response is dominated by SEUs, with most of them causing an incorrect hit/miss response. Parity protects the lock, and least-recently filled bits as well as the stored line address. To save the power and wire density impact of delivering all tags to external parity check circuitry, only the tag that hits is checked. Because it is very likely that an SEU will change a tag to a value that is not in the test, and parity is only read on tags that hit, most tag SEUs do not cause a parity error.

Tag SEUs can thus *strand* a line by moving its address outside the program space. In a WB cache, the parity error will be detected on eviction, but the line cannot be written back to the correct memory location. However, the effect in a write-through cache is benign, as the line will be reloaded when the cache misses an access to it. In the event that the post-SEU address resides in an active program, the cache will register a hit. Thus, when the upset tag location is finally read, a parity error will

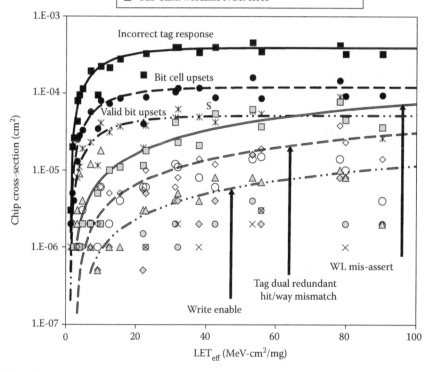

FIGURE 6.12 Tag SRAM array and control error cross-sections with the beam incident in the bit line direction. Parity errors are labeled *Bit cell upsets*, since they were confirmed by checking the actual data. Saturation cross-section is similar for the word line direction, but with reduced low linear energy transfer cross-section. The small tag array sizes allow control errors to essentially double the cross-section.

be triggered. Because of their redundancy, an SEU that flips one of the valid bits effectively invalidates the entry. However, the valid bits are split between the dual redundant comparator circuits. Thus, one may hit, if that valid bit is still correct, while the other may miss. The resulting response is a dual redundant hit/way select mismatch error. In this manner, many of the SEE-induced cache errors observed in the tests are correlated.

6.2.5.2.2   Tag Peripheral Circuit Errors

The SET-induced peripheral circuit errors have very different Weibull fits than SEU in the arrays, exhibiting softer saturation and higher threshold LET. Note, however, that the threshold LET is still low enough to be susceptible to neutron-induced upsets in a terrestrial environment—the primary mechanism for neutron upset is secondary ion generation, with the likely Mg ions generating up to 20 fC/μm of track length. At LET_{eff} = 50 MeV ·cm^2/mg, the WL errors have a cross-section about one-tenth of the tag bit SEU. Note that the relatively small, low power arrays have low threshold LET for a number of SET-induced control logic errors. The key result is that for tag arrays, such peripheral errors can contribute significant errors. The cross-sectional data provide quantitative insights as to the SET susceptibility of specific peripheral circuits. For instance, the global BL precharge EC detects few errors, and none at low LET. Moreover, many errors are strongly correlated. This was expected because many are redundant, as mentioned in Section 6.2.3.2. Consequently, depending on the IC hardness goals, some, perhaps most, of the checking circuits can be deleted. By focusing on the most likely errors, that is, WL assertion and write enable should provide sufficient reliability. By using static comparators, which are slower, some dual redundant mismatches can be eliminated, albeit at a negative operating frequency impact.

6.2.5.3   Data Array Results

The data array results in both the WL and BL directions are shown in Figure 6.13. The parity errors coincident with data load mismatches dominate. This can be expected based on the lower relative area and complexity of the peripheral circuits in the data array. The larger data arrays produce a larger cross-sectional gap between the array (SRAM cell) SEU and those due to SETs, because their relative peripheral circuit area is less than in the tag arrays. For the data arrays, no global BL precharge suppression errors were recorded, but incorrect assertion errors were. We surmise these were due to SETs in the relatively small driving circuitry. The precharge devices on these large busses are large and thus their driver cross-section is very small. As in the tag arrays, the most likely SET-induced error was WL mis-assertions. In an actual microprocessor, the root cause of such an error can be an SEU, because any address error will create a WL mis-assertion, and a latch containing the address can be upset at any pipeline stage. However, because the addresses into the cache in the test chip were self-correcting TMR logic, it can be concluded that these errors were within the cache arrays, most probably the decoders. Figure 6.14 compares the data array cross-sections for the angled beam oriented along the array WLs and along the BLs. The slightly lower WL direction SEU cross-sections can be attributed to smaller MBU extent, because cells are further apart in that direction. The Weibull fits show a small, but probably not meaningful difference for the SET-induced WL and write errors.

In most, but not all of the data array mismatches, a parity error was generated. With an MBU span of two, no upsets should escape the parity check. The root cause of these errors was traced to an interaction between the tag and data circuits. In these nonparity error data mismatch cases, tag way/hit mismatches were detected. Because the dual redundant comparators drive the left and right data arrays, respectively, different ways were selected in the separate arrays—each had correct parity. Note that in the case of a single comparator, if its result is erroneous, the chosen data would be returned

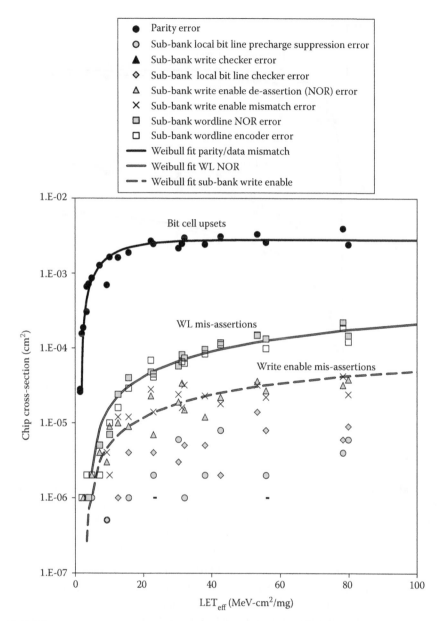

FIGURE 6.13 Recent radiation hardening by design cache data array and control error cross-sections with the beam incident in the bit line direction. Word line write enable mis-assertions dominate the control errors, although nearly all checkers did detect errors.

from the cache with no error reported, that is, an SDC would potentially occur. In the RHBD cache here, this error is caught by the way/hit comparisons of the dual redundant circuits and the SDC avoided. Only one clock spine error was reported by the device under test during a store and lookup test using Au, with an angle of 53°, the

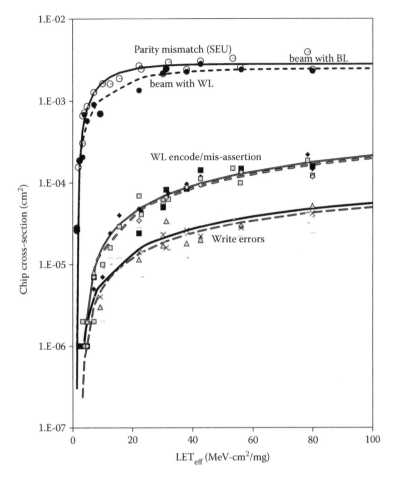

FIGURE 6.14 Recent radiation hardening by design Cache data array and control error cross-sections with the beam incident in the bit line direction (solid lines) and the word line direction (dashed lines). The higher cross-section in the bit line direction is due to more words being affected since the spacing is closer and there are no intervening N-wells. Cross-sections for control errors have little or no dependency on the beam direction.

beam along WL direction at LET_{eff} = 152 MeV·cm²/mg. Because the clock assertion is gated to control whether a subbank should *fire* and subbanks may be writing, this requires action to avoid potential state corruption. Importantly, the ion beam experiments showed that all array errors are caught—no cache state was ever upset without triggering an error.

6.2.6 PERFORMANCE IMPLICATIONS

The strategy employed for the RHBD write-through cache is to invalidate the contents whenever one of the ECs detects a SET-induced signaling error, so it cannot affect the processor architectural state. With this approach, a memory scrub is just a read of each location. Although it is likely that many of these invalidations are unnecessary, this approach avoids error *escapes*. In general, SEEs are rare events compared to the clock rate, even in the worst-case orbital conditions. For example, in geostationary earth orbit, the average time interval between particles that hit the cache and having an LET greater than 1 MeV·cm²/mg is calculated to be 3000 seconds or 50 minutes. Operating at a clock rate of 1 GHz, with a 10 core clock miss penalty and a 95% cache hit rate, the performance impact is well below 0.0001%.

6.3 ERROR DETECTION AND CORRECTION

The most commonly used ECC is based on Hsiao codes [37] which add r check bits for every group of k data bits forming a $k + r$ bit code word. A single error correction, double error detection code requires all legal code words to be at least four bit changes apart, that is, they have a Hamming distance of four. Codes capable of correcting multiple bit errors are starting to be used in L2 and L3 caches as they protect not only against MBUs, but also against sporadic bit failures, for example, bits that pass IC testing but become unstable over time. Before the data is written into the memory, the check bits are calculated and subsequently stored with the data. When data is read, the check bits are recalculated and exclusive ORed with the stored check bits to generate a syndrome code that is also r bits in length. If all the syndrome code bits are zero, then no single or double bit error has occurred and the data is assumed to be correct. Note that this may not be true—if enough bits have been upset to transform one legal code word into another, then an SDC has occurred. The syndrome is decoded to point to the incorrect bit, which is inverted. The primary cost of EDAC is that it generally requires a read, modify, and write cycle to insert data into a code word. L2 caches nearly always have such a buffer, but small embedded microprocessors rarely do.

L1 caches generally support byte writes for a *store byte* instruction commonly used in string operations. The hardened SPARC processor design in Gaisler [38] uses a (32, 7) BCH code for multibit error correction. This design is very slow, but even then double writes incur an extra clock cycle. In a high performance microprocessor L1 cache, this latency impact is unacceptable, because a pipeline stage must be added to the load/store operations. Moreover, this process must be implemented in a relatively sophisticated write buffer. An alternative is to add sufficient check bits to each byte so that each code word of data, consisting of the data byte and the check bits, supports EDAC. This implies greater than 50% memory overhead.

6.3.1 MEMORY SCRUBBING

EDAC and other parity-based schemes, with the proper column interleaving, can provide excellent SEU mitigation. However, they must be protected from upsets in the same code word due to multiple particles. The mean time to failure is given by

$$\text{MTTF} = \cfrac{t}{1 - \left[e^{-\lambda(M/C)t} + \lambda \dfrac{M}{C} e^{-\lambda(M/C)t} \right]^{BC}} \tag{6.6}$$

where M is the memory bank size, t is the interval between scrubbing operations, C is the number of code words in each bank, B is the number of memory banks, and λ is the bit failure rate. Even in orbital applications, low scrub rates are required—on the order of hours between scrub operations can be sufficient. A write-through cache can be periodically flushed because as mentioned, periodic cache invalidation is equivalent to a scrub, and is supported in a single clock cycle in the RHBD design presented here. The detection scheme here limits the invalidation performance impact, while mitigating this SDC risk.

6.4   REGISTER FILE HARDENING AND DESIGN

Because of the high performance required, microprocessor RFs commonly use parity protection, which is fast to generate, but can only allow soft-error detection. The Itanium server microprocessor incorporated parity in a unique low overhead scheme that does not add to the critical timing path, albeit at the cost of a few unprotected cycles [39]. A key problem in RF soft-error protection is the lack of time in the pipeline, that is, critical timing paths, for EDAC calculation in the stages leading to and from the RF. Mohr presented a sufficiently fast XY parity-based EDAC scheme to allow its use in RFs [36]. In addition, EDAC does not protect against erroneous data or operations caused by SETs, either in the RF or in the arithmetic logic unit (ALU)/bypass circuitry that produces and consumes data residing in the RF. Analysis of synthesized RFs on a 90-nm CMOS process showed that adding EDAC for each nibble incurred at minimum a 34% speed impact while saving only 45% the area required for TMR. For large 32- and 64-bit registers encoded as a single ECC code word, an extra pipeline stage was necessary [40]. This is unacceptable in a high performance processor.

6.4.1   RADIATION HARDENING BY DESIGN DUAL MODULAR REDUNDANT REGISTER FILES

The RHBD RF design described in this chapter combines microarchitectural and circuit level approaches, starting with dual mode redundancy (DMR) for compatibility with a DMR data path. DMR decoders generate redundant read and write WL signals [41]. Full SEE protection is provided by DMR combined with large critical node separation through bit interleaving and parity. The DMR data path is not essential; however, Ditlow, in work contemporaneous with design presented in this section, showed that DMR storage allows excellent RF SEU protection [42]. However, no SET protection was provided by Ditlow's design.

Like the RHBD cache, in this RF design, parity provides error detection. DMR allows one copy to provide clean data for SEU corrections; interleaving prevents multicell upsets (MCUs) from causing uncorrectable errors; and again, inadvertent operations are prevented by circuit approaches. The parity protection is in 5-bit groups to provide the greatest possible node spacing to mitigate MNC. A standard load-store CPU has two read ports and one write port, supporting the two ALU operands and single result, respectively. This design adds a third, that is, the Rt/Rd read port used to read a copy of the destination register contents about to be overwritten in the next cycle WB pipeline stage (see Figure 6.15). This extra port allows overwritten RF data to be restored when an error condition is detected. There are two basic ECs: in the first the DMR data are compared; the store is cancelled due to a detected data path error (SEU or SET) that is detected as the WB is in progress (the WB checker is

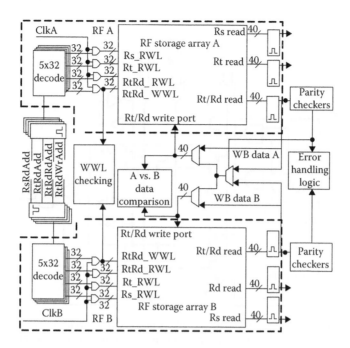

FIGURE 6.15 DMR register file configuration showing read and write port connections, including feedback to the Rt/Rd write port for error correction using data in the copy with correct parity to overwrite data in the other copy with a parity error. Correction is by nibble, so bits in both register file entries can be upset so long as both copies of any nibble are not. Mismatches in the A and B write-back data are caught on the input to the RF, stalling the processor to restart with a clean architectural state. (L. Clark et al., "A dual mode redundant approach for microprocessor soft error hardness," *IEEE Trans. Nuc. Science*, 58[6], © 2011 IEEE.)

labeled *A vs. B data comparison* in Figure 6.15). The second EC detects write word line (WWL) mis-assertions by the *WWL checking* circuit, shown in Figure 6.16 and described in detail in Section 6.4.1.1. DMR does not indicate which copy is erroneous, so the original (pre-WB pipeline stage) data are returned to the RF and then the instruction with nonmatching results is restarted from the beginning of the pipeline after the RF is checked for SEUs and those are repaired.

6.4.1.1 Circuit Design

Referring to Figure 6.15, the checking circuits are labeled *A versus B data comparison*, which check that the WB data match. The same checking mechanism is used to detect WWL mismatch errors that occur when one of the two WWL copies is asserted, but not the other. This logic is like that of the cache ECs described previously (see Figure 6.16). Clock chop circuits delay the turn on edge to allow the checked state to settle after the rising clock edge controlling the state transitions. The same delays are used to generate a pulsed precharge, which eliminates the need for a dynamic to static converter latch. The D2 domino ERRA and ERRB signals are discharged for any transient mismatch between the checked inputs. The full keeper

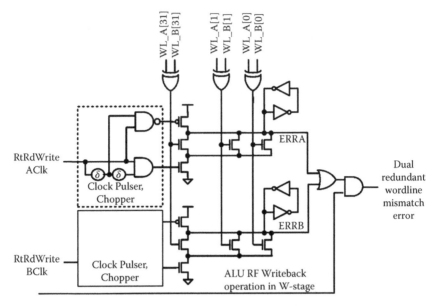

FIGURE 6.16 The dual redundant register file write word line checking circuit. This circuit follows the approach used in the recent radiation hardening by design cache. Pulsed clocks, which create the checking window, require a full keeper on dynamic nodes ERRA and ERRB.

(rather than the standard PMOS only keeper in Figure 6.9) holds the state even if the mismatch transient is short, regardless of the clock period.

The RF storage cell design is shown in Figure 6.17. The cell differs from that in a conventional dynamic RF in that a write operation requires simultaneous assertion of two redundant WWLs—this mitigates the possibility of inadvertent writes to RF locations due to control logic SETs. The aforementioned checker reports when a WWL deassertion aborts an intended write. When implemented with a DMR control and data pipeline, one WWL, WWLa, is controlled by the A pipeline copy, while WWLb is controlled by B pipeline circuits. Consequently, an SEE manifest as a control error that propagates to the RF from one copy cannot affect the architectural state in the RF. Read out is conventional, discharging the dynamic read bit line (RBL) if the cell state is a logic "1." Two of the read ports are connected to one side of the cell, and one to the other, to increase the minimum storage node capacitance, which would dominate the cell SEU hardness. When an error occurs, either due to a mismatch of the DMR input data or WWLs, the register that is the target of the write is restored to its prior state—putting back the correct architectural state. Similarly, if one of the WWLs is truncated, which may partially write both copies, the entire operation is started over from the previously correct architectural state. Soft errors manifest as read word line (RWL) errors, if they affect the data read, are corrected similarly.

6.4.1.2 Error Correction

The design is intended to use software-controlled repair mechanisms [41], but the key aspects could be handled by hardware. Basically, once an error is detected, any

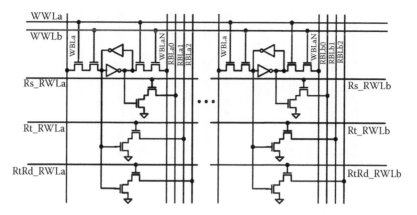

FIGURE 6.17 DMR register file cell circuit showing the read and write ports. The inverting one port allows added capacitance on both sides of the storage cell to increase Q_{crit}.

SEU can be corrected in 64 clock cycles, as 32 two-cycle operations. Each A and B 32-b register copy is read simultaneously; if they mismatch, the A or B nibble with the correct parity overwrites the copy with incorrect parity. Hardware multiplexing paths between the copies carry out these operations. Only accumulated errors affecting the same parity group in both copies result in an uncorrectable error, but due to large physical separation between cells in the same parity nibble these are almost certainly due to separate SEU events, that is, error accumulation.

Error accumulation is avoided by background scrub operations. Besides allowing potentially incorrect data to be backed out when a mismatch between the A and B data copies is detected, the Rt/Rd read port also allows opportunistic scrubbing. When an instruction will not write to the RF in the WB pipeline stage, it is not a load instruction, the Rt/Rd read port is opportunistically used to read the next register pointed to by a scrub pointer. The read-out value is automatically checked for parity errors. This allows all registers to be sequentially read in a rotating fashion as normal operations occur. The scrubs opportunistically use the Rt/Rd port, when it is not required by an instruction writing back to the RF.

6.4.1.3 Physical Design

The RF layout showing the bit interleaving is shown in Figure 6.18. The decoders are spatially separated, to prevent SEEs from corrupting both copies simultaneously. Like the cache design composing Section 6.2, the bits of each parity group are interleaved to maximize the separation and thus, likelihood of MNC affecting 2 bits in the same parity group. The interleaving allows a spacing between such critical nodes of 16 standard cell heights (greater than 32 μm).

6.4.1.4 Single Event Effect Testing

Two basic broad beam proton and ion tests were performed, static RF and dynamic RF/ALU/bypass tests. The former exercises the RF memory statically, that is, without the data path in operation, to determine the RF SEU and SET characteristics

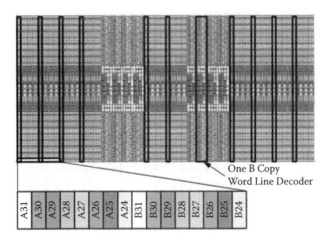

One B Copy
Word Line Decoder

A31 A30 A29 A28 A27 A26 A25 A24 B31 B30 B29 B28 B27 B26 B25 B24

FIGURE 6.18 DMR register file physical design. Interleaving of the register file columns, where A indicates the A copy, B indicates the B copy, and the number is the bit number in the word.

independent of the data path errors. The latter have shown that the DMR RF, when used with a DMR data path and control logic, can effectively mitigate all soft errors at speed, where SETs are prevalent. As in the cache testing, fully self-correcting TMR circuitry is used to test the RF in the test chips.

In the static tests, the RF is written with all 0's, all 1's, or a checkerboard pattern. The RF is then read continuously by the test engine using only a single read port. In the event of an error, the upset stored data are corrected and the test continues. Each location is read a second time if there is an error. A SET in the RF dynamic RBL circuits, address, or decoding, is detected if the read data is initially incorrect, but is correct on the second read. Both RF (A and B) copies are tested, which reflects the actual DMR processor operation, where each redundant data path/control copy receives unchecked and uncorrected RF data.

The RF WB checking circuits are used in the dynamic tests to characterize the data path checking functionality. In these tests, a DMR Kogge–Stone adder, complete with input and output bypass paths as used in a standard microprocessor performs arithmetic operations on the RF data. The types of errors generated by the data path can also be examined in this mode. Thus, these tests allow SETs to corrupt many bits in one copy, for example, due to transients on the adder prefix circuitry corrupting the most significant bits—propagating carry errors, or incorrect function selection or data bypassing due to control SETs in one DMR pipeline copy. When an error is detected in these tests, the check type is logged. Moreover, in these tests, parity checking is turned off, so as to propagate more errors into the data path. Referring to Figure 6.15, these types of errors create a *WB error* indicating the A and B WB data mismatched (A vs. B data comparison in Figure 6.15), or a *WWL error*, which indicates that the WWL checker found a WWL assertion discrepancy (WWL checking in Figure 6.15) for example, due to different A and B addresses. For these errors, the test engine dumps the RF contents, allowing accumulated RF SEUs to be examined. Finally, the RF is reloaded with correct contents and the dynamic test resumes.

6.4.1.4.1   Heavy Ion Testing Results

Boron, oxygen, neon, argon, and copper broad beam ions with nominal (normal incidence) LET of 0.89, 2.19, 3.49, 9.74, and 21.17 MeV·cm²/mg, respectively at angles ranging from 0 (normal) to 70° were used to test the RF. Testing was primarily performed at 100 MHz and 200 MHz at V_{DD} = 1, 1.2, and, 1.4 V. At angles, the beam was aimed across the words, the direction for which MBUs would be most likely. During heavy ion testing, the PLL clock source was shielded. The RF bit cross-section (Figure 6.19) shows a low threshold LET of approximately 1 MeV·cm²/mg, presumably due to near minimum sized PMOS transistors. However, it is perceptibly higher than the SRAM cells in the cache. We attribute this to the significantly greater capacitance of the multiple read ports, which as mentioned were distributed to both storage nodes. Over 90% of the SEUs were single bits. The largest MNC extent was 2 bits. Over 98% of the static test SETs were detected with logic "1" stored, clearly indicating that the dynamic RBLs dominate the RF SET cross-section.

Chip-level RF cross-sections measured with dynamic testing, for argon ions with nominal (normal incidence) LET = 12.89 MeV·cm²/mg, at 40°beam angle, is shown in Figure 6.20. WB error and WL error refer to how the initial error was detected; SEU refers to other errors inside the RF when the contents were dumped; scrubbing was not run, but writes naturally update some RF contents. The bar labeled *WB error* indicates that an A/B copy mismatch detection triggered an RF dump. The SEU bars indicate only RF errors that were not the original cause of the detected error, that is, accumulated upsets. WB error MBUs are shown, indicating SETs in the data path are detected. Their prevalence demonstrates the importance of protecting against data path SETs in an RHBD design. The *MBU* column shows upsets of multiple cells, where two-bit upsets predominated, as in the static tests. However, these can also include the logic fan-out-induced errors mentioned previously.

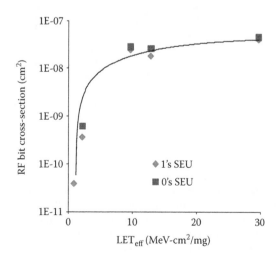

FIGURE 6.19   Measured register file bit cell heavy ion cross-section determined by static testing at V_{DD} = 1.2 V. (L. Clark et al., "A dual mode redundant approach for microprocessor soft error hardness," *IEEE Trans. Nuc. Science*, 58[6], © 2011 IEEE.)

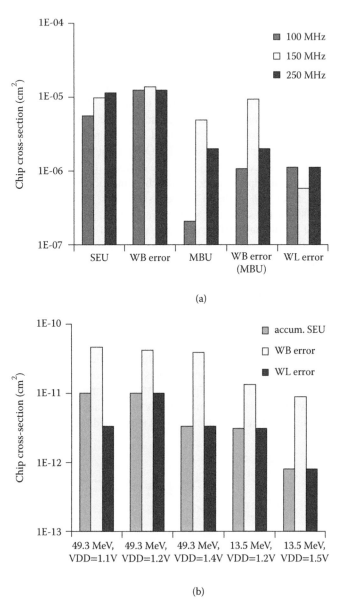

FIGURE 6.20 (a) Measured register file and ALU/bypass circuit heavy ion cross-section from dynamic testing exercising both at V_{DD} = 1.2 V and linear energy transfer = 12.89 MeV/cm²/mg (40° angle). (b) Measured register file and ALU/bypass circuit proton cross-section from dynamic testing exercising both at multiple proton energies and V_{DD}. write-back error and word line error refer to how the initial error was detected; Accumulated SEU refers to other (SEU) errors inside the register file when the contents were dumped. (L. Clark et al., "A dual mode redundant approach for microprocessor soft error hardness," *IEEE Trans. Nuc. Science*, 58[6], © 2011 IEEE.)

6.4.1.4.2 *Proton Testing Results*

Proton beam energies of 49.3 and 13.5 MeV were used. Although tests with the PLL were performed, most of the testing was performed at 60 MHz using the PLL bypass mode, because PLL upsets can cause test chip malfunctions due to bad clocks and the PLL is difficult to shield from protons. For most RF tests, the total fluence was $5(10^{11})$ protons/cm^2. Flux was $2.3(10^8)$ to $8.9(10^8)$ protons/cm^2/s. For the 49.3 MeV proton energy, the measured cell (per bit) cross-sections for static SEU testing of the RF were $3.71(10^{-14})$ cm^2, $2.68(10^{-14})$ cm^2, and $1.12(10^{-14})$ cm^2 at $V_{DD} = 1$, 1.2, and 1.4 V, respectively. Similar response was measured for all 0's and all 1's patterns, which we attribute to the well-balanced RF cell storage node capacitance. In static testing, SETs were observed only for the read all 1's case but with a very low cross-section. This is also as expected—N-type diffusions predominate on the BLs, so they are generally pulled in one direction by a SET. For the 13.5 MeV proton beam energy, the measured cell cross-sections for static SEU testing of the RF were $1.07(10^{-14})$ cm^2, $8.98(10^{-15})$ cm^2, and $1.72(10^{-15})$ cm^2 at $V_{DD} = 1$, 1.2, and 1.4 V, respectively. MCUs were rare and the longest upset extent was 2 bits. Cross-sections V_{DD} supply voltages and proton energies measured in the dynamic tests comprise Figure 6.20b. A modestly reduced cross-section at higher V_{DD} and lower energy are evident.

6.5 CONCLUSIONS

This chapter has focused on novel error detection techniques to mitigate soft errors in high speed microprocessor memories such as L1 caches and RFs. A write-through policy allows the former to be invalidated when errors are detected, and the likelihood of errors due to cache control logic was shown to be significant, particularly in the tag memory arrays. The cache approaches have been experimentally proven to be effective, although a reduced EC set may be adequate, reducing the area overhead below the 15% achieved here.

The RF testing data confirm that DMR hardening techniques are effective for mitigating errors in these timing critical memories. SEU MCU extent was far below the critical node spacing achieve by column interleaving. While many bit errors (representing a potentially large equivalent MCU extent) were detected in the dynamic tests, the many bit errors occurred only in one, for example, the A but not B, copy—clearly indicating data path logic error propagation. Regardless, these operations are restarted. The DMR RF provides excellent SEU immunity, and is consistent with commercial designs that have split the RF for performance, rather than for hardening reasons, because the RF read port overhead often dominates that of the storage itself.

Adding ECs to detect errors introduced into or produced by the memory periphery circuits such as sense, write, prechargers, and the decoders has been shown to be effective in mitigating such SET and pipeline latch SEU errors. The EC circuits themselves do not contribute to the circuit critical paths substantially, although mechanisms must be provided to undo some of the side effects of inadvertent operations, for example, the RF writes.

REFERENCES

1. K. Label et al., "Single event effect proton and heavy ion test results for candidate spacecraft electronics," *Proc. Radiation Effects Data Workshop*, 1994, pp. 64–71.
2. D. Rea et al., "PowerPC RAD750-A microprocessor for now and the future," *Proc. IEEE Aerospace Conf.*, Big Sky, MT, 2005, pp. 1–5.
3. N. Haddad et al., "Second generation (200MHz) RAD750 microprocessor radiation evaluation," *Proc. RADECS*, 12th European Conference, Sevilla, Spain, 2011, pp. 877–880.
4. R. Lacoe, J. Osborne, R. Koga, and D. Mayer, "Application of hardness-by-design methodology to radiation-tolerant ASIC technologies," *IEEE Trans. Nuc. Sci.*, 47(6):2334–2341, December 2000.
5. C. Hafer et al., "LEON 3FT processor radiation effects data," *Proc. Radiation Effects Data Workshop*, Quebec, Canada, 2009, pp. 148–151.
6. S. Guertin, C. Hafer, S. Griffith, "Investigation of low cross-section events in the RHBD/FT UT699 Leon 3FT," *Proc. Radiation Effects Data Workshop*, Las Vegas, NV, 2011, pp. 1–8.
7. F. Sturesson, J. Gaisler, R. Ginosar, T. Liran, "Radiation characterization of a dual core LEON3-FT processor," *Proc. RADECS*, Sevile, Spain, 2011, pp. 938–944.
8. L. Clark et al., "An embedded 32-b microprocessor core for low-power and high-performance applications," *IEEE J. Solid-state Circuits*, 36(11): pp. 1599–1608, November 2001.
9. F. Ricci et al., "A 1.5 GHz 90-nm embedded microprocessor core," *VLSI Cir. Symp. Tech. Dig.*, pp. 12–15, June 2005.
10. J. Haigh et al., "A low-power 2.5-GHz 90-nm level 1 cache and memory management unit," *IEEE Journal of Solid-state Circuits*, 40(5):, pp. 1190–1199, May 2005.
11. S. Yang et al., "A 32nm high-k metal gate application processor with GHz multi-core CPU," *ISSCC Tech. Dig.,* San Francisco, CA, 2012, pp. 214–215.
12. D. Bradley, P. Mahoney, B. Stackhouse, "The 16 kB single-cycle read access cache on a next-generation 64 b Itanium microprocessor," *ISSCC Technical Digest*, San Francisco, CA, 2002, pp. 110–111.
13. C. Hsieh, P. Murley, and R. O'Brien, "A Field-funneling effect on the collection of alpha-particle-generated carriers in silicon devices," *IEEE Elec. Dev. Let.*, EDL-2(4): pp. 103–105, 1981.
14. F. Sexton et al., "SEU simulation and testing of resistor hardened D-latches in the SA3300 microprocessor," *IEEE Trans. Nuc. Sci.*, 38(6): pp. 1521–1528, December 1991.
15. P. E. Dodd and F. W. Sexton, "Critical charge concepts for CMOS SRAMs," *IEEE Trans. Nucl. Sci.*, 42: pp. 1764–1771, 1995.
16. M. Gadlage et al., "Single event transient pulsewidths in digital microcircuits," *IEEE Trans. Nuc. Sci.*, 51: pp. 3285–3290, December 2004.
17. J. Benedetto et al., "Heavy ion induced digital single-event transients in deep submicron processes," *IEEE Trans. Nuc. Sci.*, 51(6): pp. 3480–3485, December 2004.
18. D. Kobayashi, T. Makino, and K. Hirose, "Analytical expression for temporal width characterization of radiation-induced pulse noises in SOI CMOS logic gates," *Proc. IRPS*, Montreal, Quebec, Canada, pp. 165–169, 2009.
19. A. Holmes and L. Adams, *Handbook of Radiation Effects*, Oxford University Press, New York, 2002.
20. D. Heidel et al., "Low energy proton single-event-upset test results on 65 nm SOI SRAM," *IEEE Trans. Nuc. Sci.*, 55(6): pp. 3394–3400, December 2008.
21. T. Calin, M. Nicolaidis, and R. Velazco, "Upset hardened memory design for submicron CMOS technology," *IEEE Trans. Nuc. Sci.*, 43(6): pp. 2874–2878, December 1996.

22. R. Koga et al., "Single ion induced multiple-bit upset in IDT 256K SRAMs," *Proc. RADECS*, St Malo, United Kingdom, pp. 485–489, 1993.

23. T. Karnik, P. Hazucha, and J. Patel, "Characterization of soft errors caused by single event upsets in CMOS processes," *IEEE Trans. Secure and Dependable Computing*, 1(2): pp. 128–144, April 2004.

24. G. Gasiot, D. Giot, and P. Roche, "Multiple cell upsets as the key contribution to the total SER of 65 nm CMOS SRAMs and its dependence on well engineering," *IEEE Trans. Nucl. Sci.*, 54(6): pp. 2468–2473, December 2007.

25. P. McDonald, W. Stapor, A. Campbell, and L. Massengill, "Non-random single event upset trends," *IEEE Trans. Nuc. Sci*, 36(6): pp. 2324–2329, December 1989.

26. L. Jacunski et al., "SEU immunity: The effects of scaling on the peripheral circuits of SRAMs," *IEEE Trans. Nuc. Sci*, 41(6): pp. 2324–2329, December 1989.

27. D. Mavis et al., "Multiple bit upsets and error mitigation in ultra-deep submicron SRAMs," *IEEE Trans. Nuc. Sci.*, 55(6): pp. 3288–3294, December 2008.

28. Q. Zhou and K. Mohanram, "Gate sizing to radiation harden combinational logic," *IEEE Trans. CAD*, 25: pp. 155–166, January 2006.

29. K. Mohr and L. Clark, "Experimental characterization and application of circuit architecture level single event transient mitigation," *IRPS Proc.*, pp. 312–317, April 2007.

30. J. Montanaro et al., "A 160 MHz 32 b 0.5 W CMOS RISC microprocessor," *IEEE J. Solid-state Circuits*, 31(11): pp. 1703–1714.

31. K. Warren et al., "Heavy ion testing and single event upset rate prediction considerations for a DICE flip-flop," *IEEE Trans. Nuc. Sci.*, 56(6): pp. 3130–3137, December 2009.

32. N. Seifert et al., "Radiation induced clock jitter and race," *IRPS Proc.*, pp. 215–222, April 2005.

33. S. Chellapa, L. Clark, and K. Holbert, "A 90-nm radiation hardened clock spine," *IEEE Trans. Nuc. Science,* 59(4): pp. 1020–1026, 2012.

34. X. Yao, D. Patterson, K. Holbert, and L. Clark, "A 90 nm bulk CMOS radiation hardened by design cache memory," *IEEE Trans. Nuc. Science*, 57(4): pp 2089–2097, August 2010.

35. X. Yao, L. Clark, D. Patterson, and K. Holbert, "Single event transient mitigation in cache memory using transient error checking circuits," *Proc. CICC*, San Jose, CA, pp. 1–4, 2010.

36. K. Mohr et al., "A radiation hardened by design register file with low latency and area cost error detection and correction," *IEEE Trans. Nuc. Sci.*, 54(4): pp. 1335–1342, August 2007.

37. M. Hsiao, "A class of optimal minimum odd-weight-column SEC-DEC codes," *IBM J. Res. Develop.*, 14(4): pp. 395–401, July 1970.

38. Gaisler, J., "A portable and fault-tolerant microprocessor based on the SPARC v8 architecture," *Proc. Dependable Systems and Networks*, 2002, pp. 409–415.

39. E. Fetzer, D. Dahle, C. Little, and K. Safford, "The parity protected, multithreaded register files on the 90-nm Itanium microprocessor," *IEEE J. Solid-State Circuits*, 41(1): pp. 246–255, January 2006.

40. R. Naseer, R. Bhatt, and J. Draper, "Analysis of soft error mitigation techniques for register files in IBM Cu-08 90nm technology," *Proc. IEEE Int. Midwest Symp. Circ. and Sys.*, San Juan, PR, pp. 515–519, 2006.

41. L. Clark, D. Patterson, N. Hindman, K. Holbert, and S. Guertin, "A dual mode redundant approach for microprocessor soft error hardness," *IEEE Trans. Nuc. Science,* 58(6): pp. 3018–3025, 2011.

42. G. Ditlow et al., "A 4R2W register file for a 2.3GHz wire-speed POWER™ processor with double-pumped write operation," *ISSCC Tech. Dig.*, San Francisco, CA, 2011, pp. 256–258.

7 Design Space Exploration of Wavelength-Routed Optical Networks-on-Chip Topologies for 3D Stacked Multi- and Many-Core Processors

Luca Ramini and Davide Bertozzi

CONTENTS

7.1   INTRODUCTION

Photonic interconnect technology is considered a promising way of relieving power and bandwidth restrictions in next generation multi- and many-core integrated systems.

Optics could solve many physical problems of on-chip interconnect fabrics, including precise clock distribution, system synchronization (allowing larger synchronous zones, both on-chip and between chips), bandwidth and density of long interconnections, and reduction of power dissipation. Optics may relieve a broad range of design problems, such as crosstalk, voltage isolation, wave reflection, impedance matching, and pin inductance [1]. It may allow continued scaling of existing architectures and enable novel highly interconnected or high-bandwidth architectures.

Silicon photonics has advanced substantially in recent years and has demonstrated many of the key components for the implementation of optical networks-on-chip (ONoCs) in an integrated CMOS process [2].

Such components include power-efficient laser sources, low-loss waveguides, high-bandwidth modulators, broadband photonic switches, and high-sensitivity photodetectors. The improvement of the quality metrics of these components, as well as the integration route with CMOS manufacturing processes, is being relentlessly pursued.

However, despite the arguments in favor of optics for interconnects on the silicon chip, and the success of technology platforms fostering a fables silicon photonics ecosystem [3], ONoCs are fundamentally still at the stage of a promising research concept.

At least three reasons can be identified. First, the adoption cost of this technology is still very high, far away from that of the inexpensive on-chip electronic interconnects. This implies that the new interconnect technology will become practically viable only when it will be proven to deliver out-of-reach performance or power figures in the context of compelling use cases. Second, technology maturity is currently lagging behind actual industrial standards (e.g., due to thermal sensitivity concerns), and again only compelling cases for silicon nanophotonic links can foster a larger investment on technology development. Finally, the availability of mature optical components is not currently supported by mature cross-layer design methods and tools for system design. System designers should be equipped with the needed methodologies and toolflows to do design with the new interconnect technology.

This chapter focuses on the latter challenge illustrated previously, and reports on the latest developments in the design methodologies for ONoC design. Therefore, the focus will not be on the technology layer, but rather on cross-layer design aspects with photonic interconnect technology. Similar to electronic interconnect fabrics, also optical interconnection networks suffer from the design predictability gap, which will be quantified in this chapter by pointing out the difference between abstract projections and postplace and route measurements of insertion losses across optical paths. Such deviation may lead to underestimate the main source of ONoC static power, namely power consumption of optical laser sources. This concern affects ONoC design from the ground up, as its consideration is of the utmost importance to properly select a network connectivity pattern (i.e., a topology) that can preserve its quality metrics across the design hierarchy, up to the final technology mapping.

The main cause for the design predictability gap is the effect of place and route constraints on topology layout. In fact, placement of optical components on the 2D silicon-on-insulator (SOI) surface, and the consequent routing of waveguides to connect them together, should account for the actual position of initiator and target network interfaces. Many ONoC topologies make unrealistic assumptions on such positioning, hence resulting in inaccurate floorplans. The main source of such inaccuracy consists of the actual number of waveguide intersections, which has a direct implication on the optical power loss in the network. Due to the wiring intricacy of actual layouts, the number of crossings might in the end turn out to be much higher than that projected by means of an extrapolation from the logic topology. Especially in 3D stacked architectures, initial interface positioning stems from a comprehensive floorplanning strategy of the system as a whole in an attempt to save area, therefore, there are very few degrees of freedom with respect to the initial assignment.

For this reason, this chapter considers a 3D stacked multicore architecture as the baseline experimental setting, and reports on the engineering effort of an optical layer capable of delivering both intracluster and off-chip connectivity. Without lack of generality, wavelength routing is selected as the reference routing methodology. The chapter will highlight the robustness of the main ONoC topologies with respect to the design predictability gap, which may lead to quality metrics degradation, but also to practical infeasibility once the actual implementation constraints are considered.

The set of topologies under test is selected also to shed light on the main properties of two fundamental categories of connectivity patterns: filter-based topologies as opposed to ring structures.

7.2 SILICON PHOTONICS AS A TECHNOLOGY ENABLER

Silicon is a well-known material used in complementary metal–oxide–semiconductor (CMOS) microelectronic chips. Silicon photonics offers the compatibility with standard CMOS fabrication processes, enabling dense integration with advanced microelectronics. The capability of silicon photonic devices to be integrated into complex platforms, coupled with decades of high-quality development driven by the microprocessor industry, allows their low-cost and mass-volume production.

Silicon photonics provides also an excellent high index contrast between the refractive index of the core (typically 3.5 for crystalline silicon) and the above cladding (typically 1.5 for silicon dioxide).

This high index contrast generates higher optical modes confinement, so that the optical signal can be easily guided by devices with subwavelength dimensions.

Hereafter an overview of all silicon photonic devices of interest for ONoC implementation is presented, starting from optical links up to laser sources and photodetectors [4].

7.2.1 OPTICAL LINKS

The optical link is the fundamental building block that must be used to guide the high-speed optical signals from the photonic source up to the receiver. The optical link is commonly referred to as *waveguide* in the optical domain.

Recently, *submicrometer crystalline silicon waveguides* [5] have been an excellent option for optical links. Such a structure is able to propagate parallel wavelengths with terabit-per-second data rates throughout the whole chip. Thanks to these appealing properties, it is possible to further build straight, bend, and crossing waveguides, as well as couplers, thus providing all the basic structures for *optical communication channels*.

From experimental characterization, it has been demonstrated that crystalline silicon waveguides are able to deliver data rates up to 1.28 Terabit/s, including 32 wavelengths modulated at 40 Gbit/s each, through a communication link of 5 cm [5].

Waveguide crossings (i.e., intersections of two waveguides) represent the major source of optical power degradation across optical paths although they cannot be really avoided on a single plane chip. Nonnegligible attenuations are incurred across optical paths also in terms of propagation loss (in straight waveguides), or bending loss (in bending waveguides).

In the open literature, two-dimensional tapers have been proposed in an attempt to minimize the crossing loss across optical paths. Among the most relevant ones, it is worth mentioning the standard *elliptical taper* [6] and the *multimode interference (MMI) taper* [7].

In contrast to submicrometer crystalline silicon waveguides, *deposited silicon nitride waveguides* offer many advantages for integrated photonics. Unlike crystalline silicon, the antagonist silicon nitride can be deposited in multiple layers, similar to electronic wires.

The latter case has the capability of eliminating in-plane waveguide crossing losses, once vertical optical couplers are in place [8].

Experimental results show that the transmission of high-speed optical data through a deposited silicon nitride waveguide can achieve 1.28 Terabit/s (as usual including 32 wavelengths modulated at 40 Gbit/s each) throughout a 4.3 cm silicon nitride waveguide [9].

7.2.2 MODULATORS

The *silicon electrooptic modulator* is an essential device for photonically enabled on-chip links, as it performs high-speed conversion of an electrical signal into an optical one.

It encodes data on a single wavelength that can be then combined with additional optical signals through wavelength division multiplexing (WDM) on the same physical medium, thus resulting in a cohesive wavelength parallel optical signal. *Crystalline silicon microring resonator electrooptic modulators* are the most recently used devices among those presented in the open literature. They consist of a microring resonator (MRR) configured as p-doped-intrinsic-n-doped carrier injection device. The standard operation of these devices relies on nonreturn-to-zero, and on on-off-keyed modulation signals.

To achieve high-modulation rates that are typically limited by carrier lifetimes, modulators are driven using a particular mechanism called preemphasis method [10]. The *electrooptic modulator has also been proposed for polycrystalline silicon* [11]. The grain boundaries inherent in the material result in increased optical loss due to scattering and absorption, which end up reducing free-carrier lifetime, and may increase the intrinsic speed of the modulator accordingly. Unlike crystalline silicon, polycrystalline silicon can be also deposited and stacked with other silicon photonic materials for multilayer integration. Finally, modulators can be also embedded across arrays (*silicon electrooptic modulators arrays*), so to deliver a major bandwidth boost along the communication channel.

Hence, at the output stage of each array, the optical data stream contains multiple wavelengths ready to be transmitted throughout the interconnection network, ending up at the receiver stage.

7.2.3 Photonic Switching Elements and Optical Routers for Optical Networks-on-Chip

Broadband *photonic switching elements* (PSEs) with one or two inputs and two outputs are the fundamental building blocks of an ONoC. The former case (one input and two outputs) consists of a MRR positioned adjacent to a waveguide intersection.

Alternatively, a parallel switching element denoted as 1×2 comb switch has been also presented in the recent literature [12]. Simultaneous switching of 20 continuous-wave wavelength channels with nanosecond transition times has been demonstrated by using the comb-switching technique. 2×2 PSEs instead (two inputs and two outputs) consist of a waveguide intersection and two ring resonators. The switching function is achieved through resonance modulation via carrier injection into the ring.

The fundamental switching elements introduced earlier (1×2 PSE, 2×2 PSE) are typically composed to derive higher order switching structures.

A 4×4 nonblocking nanophotonic switching node [6] is a clear example thereof. This optical router may include either eight 1×2 PSEs or a mixture of them, so that each input port is capable of reaching all three output ports (because self-communication is not allowed), thus enabling nonblocking functionality.

A 5×5 Cygnus [13] is another example of *strictly nonblocking router* for ONoCs. It consists of a switching fabric, and a control unit that uses electrical signals to configure the switching fabric according to the routing requirement of each packet. The switching fabric is built from the parallel and crossing switching elements. Cygnus uses only 16 MRRs, 6 waveguides and 2 terminators.

A 4×4 optical turnaround-router is an optical router, as always nonblocking, which has been customized for *FONoCs* (*Fat Tree-based Optical NoC topologies* [14]). It combines a mix of 1×2 and 2×2 PSEs, and is conceived to implement the turnaround routing algorithm typically used by fat-tree topologies.

All the discussed optical routers can be used to build any *space-routed ONoC*, such as mesh, torus topologies, as well as *FONoCs*.

Space routing means that reserved optical connections between any two initiator–target pairs should be set up before the actual communication takes place. For this purpose, a dual electronic NoC can be used to convey path programming packets. With this paradigm, the entire optical bandwidth along the reserved optical path is allocated for end-to-end communication.

A different approach is taken by the 4×4 λ-router [15], *the milestone switching fabric for wavelength-routed ONoC topologies*. Here, the network routing function is statically determined based on the wavelength of the optical signals.

For a given initiator, signals modulated on different wavelengths will be routed differently in the network, and will reach different destinations. Topologies are designed in such a way that signals with the same wavelength originating from different initiators will never interfere with each other. The appealing property of these topologies is that they enable contention-free communication, hence there is neither path setup nor contention resolution phase before optical packet transmission. This is achieved at the cost of penalizing the bandwidth of each communication stream, although a limited amount of wavelength parallelism is still feasible [16]. Alternatively, spatial division multiplexing can be used.

In the λ router case, with six 2×2 optical filters tuned on four different wavelengths, it is possible to realize four filtering stages, resulting in a 4×4 multistage optical network.

Hence, increasing the total number of wavelengths (and in turn the corresponding number of stages), and replicating the number of the 2×2 optical filters, it is possible to derive topologies of arbitrary size.

Other switching structures have been proposed that follow the wavelength routing paradigm, such as the 4×4 generalized wavelength-routed optical router (GWOR) [17]. This optical structure is capable of enabling 12 contention-free optical paths (three from each input port) thanks to its four 2×2 optical filters and three wavelengths. It is more suitable to connect initiators and targets distributed across the four cardinal points.

7.2.4 Photodetectors

At the destination front end, a photodetector is necessary to convert the incoming optical signal into an electrical one. As usual, before sensing the optical signal, a MRR is needed to filter the wavelength-parallel signal, hence treating each component separately.

Recently, developments in integrating germanium photodetectors with crystalline silicon waveguide have enabled to manufacture many high-performance and CMOS-compatible devices [18,19], aiming at high bandwidth (40 GHz), high responsivity (1 A/W), quantum efficiency above 90%, low capacitance (around 2 fF), and finally a dark current below 200 nA.

Another emerging methodology used in the design of photodetectors consists of adopting silicon with crystal defects as the absorbing material. The latest efforts in this field have yielded silicon photodetectors with bandwidth and responsivity higher than 35 GHz and 10 A/W respectively [20].

Similar to modulators, photodetectors can be structured into photodetector arrays. This strategy is very common when a parallel data stream comprised by multiple wavelengths has to be received at the destination stage of a given ONoC architecture.

7.2.5 Laser Sources

For on-chip application, *laser sources* can be implemented either on-chip or off-chip, depending on the power and bandwidth requirements of the system at hand, and their trade-offs. Recent emerging technologies continue to mature, and high-quality on-chip lasers compatible with CMOS processes start to appear. Other solutions have been yielded more recently, such as *electrically pumped hybrid silicon lasers and electrically pumped rare-earth-ion lasers on silicon* [21].

Alternative solutions leverage on III-V compound semiconductors to produce off-chip laser sources where the light is emitted by the external source and then brought on-chip using couplers. For instance, *quantum dot lasers*, based on III-V compound semiconductors are typically used in WDM applications as they are able to deliver many narrow-spectrum peaks across the frequency range of interest. Opportunely coupled with quantum dot semiconductor amplifiers, these lasers are able to provide several wavelengths within a low relative intensity noise, so that light will be modulated, transmitted, and received with error-free performance.

All devices presented earlier are key enablers to materialize ONoC consisting of multiple optical routers (broadband active switches or passive filters, depending on the routing methodology), that are properly interconnected with each other using silicon waveguides, that in turn may be straight, bent, or crossed depending on the topology requirements. As mentioned earlier, nowadays all devices necessary to build an entire on-chip optical communication infrastructure are viable for integration on a silicon chip, thus paving the way for the assessment of the ONoC paradigm.

Finally, it is worth observing that *3D stacked integrated systems* represent the most likely target for the exploitation of optical interconnect technology. The key reason is that it is a cost-effective solution for the integration of layers manufactured with different technologies, that this way do not need to be made compatible with one another, except for the obvious alignment and interlayer communication requirements. Across the same vertically integrated environment, we can accommodate processing, memory, and optical layers, thus resulting in a successful strategy to improve bandwidth scalability in next generation high-performance multi- and many-core systems.

7.3 NEED FOR PATHFINDING

As depicted in Figure 7.1, there are three fundamental groups of researchers that are currently involved in the pathfinding effort from the elegant ONoC concept to an actual technology of practical relevance.

The first group is focused on the characterization and optimization of *silicon photonic devices,* and on their monolithic integration with mainstream CMOS manufacturing processes. Here the landscape is actually far from consolidating.

In fact, there are two main paths toward an integrated platform. Hybrid/heterogeneous designs [22,23] enable each component to be custom-tailored, but suffer from large packaging parasitics, increased manufacturing costs due to requisite process flows, and nonmature 3D integration or microbump packaging. Monolithic integration mitigates integration overheads, but has not penetrated deeply scaled technologies due to necessary process customizations [24]. The first monolithic integration of photonic devices and electronic–photonic operation in sub-100 nm (45-nm SOI process with zero foundry changes) is demonstrated in Reference 25.

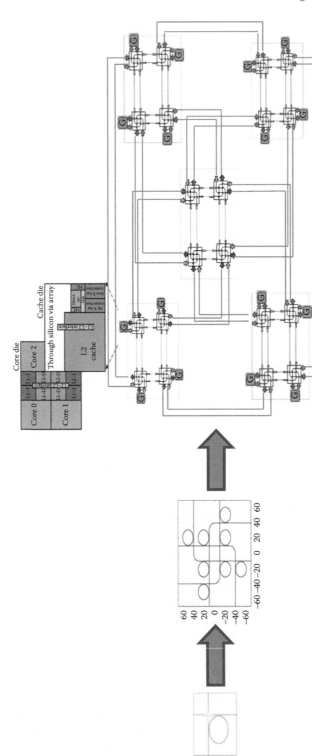

FIGURE 7.1   The pathfinding requirement.

A relevant gap separates baseline silicon photonic devices with *on-chip communication architectures,* which combine such devices together to materialize higher-order switching structures, complete communication channels, network topologies, routing and flow control methodologies, layout constraints aware physical designs. Network interfaces are an essential part of an ONoC architecture: their focus goes well beyond that of the low-level circuitry for domain conversion, but also covers typical networking topics such as proper buffer configuration and sizing, message-dependent deadlock avoidance, clock (re-) synchronization, end-to-end flow control, and so on.

The last group of researchers is instead involved in the redesign of an entire system to take the maximum advantage of the new interconnect technology. At this level, interconnect fabric design is tightly intertwined with system-level design issues such as memory organization, cache coherence protocol, application-perceived performance, number of memory controllers, and so on. Complex and scalable optical interconnects such as torus, square roots in J. Chan et al. [26], hierarchical wavelength-routed optical ring architectures in S. Koohi et al. [27], as well as Corona and Firefly frameworks [28,29] have been recently reported in this context.

Although many valuable research works have been reported in these emerging research fields, very few structured and coherent methodologies have been proposed so far (C. Batten et al. [30] is a nice example) to bridge the gap between the above abstraction layers for designing ONoC architectures. Such a pathfinding effort should address two relevant gaps that are highlighted in Figure 7.1. The first one exists between silicon photonic devices and on-chip communication architectures, which could be referred to as the *physical gap.* A design methodology addressing this gap should for instance deal with the deviation of physical topologies with respect to their logic schemes, take placement and routing constraints into close account for topology assessment and selection, engineer and optimize the electronic side of the network interface architecture, consider aggressive electrical baselines for the sake of crossbenchmarking, carefully include static power overhead in the overall power budget, or devise smart selection policies of network routes in hybrid electro-optical interconnect fabrics.

The second gap separates on-chip communication architectures with system-level design frameworks, and could be referred to as the *systemability concern.* Here, the focus is on the codesign of the ONoC architecture with the requirements dictated by the target system, and on future generations of such systems. Therefore, a design methodology addressing this gap should deal with scalability methods to hundreds of cores, cache coherence signaling, tuning ONoC parameters to system requirements, interconnect, and memory hierarchy codesign. The precondition to address the systemability gap in a trustworthy way is to enforce technology- and layout-aware decision making across all the layers of the design hierarchy, that is, to bridge the physical gap. Without that, conclusions of system-level studies about the potentials of optical interconnect technology turn out to be overly optimistic. The focus of this chapter is therefore on the characterization of the physical gap for the most relevant wavelength-routed ONoC (WRONoC) topologies, and on the evaluation of their robustness to the insertion loss degradation that takes place when refining the logic topology into the physical one. It will be showed that the predictability of a

topology does not only depend on the careful back-annotation of physical parameters into its logic scheme, but also on taking the implications of place and route
constraints on the actual topology layout into close account.

7.4   PREDICTABILITY–CRITICAL OPTICAL NETWORKS-ON-CHIP TOPOLOGIES

Several connectivity patterns have been proposed for both space-routed and
WRONoCs, however their properties are often discussed with reference to the logic
scheme only. For space-routed ONoCs, which reserve the entire bandwidth of an optical path to a specific communication flow, this practice is likely to deliver accurate
results just at the same as topologies are typically inspired by the regular ones in use
in general purpose electronic NoCs (e.g., mesh, torus, spidergon) [26], see Figure 7.2a.
This is a consequence of the good matching between the wavelength parallelism of
space-routed ONoCs and the bit parallelism of electronic NoCs.

In contrast, WRONoCs are fundamentally different from electronic and space-
routed optical networks, as they inherently deliver full connectivity (i.e., contention-
free and potentially parallel communication between every master and every
slave), while sharing the bandwidth of optical paths between multiple communication flows. WRONoC topologies are typically built up with optical add-drop filters
[31]. On one hand, they split multiwavelength signals into their basic components,
each one routed to a different destination. On the other hand, they aggregate signals from different masters (featuring different wavelengths) for each destination
(see Figure 7.2b).

Although topologies differentiate themselves for the amount of physical resources
(waveguides, optical filters) and number of wavelengths they require, they can be
ultimately viewed as multistage interconnection networks (MINs), a connectivity
pattern that maps inefficiently onto a 2D floorplan. In some cases, such networks
are effectively folded to match specific placement constraints of masters and slaves,
like for GWOR topologies (see Figure 7.2c). Unfortunately, the implications of different placement assumptions on topology properties are hard to predict. Clearly,
designers that intend to leverage on the contention-free communication paradigm
of WRONoCs have to deal with the *design predictability gap* of their topologies,
assessing the deviation of topology layouts from their logical views, due to the *placement constraints* of network initiators and targets on the system at hand.

A typical experimental setting consists of deriving such constraints from the
floorplan of a clusterized multicore processor with an optical layer vertically stacked
on top of it, providing intercluster as well as processor-memory communication. In
practice, they concern the actual position of optical network interfaces and photonically integrated memory controllers on the optical layer. The key sources of nonide-
ality are the additional number of waveguide crossings, which derives from mapping
the connectivity pattern on a 2D layout, and the propagation loss arising from long
waveguides. Consequently, the ultimate implications of topology actual layouts on
network total power consumption can be derived, as optical power losses need to be
compensated by a higher power output by laser sources, in an attempt to feed photodetectors with enough optical power for reliable detection.

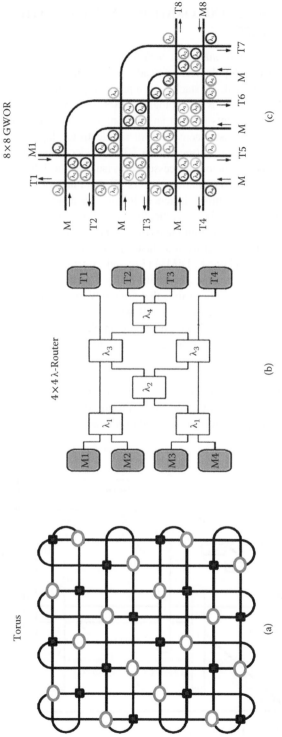

FIGURE 7.2 (a) Space-routed torus, (b) Wavelength-Routed λ-router, and (c) Generic Wavelength-Routed Optical Router (GWOR).

7.5   DESIGN SPACE EXPLORATION OF WAVELENGTH-ROUTED TOPOLOGIES

To assess the role of *placement constraints* in the comparison framework of wavelength-routed network topologies, we need to make specific assumptions on the target system. As mentioned, 3D stacking is a promising scenario for the cost-effective integration of heterogeneous technologies. ONoCs are very likely to be integrated with electronic components in such an environment. Therefore, we focus on a 3D stacked multicore processor, consisting of an electronic layer and an optical one vertically stacked on top of it (see Figure 7.3). The electronic layer is composed by an array fabric of homogeneous processor cores (such as the Tilera architectures [5]), which are grouped into computation clusters. Without lack of generality, we assume that cores are aggregated into 4 clusters of 16 cores each, and that each cluster has its own gateway to the optical layer (vertically stacked with a corresponding hub in the optical layer). This latter accommodates three kinds of communications: (1) between any two pairs of clusters; (2) from a cluster to a memory controller of an off-chip photonically integrated DRAM DIMM [32]; and (3) from a memory controller back to a cluster. Finally, we assume a core size of 1×1 mm and a die size of 8×8 mm. In this context, placement constraints that are likely to be enforced during the place and route of network topologies are as follows.

First, hubs (i.e., the optical components of an electro-optical and optoelectronic network interface) should be positioned in the middle of each cluster, hence they end up being placed along a square in the middle of the optical layer (see *Hi* blocks in Figure 7.3).

Second, we assume 4 memory controllers located pairwise at the opposite extremes of the chip (see *Mi* blocks in Figure 7.3), as proposed in conventional chip multiprocessor architectures [33]. Overall, we need to connect eight initiators (four hubs, four

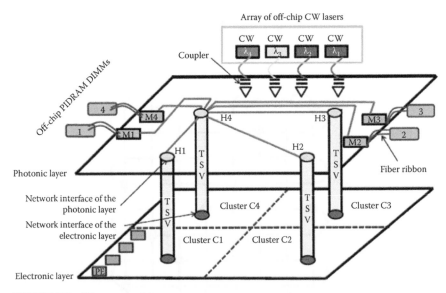

FIGURE 7.3   3D target architecture.

memory controllers) with eight targets (the target interface of the same four hubs and four controllers). For this purpose, we leverage on a WRONoC topology. Global connectivity is clearly the easiest architectural solution that can be exploited to connect initiators and targets with each other. Therefore, this approach will initially be addressed. To provide global connectivity, an 8×8 WRONoC topology is required. Ultimately, the most relevant topology logic schemes that can address this connectivity requirement are discussed, followed by their implementation in the target system.

7.5.1 GLOBAL CONNECTIVITY

Figure 7.4a shows the first topology under test: an 8×8 λ-router, which was proposed by Scandurra and Connor [31]. To interconnect eight initiators with eight targets, the network utilizes eight stages of four and three add-drop filters. The topology reflects the connectivity pattern of unidirectional MINs in the electronic domain, where however the interstage pattern is custom-tailored for the routing methodology of WRONoCs. Unfortunately, the attractive logic scheme of this topology does not match the actual placement constraints of most real-life systems, where it is almost impossible to find all initiators on one side and all targets on the other side of the chip. At the same time, many cores are both initiators and targets of on-chip communication transactions, hence the physical implementation of this topology implies some degree of folding.

Due to the lack of consolidated automatic place and route tools for ONoCs, the connectivity pattern of all topologies were manually laid out, hence coming up with full custom design solutions. Figure 7.4d depicts the actual layout of the 8×8 λ-router topology after manual placement of MRRs, and routing of waveguides. Later validation experiments with prototype place and route tools delivered lower absolute numbers (waveguide crossings, insertion loss), but the same relative trends presented in this chapter [34]. In Figure 7.4, the central boxes in the layouts represent hubs, whereas the side ones represent memory controllers. For each box, the internal splitting lines denote initiator and target sections of network interfaces.

The followed layout design guidelines were inspired by and similar to those used to lay out fat-tree topologies in electronic NoCs [35]. Moreover, the criteria were a uniform exploitation of the floorplan space and the equalization of waveguide length. In spite of this, the difference between the logic and the physical topologies is still apparent. The ultimate effect is an increase of the total amount of insertion losses (dominated by waveguide crossings), which in turn leads to an increased requirement on the laser power needed by the optical signals to stay above the minimum detection threshold of the photodiodes.

To find the best solution for global connectivity, the previous topology is compared next with the 8×8 GWOR [17] and an optimized crossbar variant, hereafter referred to as the 8×8 folded crossbar.

According to the wavelength assignment proposed in X. Tan et al. [17], the 8×8 GWOR (see Figure 7.4b) is constructed starting from its lower basic cell, the 4×4 GWOR. This latter is well suited for those cases where initiators and targets are distributed across the cardinal points. In fact, the topology consists of four waveguides that intersect each other, with MRRs placed pairwise at each intersection.

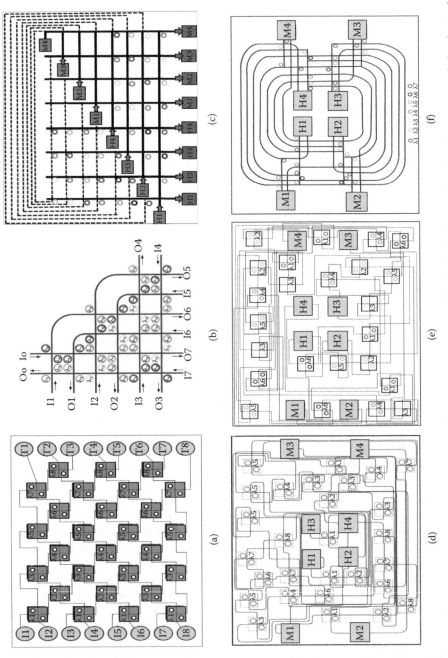

FIGURE 7.4 Logic topologies (a)–(c) versus corresponding physical topologies (d)–(f). Circles denote the micro ring resonators belonging to the interconnection network, whereas external boxes indicate initiators and targets (i.e., the inputs and outputs of the network).

Unfortunately, the scaled pattern to an 8 × 8 network keeps making the same physical placement assumptions, which is not realistic, as it is very unlikely that all cores are placed around a central fence including the optical interconnect fabric.

Moreover, it is worth recalling that, unlike the previous topology, this one saves some resources by avoiding the useless self-communication capability. The physical view of the 8 × 8 GWOR is illustrated in Figure 7.4e, and confirms that the placement constraints of the target system are unnatural for the GWOR connectivity pattern, which ends up in a circuitous wiring that makes the original pattern hardly recognizable.

Again, the waveguide crossings arising as an effect of the 2D surface mapping are apparent. Finally, an 8 × 8 optical crossbar is considered. This topology places MRRs at each intersection of a point matrix, thus establishing connections between a given initiator and the desired target. Although considered quite inefficient in abstract analysis frameworks, the topology lends itself to an interesting optimization that applies directly to its logic scheme. In the original topology, every initiator delivers optical signals to targets in a given order. By changing this order for every initiator (see Figure 7.4c), then a waveguide length overhead is apparently generated (see the wraparound links). However, this is only an apparent effect of the logic scheme, as the actual layout is in contrast facilitated: every initiator can in fact drive an optical waveguide that is part of a ring-like topology that dispatches optical packets to the possible destinations (Figure 7.4f). Moreover, the logic scheme of this topology makes only use of 1 × 2 optical filters with respect to other topologies, which instead basically use 2 × 2 optical filters. Clearly, the layout is much more regular than the 8 × 8 λ-router and the 8 × 8 GWOR, and MRRs are clearly positioned close to communication targets for wavelength-selective ejection of optical signals. In previous comparison frameworks, such layout-level details are typically omitted, and considered quality metrics include mainly the number of MRRs, and the maximum number of waveguide crossings on the logic scheme. From this viewpoint, the 8 × 8 folded crossbar features the lowest number of MRRs (44), as opposed to 48 for the 8 × 8 GWOR, and 56 for the 8 × 8 λ-router. Unfortunately, this analysis methodology is only partially informative and even misleading. The ranking is in fact exactly the opposite when the number of waveguide *logic* crossings is considered: the λ-router would exhibit 7 crossings as opposed to the 10 ones of GWOR and the 15 ones of the optical crossbar. Clearly, such abstract and even conflicting considerations alone are not enough to derive trustworthy conclusions, especially when the most power-efficient solution is searched for. For this reason, physical layer and layout analysis are required to assess the real trade-offs. Finally, it is important to keep in mind that optical crossbar and GWOR utilize seven distinct wavelengths (i.e., seven continuous-wave laser sources) by construction to deliver full and global connectivity versus the eight ones of an 8 × 8 λ-router.

7.5.2 RELATIVE TOPOLOGY COMPARISON

Insertion loss is the most important physical metric that must be quantified to determine the laser power that guarantees a predefined bit error rate at receivers. In fact, once *ILmax* is obtained (the maximum insertion loss across all wavelengths and optical paths), and the detector sensitivity is known (S), it is possible to evaluate the lower limit of optical laser power (P) to reliably detect the corresponding photonic signal at the receiver end. A simplifying assumption can be made at this point: the worst-case ILmax can be quantified across the entire global network, and then the practical assumption can be made that such a worst-case ILmax dictates the power output by all laser sources. This study assumes the loss parameters summarized in Table 7.1. A Simulink® simulation framework is used to assess physical metrics of ONoCs by modeling every single path of a given topology while accounting for the above loss parameters. Finally, the corresponding insertion loss is obtained as a sum of all components losses such as PSEs, straight, bend, crossing waveguides, and drop-into ring losses encountered in the path under test. The topology models assume die sizes of 8 × 8 mm. Figure 7.5 shows insertion loss deviations between logic and physical ONoCs for all topologies considered in this comparison framework, when the standard elliptical tapers are assumed at every waveguide crossing [6]. The insertion loss critical path is more than six times worse in two physical networks out of three with respect to the corresponding logic schemes. Especially GWOR suffers from 72 waveguide crossings against the 10 expected ones, whereas λ-router reports 64 crossings versus 7. Surprisingly, the *folded crossbar* maps more efficiently to the target placement constraints, although it is frequently discarded in abstract analysis frameworks, which only consider the logic schemes and the abstract properties. The physical implementation is so efficient (i.e., only very few additional

TABLE 7.1

Loss Parameters

Parameters	Value
Propagation loss from J. Chan et al. [26]	1.5 dB/cm
Bending loss from J. Chan et al. [26]	0.005 dB
Crossing loss (optimized by elliptical taper) from 2D FDTD	0.52 dB
Drop loss (optimized by elliptical taper) from 2D FDTD	0.013 dB

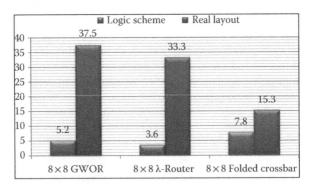

FIGURE 7.5   Maximum insertion loss across topologies.

crossings from layout constraints) to offset the inherently higher number of wave-guide crossings of the logic scheme. More in details, propagation loss is a significant contribution in the folded crossbar topology, indicating that the critical path now is both waveguide- and crossing-dominated. Due to its long optical waveguide of 25.5 mm (worst case) and 22 crossings, the folded crossbar achieves 15.3 dB optical power loss on the critical path. On the contrary, for λ-router (33.3 dB) and GWOR (37.5 dB) only crossing losses have been computed as their contribution is dominant in the breakdown. Obviously, had we accounted for their propagation losses too, their current huge gap with the crossbar would have become even worse, without providing any significant novelty to the discussed results.

By using the critical path insertion loss, it is then possible to derive the needed laser power to get a bit error rate of BER=10^{-12} at optical receivers with a sensitiv-ity of -17 dBm [36]. By also considering the contribution of modulator [37] and detector power [36], and of thermal tuning [26] listed in Table 7.2, the total power results in Figure 7.6 can be derived. The total power of GWOR is larger than that of other topologies, even if the λ-router features one laser source more than GWOR and the folded crossbar to provide the same (full) connectivity. More precisely, the total power of the λ-router topology is 2.47 times lower than the GWOR one. The folded crossbar turns out to be the most power efficient solution. It consumes only 276 mW, almost two orders of magnitude less than GWOR. These results certainly indicate that GWOR and λ-router are unfeasible for the placement constraints of the target 3D system, at least when they target global connectivity.

TABLE 7.2
Device Parameters

Device	Features
Laser	CW (continuous-wave)
	Laser efficiency
	PLE = 20%
	Coupling laser-link
	PCW = 90%
Modulator	Silicon-disk
	Launch-efficiency $\beta = 20\%$
	Dyn. dissipation = 3 fj/bit
	Static power = 30 μW
	$V_{dd} = 1$ V
	Modulator-power depends on ILmax [37]
Detector	CMOS (45 nm)
	Hybrid silicon receiver
	Sensitivity, S = -17 dBm
	(BER = 10^{-12} at 10 Gbit/s)
	Power = 3.95 mW [36]
Photonic-switching-elements	Thermal tuning = 20 μW/ring [26]

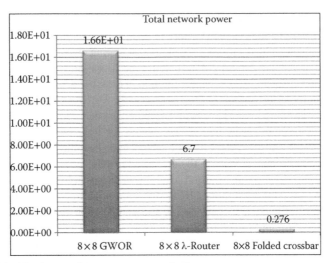

FIGURE 7.6   Total network power across topologies.

7.5.3 COMPARISON WITH AN OPTICAL RING TOPOLOGY

The layout of the folded crossbar is very similar to a ring topology. This latter is in fact the simplest connectivity solution among all network topologies presented in the open literature, and apparently the less sensitive one to place and route constraints. The only one way to assess whether the folded crossbar is the most efficient solution for the target system consists of comparing it with an actual ring topology. To correctly compare these topologies, a ring is designed assuming the same number of wavelengths utilized in the crossbar, that is, seven. The use of multiple ring waveguides (i.e., spatial division multiplexing) is the only way to meet this requirement. Figure 7.7b depicts the real layout of the ring topology after manual placement of the 3D stacked optical layer under test with the given physical constraints (i.e., hubs in the middle and memory controllers positioned at the opposite extremes). Clearly, this topology better fits the target constraints. Essentially, it works like a bus, in which multiple waveguides are contained into it. In this case, seven parallel waveguides are needed to deliver full and contention-free communication parallelism. Figure 7.8a illustrates the postlayout insertion loss critical path comparison between the 7-way ring and the 8 × 8 folded crossbar.

The 7-way ring achieves 7.75 dB insertion loss against 15.3 dB of the crossbar on the critical path, thus resulting 50% more power efficient. The key reason lies in the fact that the 7-way ring provides less wiring length on the critical path (2 vs. 2.55 cm in the crossbar). Moreover, the crossbar has 22 waveguide crossings (localized in the optical network), while only 9 crossings are there in the 7-way ring. Even if in the ring topology, there are no intersections in principle, they are actually needed at initiator interfaces to connect to the parallel ring waveguides that are furthest away from the injection point, as shown in Figure 7.7a. In contrast, such crossings may not appear at target interfaces, since the output signal of photodetectors might directly leave the optical plane by means of through-silicon vias (TSVs) in the best case.

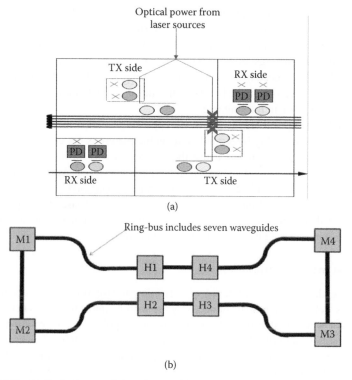

FIGURE 7.7 (a) Optical network interface architecture. (b) Manual layout of an optical ring topology.

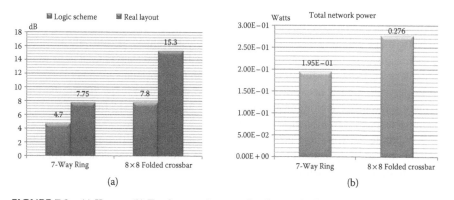

FIGURE 7.8 (a) ILmax. (b) Total network power for the topologies under test.

Notice that the logic scheme of any ring topology is characterized by such obstacles, which may degrade the insertion loss, and as a consequence the total power. In this case, ILmax of the ring logic scheme is about 4.7 dB, while it gets almost doubled when postlayout results are considered. Here, the wiring length contribution becomes dominant through its propagation loss.

The total power consumption of the two topologies is shown in Figure 7.8b. Thanks to the lower insertion loss, the 7-way ring topology results more efficient than the 8 × 8 folded crossbar by about 30%. This latter is heavily penalized by the larger number of crossings, and the higher wiring length. In this case, the insertion loss gap of 50% is reduced to 30% in terms of total power, due to the relevant contribution of optical receivers in the breakdown. They contribute for 63% in the crossbar topology and for 89% in the ring one.

Ultimately, the optical ring is clearly an appealing solution for such a small scale system. However, in highly integrated systems the picture is not entirely clear, as the larger connectivity requirements cause the number of ring waveguides to increase, hence worsening the crossing concerns at initiator interfaces. At the same time, in larger dies the critical path becomes propagation loss dominated, thus raising another concern for optical rings. Finally, the quality metrics of filter-based topologies will be effectively scaled up in future systems as computer-aided design (CAD) tools for automatic place and route of ONoC topologies become available.

7.5.4   NETWORK PARTITIONING

To increase the level of confidence of this comparative framework, optimization techniques well beyond global connectivity are worth exploring such as network partitioning for wavelength reuse (all topologies) and topology transformations for the optical crossbar and GWOR topologies for more flexible and/or efficient place and route. This work does not blindly apply topologies under test to the master/slave connectivity problem of the target system, as the previous section has demonstrated that even at such a small system scale, a typical global topology for all communication actors becomes easily infeasible: too many waveguide crossings arise in an attempt to accommodate the connectivity pattern onto the 2D floorplan. Of course, should the routing methodology change (e.g., arbitration of waveguides), the picture would change again and deserve further analysis.

Network partitioning is considered in this analysis framework, not only as a means of increasing design predictability but also of enabling wavelength (and laser source) reuse across partitions, similar to what is done in telecommunication networks.

In particular, each network partition can be devoted to a specific traffic class, namely intercluster communications, memory access requests from clusters, and memory responses from memory controllers.

A separate network, and associated topology, can be inferred in each partition. Finally, this strategy enables to cut down on the number of wavelengths from eight to just four, thanks to their reuse.

7.5.5   LOGIC TOPOLOGIES

This section illustrates the logic schemes of WRONoC topologies under test, considering that each network partition will have to interconnect at most four masters with four slaves. We consider the most relevant schemes that have been proposed so far in the open literature (the same used for global connectivity, although scaled down), in addition to engineering an ad-hoc topology for the 3D stacked system at hand.

As mentioned in the previous paragraph, the 8 × 8 GWOR is a scalable and non-blocking wavelength-routed optical router. The basic cell of the former solution is represented by the 4 × 4 GWOR that has four bidirectional ports located on the cardinal points. Furthermore, two horizontal and two vertical waveguides are used, which intersect each other to form a basic check shape, and MRRs are placed pairwise on waveguide intersections. The proposed topology does not support self-communication, hence its use for the memory request and response networks requires its extension to a 5 × 5 configuration. This is possible, as its wavelength assignment enables any size of GWOR architecture. As it can be seen in Figure 7.9a, 5 × 5 GWOR is constructed starting from its lower basic cell (4 × 4 GWOR). With respect to the baseline scheme, three more MRRs need to be inserted to work around the lack of self-communication, and enable each master to be connected with all of the four slaves. At the same time, one input goes unused, therefore redundant MRRs are removed.

On the contrary, Figure 7.9c illustrates the 4 × 4 λ-router. To interconnect four masters with four slaves, the network makes use of four stages of two and one add-drop optical filters. This network is obtained by scaling down the previously proposed 8 × 8 λ-router, or vice versa it could be seen as the preliminary cell to build any size of the λ-router network. With respect to the original scheme, the native 2 × 2 add-drop filters are replaced with 2 × 2 photonic switching elements, the only difference being an easier physical design thanks to the orthogonally intersected waveguides. Figure 7.9b shows the scaled-down version of the 8 × 8 folded crossbar, that is customized for connecting four initiators with four targets.

Finally, a custom-tailored solution for processor-memory communication is described, namely the Snake topology. The pattern (Figure 7.9d) is also flexible, as a different number of initiators and targets can be easily accommodated. In the 4 × 4 Snake, six wavelength filters (2 × 2 PSEs) are tuned to different wavelengths, and their number scales up from the upper side to the lower one. Four main optical links connect the slaves, while enabling some placement flexibility.

With respect to the λ-router, such a topology first breaks the monodimensional assumption, and also grows vertically instead of horizontally. In the end, the Snake is easily capable of providing asymmetric solutions (e.g., 4 × 8, 8 × 4), much more easily than λ-router (although this is in principle still feasible with this topology as well).

Finally, the Snake topology is conceived to map efficiently to the placement constraints of the target system, and should be viewed as a custom-tailored solution for the system at hand to ease reachability of memory controllers. In practice, communication from the center of the chip to its upper or lower side is facilitated.

For the sake of comparison, all topologies are constrained to use the same number of wavelengths and laser sources, and to instantiate physical resources accordingly. Therefore, all topologies have to deliver the same communication bandwidth.

Ultimately, as it was done for global connectivity, an optical ring topology is designed to connect four masters with four slaves, and then added in the comparison framework. In practice, rings are devoted to specific message classes in this instance (hereafter referred to as ORNoC).

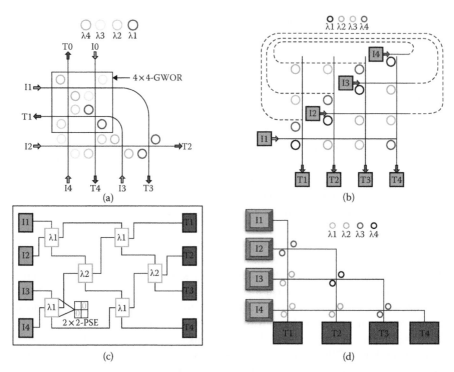

FIGURE 7.9 Network partitioning logic schemes (a)–(d). Circles denote the micro ring resonators belonging to the interconnection network, whereas external boxes indicate initiators and targets (i.e., the inputs and outputs of the network).

7.5.6 PHYSICAL TOPOLOGIES

This section deals with the problem of assigning topologies to network partitions and of laying them out. For the intercluster ONoC, the choice is straightforward: 4×4 GWOR delivers the needed connectivity in a scenario where its physical placement assumptions are perfectly satisfied. At the same time, it features the lowest number of MRRs. Therefore, the problem can be restricted to identifying the topologies that are better suited for processor-memory communication, and to lay them out twice: for the memory request network (from hubs to memory controllers), and the memory response one (from controllers to hubs). The fundamental difference lies in the flipped position of masters and slaves, which makes them asymmetric.

Manual layouts have been drawn with similar criteria to those used for global connectivity. The only approximation lies in the lack of a distribution network of optical power. It is as though we were neglecting the top-level clock tree in the layout of an electronic network. After topology place and route, the difference between logic and physical topologies is still apparent although the use of network partitioning mitigates this effect to a significant extent.

The 5×5 GWOR (Figure 7.10a) suffers from the different placement position of network interfaces with respect to the logic scheme, to such an extent that the critical path increases from 4 crossings to 31, whereas the total number of MRRs achieves

40 (8 in the intercluster network, plus both 16 for the memory request network and the response one).

Despite a higher worst-case number of crossings in the logic scheme (6), the layout of the 4 × 4 folded crossbar in Figure 7.10b results only in 21 crossings, with the same number of MRRs reported by GWOR (40).

The layouts of the 4 × 4 λ router (Figure 7.10c), ORNoC (Figure 7.10d), and 4 × 4 Snake (Figure 7.10e) are clearly less intricate than the previous ones, hence potentially resulting in lower insertion loss critical paths. More precisely, λ-router reports eight crossings, whereas Snake only six.

By using the wavelength assignment in S. Le Beux et al. [38], and a convenient ordering of nodes along waveguides, ORNoC turns out to exhibit three crossings on the critical path, all localized close to network interfaces for the sake of waveguide reachability.

The key properties of the topologies under test, measured after their physical design and inclusive of the three partitions, are the following:

- Although all topologies natively use four wavelengths, a spatial division multiplexing over four waveguides has to be used for ORNoC to achieve the same goal.
- Snake and λ-router solutions make use of 32 MRRs (24 in the request and response networks vs. 8 in the intercluster one) against 40 of the ring one.

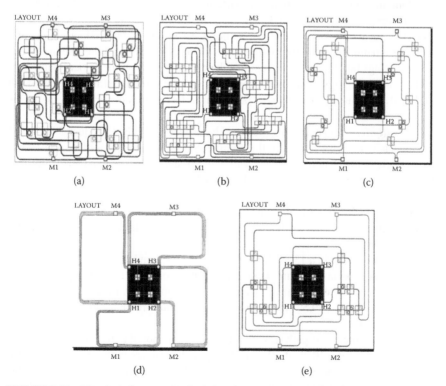

FIGURE 7.10 Topology layouts: (a) Generic Wavelength-Routed Optical Router (GWOR), (b) Folded Crossbar, (c) λ-router, (d) Ring, and (e) Snake.

The key reason lies in the fact that each optical network interface in the ring needs four MRRs to inject modulated wavelengths into their waveguides, in addition to eight MRRs needed in the intercluster network.

All other topologies instead do not have any injection filters, as they get a branch of the light distribution network, which directly enters the network. In the ring, the injection waveguide needs to be bridged to the ring waveguides. Extraction filters at receivers are common for all topologies, hence are not considered in the count.

7.5.7 Power Efficiency of Topologies

Figure 7.11a shows the worst-case insertion loss across all topologies considered in this comparison, assuming two kinds of tapers for the optimization of waveguide crossings: the standard elliptical taper and the MMI one.

However, the feasibility of the MMI taper should not be taken for granted, as it depends on the maturity of the manufacturing process, and on the device size. In fact, it has a larger area footprint with respect to the elliptical taper, therefore, it might be suitable for layout-induced waveguide crossings, but might be unfeasible for the internal crossings of photonic switching elements, where ring resonators should be placed close to the waveguides for the sake of efficient coupling. The insertion loss parameters utilized for this optimization were derived from 2D finite-difference-time-domain (FDTD) simulations. In particular, the crossing loss is about 0.18 dB while the drop loss exhibits 0.0087 dB.

GWOR turns out to be the worst solution, as it suffers from 31 crossings and 24 mm of wiring length on the critical path, whereas ORNoC (the best solution) has just 3 crossings but 32 mm of waveguides length as always on the critical path.

The Snake topology, with its six crossings and the same maximum guide length of GWOR, becomes competitive, as propagation losses are not very relevant at this chip size yet.

With *elliptical taper*, the overhead with respect to ORNoC is just 5%. 4 × 4 λ router has reasonable results in the comparison as it has 22 mm of wiring length and eight crossings, whereas the 4 × 4 folded crossbar is better than GWOR for two reasons: lower number of crossings (21), and a 4 mm shorter link length on the insertion loss critical path.

The effect of MMI is highly beneficial for the Snake, as it minimizes the impact of its crossings over ILmax, whereas benefits are not so relevant for the waveguide-dominated ORNoC.

This latter ends up in a 13.2% higher insertion loss than Snake. This result is very interesting, as it points out that there is actually a role also for nonring topologies in WRONoCs, in spite of their apparently higher complexity. In turn, Snake results in a 2.5%, 32.6%, and 49.5% lower insertion loss than λ-router, folded crossbar and GWOR respectively.

By using such critical path insertion losses, it was possible to derive the needed laser power to meet a BER of 10^{-12} at the optical receivers with a fixed sensitivity of -17 dBm. It is then possible to account for the power contribution of modulators, detectors, and thermal tuning, thus estimating total power for each topology. Relevant parameters are in Table 7.2.

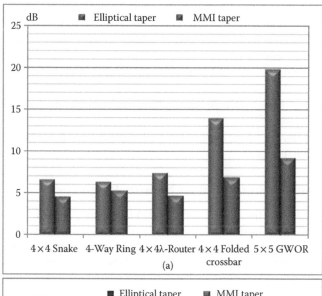

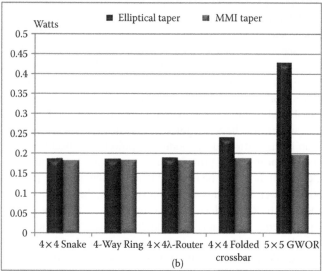

FIGURE 7.11   (a) ILmax. (b) Total network power comparison among topologies.

Figure 7.11b shows the total power across all topologies when the energy consumption of the detector is 395 fj/bit (or 3.95 mW). Power refers to the scenario where the maximum aggregate bandwidth of the network is used (around 440 Gbit/s with modulation rates of 10 Gbit/s).

As it can be seen, the total power of GWOR is higher than that of other topologies regardless of the specific taper. With elliptical tapers, GWOR is clearly infeasible under the given place and route constraints. The same holds, for the folded crossbar. The capability of the Snake topology to track power efficiency of the optical ring (the best solution) is remarkable at this system scale.

The effect of *MMI tapers* is to reduce the critical path differentiation across topologies, hence significantly bridging the gap between the best and the worst one.

Laser and modulator power are closely related to the ILmax of the topologies, however the total network power is dominated by receiver power with current technology assumptions (average 75% with elliptical taper as opposed to 90% with MMI taper), therefore, the remaining gap between topologies in Figure 7.11a maps to the total power gap of Figure 7.11b after going through an attenuation factor: just 15 mW of difference between Snake (the best) and GWOR (the worst). Of course, different laser sources (e.g., wall-plug laser efficiency) or receiver (e.g., energy) parameters may further widen the gap.

7.5.8   GLOBAL CONNECTIVITY VERSUS NETWORK PARTITIONING

This section describes the comparison between the best topologies, both ring-based and filter-based, implemented both for global connectivity and network partitioning.

Figure 7.12a shows the total power comparison between the message-class-specific rings (from partitioning optimization) and the global ring. As it can be seen, both implementations almost provide the same total power, particularly, the partitioned ring consumes 188 mW, instead the global one 195 mW (roughly 3.6% more). This marginal deviation is determined by the laser power. In particular, the global ring topology features more laser sources then the local ring (seven vs. four). Let us denote also that the receiver power results to be the most important contributor in the power breakdown.

In contrast, Figure 7.12b illustrates the total power comparison of the best filter-based topologies, both in the global connectivity and the network partitioning cases. Similar to ring-based topologies, the receiver power significantly contributes to the total power. However, for the 8 × 8 folded crossbar (the best global topology) such impact ends up being mitigated by modulator power, that becomes relevant as the 8 × 8 folded crossbar topology provides higher insertion loss than the 4 × 4 Snake (15.3 dB vs. 6.75 dB). It should be recalled that modulator power partly depends on the input optical power. By including all contributions, the 4 × 4 Snake results to be more efficient than the 8 × 8 folded crossbar by about 30% (189 vs. 276 mW).

From the proposed analysis, it is clear that filter-based topologies benefit the most from the network partitioning optimization, whereas ring topologies are in any case global structures. Also, in small-scale systems topology selection can be ultimately dictated by design simplicity considerations, as it is not difficult to engineer both filter-based and ring-based topologies with similar power figures. From this viewpoint, ring solutions are clearly appealing.

7.6   OPTICAL RING VERSUS FILTER-BASED TOPOLOGY IN SCALED SYSTEMS

As the next step, the impact of system scale and technology evolution over the illustrated trend is characterized. For this purpose, a future generation of the target system is projected.

Let us now assume 128 cores in the tile-based electronic plane, getting access to the optical layer through eight gateways (and eight corresponding hubs in the optical

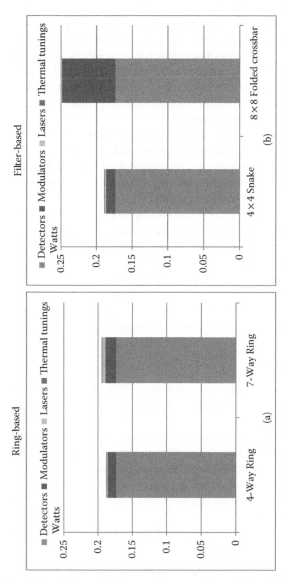

FIGURE 7.12 (a) Partitioned versus global Ring topology. (b) Best partitioned (4×4 Snake) versus best global (8×8 folded crossbar) topologies.

plane). The number of memory controllers can be kept the same, which might be feasible due to the benefits of photonics integration deeper into the DRAM DIMM [11]. Consequently, the die size grows to 16×16 mm.

We limit the comparison between ORNoC and the best filter-based topology found so far, that is, the partitioned Snake, and omit the intercluster network. Therefore, two four-waveguide ORNoCs and two separate Snake topologies (an asymmetric 8×4 network for memory requests, and a 4×8 one for responses) are considered.

MMI tapers are assumed to be mainstream in the implementation of these topologies, and detector energy is assumed to be improved up to 50 fj/bit [26], while conservatively preserving the same receiver sensitivity, a projection that is supported by the physical considerations in M. Georgas et al. [39] about silicon photonics in 3D stacked systems and receiver circuitry.

Figure 7.13a shows the insertion loss critical path breakdown of each topology.

The ring architecture (built on eight waveguides) is in fact heavily penalized by the high-wiring length over the new die size (64 vs. 48 mm of Snake), which leads to a larger amount of propagation loss regardless of the higher number of crossing losses in Snake (1.75× higher than the ring).

The total power consumption across the two topologies is shown in Figure 7.13b. Thanks to the lower insertion loss on the critical path and the higher maturity of receiver technology, the Snake results more efficient than the ring architecture by about 15%. This certainly confirms that optical ring topologies are not the most power efficient and least complex solution in all WRONoC scenarios, but that conclusions are tightly instance- and technology-specific. As the technology is still admittedly in the early stage, it is currently not possible to drop neither rings nor filter-based topologies for future ONoC topology design.

7.7 CONCLUSION

This chapter argued a comparative analysis of WRONoC topologies by considering both the properties of optical links as well as placement constraints on a target system of practical interest.

First, there is a large deviation of insertion loss between the logic scheme and the physical implementation as an effect of placement constraints.

Second, the most promising logic schemes may turn out to be the worst physical topologies, so the design predictability gap should be carefully quantified.

Third, network partitioning is an effective way of reusing wavelengths and simplifying ONoC design.

Fourth, the best topologies for global connectivity are not necessarily the best options for network partitioning.

On one hand, for small scale systems, a spatial division multiplexed ring topology is hard to beat. Even in this context, should technology evolutions improve optical receiver energy, filter-based networks could again have a role. In practice, an optical ring is ideally the best WRONoC topology, but its practical nonidealities (e.g., waveguide reachability in the injection system, worse waveguide length scalability on the critical path) make an actual comparative test with other filter-based topologies mandatory in the target system.

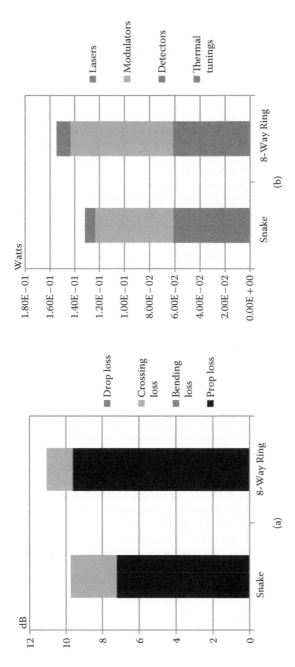

FIGURE 7.13 (a) ILmax. (b) Total network power Snake versus Ring architecture.

On the other hand, for future larger scale systems, where connectivity requirements and die size increase, spatial division multiplexing combined with the relevant role of propagation losses seriously penalizes optical ring architectures, so that filter-based topologies may become appealing. This trend will be further consolidated by the development of CAD tools for automatic place and route of filter-based topologies, which will optimize their quality metrics in layout-intricate and/or highly integrated scenarios.

REFERENCES

1. D. A. B. Miller, Rationale and challenges for optical interconnects to electronic chips, *Proc. IEEE*, 88(6), 728–749, 2000.
2. D. Marris-Morini et al., HELIOS project: Deliverable D010: State-of-the-Art on Photonics on CMOS, October 2011.
3. ePIXfab: The silicon photonics platform.
4. A. Biberman and K. Bergman, Optical interconnection networks for high-performance computing systems, *Rep. Prog. Phys.*, 75(4), 2012.
5. B. G. Lee et al., Ultrahigh bandwidth silicon photonic nanowire waveguides for on chip networks, *IEEE Photonics Technol. Lett.*, 20, 398–400, 2008.
6. N. Sherwood-Droz et al., Optical 4x4 hitless silicon router for optical networks-on-chip (NoC), *Opt. Express*, 16(20), 15915–15922, 2008.
7. G. R. Hadley, Effectivesurface-emitting lasers, *Opt. Lett.*, 20, 1483–1485, 1995.
8. A. Biberman et al., Photonic network-on-chip architectures index model for vertical-cavity using multilayer deposited silicon materials for high-performance chip multiprocessors, *ACM J. Emerg. Technol. Comput. Syst.*, 7(2), 7:1–7:25, 2011.
9. N. Ophir et al., Demonstration of 1.28 Tb/s Transmission in Next Generation Nanowires for Photonics Networks-n-Chip, 23rd Annual Meeting of the IEEE Photonic Society, pp. 560–561, Denver, CO, November 7–11, 2010.
10. S. Manipatruni, Q. Xu, B. Schmidt, J. Shakya, M. Lipson, High Speed Carrier Injection 18 Gbit/s Silicon Microring Electro-Optic Modulator, 20th Annual Meeting of the IEEE Lasers and Electro-Optics Society, Lake Buena Vista, FL: *IEEE*, 537–538, 2007.
11. K. Preston, P. Dong, B. Schmidt, M. Lipson, High-speed all optical modulation using polycrystalline silicon microring resonators, *Appl. Phys. Lett.*, 92, 151104, 2008.
12. B. G. Lee, A. Biberman, P. Dong, M. Lipson, K. Bergman, All-optical comb switch for multiwavelength message routing in silicon photonic networks, *IEEE Photonics Technol. Lett.*, 20(10), 767–769, 2008.
13. H. Gu, K. H. Mo, J. Xu, W. Zhang, A Low-Power Low-Cost Optical Router for Optical Networks-on-Chip in Multiprocessor Systems-on-Chip, IEEE Computer Society Annual Symposium on VLSI, Tampa, FL: *IEEE*, 19–24, 2009.
14. H. Gu et al., A Low-Power Fat Tree-Based Optical Network-on-Chip for Multiprocessor System-on-Chip, DATE: Conference on Design, Automation and Test in Europe, Nice, France: *IEEE*, 3–8, 2009.
15. I. O'Connor et al., Toward Reconfigurable Optical Network on Chip. ReCoSoC: Reconfigurable and Communication-Centric Systems-on-Chip, Univ. Montpellier II, France, 121–128, 2005.
16. L. Ramini, P. Grani, S. Bartolini, D. Bertozzi, Contrasting Wavelength-Routed Optical NoC Topologies for Power-Efficient 3D-Stacked Multicore Processors Using Physical-Layer Analysis, DATE: Conference on Design, Automation and Test in Europe, Grenoble, France: *IEEE*, 1589–1594, 2013.
17. X. Tan et al., On a Scalable, Non-Blocking Optical Router for Photonic Networks-on-Chip Designs, Photonics and Optoelectronics (SOPO), Wuhan, China: *IEEE*, 2011.

18. S. Assefa et al., CMOS-integrated high-speed MSM germanium waveguide photdetector, *Opt. Express*, 18(5), 4986–4999, 2010.
19. T. Yin et al., Ge n-i-p waveguide photodetectors on silicon-on-insulator substrate, *Opt. Express*, 15(21), 13965–13971, 2007.
20. M. W. Geis et al., Silicon waveguide infrared photodiodes with > 35 GHz bandwidth and phototransistors with 50 A/W-1 response, *Opt. Express*, 17, 5193–5204, 2009.
21. O. Jambois, Towards population inversion of electrically pumped er ions sensitized by Si nanostructures, *Opt. Express*, 18, 2230–2235, 2010.
22. I. Yoiung, E. Mohammed, J. Liao, A. Kern, S. Palermo, B. Block, M. Reshotko, P. Chang, Optical I/O Technology for Tera-Scale Computing, *ISSCC, Dig. Tech. Papers*, 45(1), 468–469, 2009.
23. F. Liu et al., 10 Gbps 530fJ/b Optical Transceiver Circuit in 40 nm CMOS, Symposium on VLSI Circuit, 290–291, June 2011.
24. A. Huang et al., A 10Gb/s photonic modulator and WDM MUX/DEMUX integrated with electronics in 0.13 μm SOI CMOS, Solid-State Circuits Conference ISSCC 2006, *Dig. Tech. Papers*, San Francisco, CA: *IEEE*, 922–929, 2006.
25. J. Orcutt et al., Open foundry platform for high-performance electronic-photonic integration, *Opt. Express*, 20(11), 12222–12232, 2012.
26. J. Chan et al., Architectural exploration of chip-scale photonic interconnection network designs using physical-layer analysis, *J. Lightwave Technol.*, 28(9), 1305–1315, 2009.
27. S. Koohi, M. Abdollahi, and S. Hessabi, All-Optical Wavelength-Routed NoC Based on a Novel Hierarchical Topology, NOCS'11: International Symposium on Networks-on-Chip, Pittsburgh, PA: *IEEE*, 97–104, 2011.
28. D. Vantrease et al., Corona: System Implications of Emerging Nanophotonic Technology, International Symposium on Computer Architecture, 153–164, Beijing, China, 2008.
29. Y. Pan, P. Kumar, J. Kim, G. Memik, Y. Zhang, and A. Choudhary Firefly: Illuminating Future Network-on-Chip with Nanophotonics, International Symposium on Computer Architecture, p. 429440, Austin, TX, June 2009.
30. C. Batten et al., Designing chip level nanophotonic interconnection networks, Emerging and selected topics in circuits and systems, *IEEE Journal*, 2(2), 137–153, 2012.
31. A. Scandurra and I. O'Connor, Scalable CMOS-Compatible Photonic Routing Topologies for Versatile Networks on Chip, Network on Chip Architecture, 2008.
32. Scott Beamer et al., Re-Architecting DRAM Memory Systems with Monolithically Integrated Silicon Photonics, 37th ACM/IEEE International Symposium on Computer Architecture (ISCA), 2010.
33. D. Wentzlaff et al., On-Chip Interconnection Architecture of the Tile Processor, *IEEE Micro*, 27(5), 15–31, 2007.
34. A. Boos, L. Ramini, U. Schlichtmann, D. Bertozzi, PROTON: An Automatic Place-and-Route Tool for Optical Networks-on-Chip, ICCAD'13: International Conference on Computer Aided Design, San Jose, CA: *IEEE*, 2013.
35. D. Ludovici et al., Assessing Fat-Tree Topologies for Regular Network-on-Chip Design under Nanoscale Technology Constraints, DATE'09: Conference on Design, Automation and Test in Europe, Nice, France: *IEEE*, April 2009.
36. Xuezhe Zheng et al., Ultra-efficient 10Gb/s hybrid integrated silicon photonic transmitter and receiver, Optical Society of America, 2011.
37. D. A. B. Miller, Energy Consumption in optical modulators for interconnects, Optical Society of America, 2012.
38. S. Le Beux et al., Optical Ring Network-on-Chip (ORNoC): Architecture and Design Methodology", DATE: Conference on Design, Automation and Test in Europe, Grenoble, France: *IEEE*, 788–793, 2011.
39. M. Georgas et al., A monolithically-integrated optical receiver in standard 45-nm SOI, *Solid-State Circuits*, 47, 1693–1702, 2002.

8 Quest for Energy Efficiency in Digital Signal Processing
Architectures, Algorithms, and Systems

Ramakrishnan Venkatasubramanian

CONTENTS

8.1   INTRODUCTION

Digital signal processors (DSPs) are essential for real-time processing of real-world digitized data, performing the high-speed numeric calculations necessary to enable a broad range of applications—from basic consumer electronics to sophisticated industrial instrumentation. Software programmability for maximum flexibility and easy-to-use, low-cost development tools of DSPs enable designers to build innovative features and differentiate value into their products, and get these products to market quickly and cost-effectively.

A modern-day MacBook Air operated at the energy efficiency of computers from 1991 would last only 2.5 seconds on its fully charged battery [1]. Technology scaling and architectural improvements have enabled tremendous energy efficiency in end products with the energy efficiency revolution over last three decades. The compute and memory bandwidth trends are only pointing us to improved computing and memory bandwidth requirements to enable rich and sophisticated user experience in handheld devices and the server platforms in the future.

8.2   DSP COMPUTING: COMPUTE AND
MEMORY BANDWIDTH TRENDS

In 2008, 68% of all DSPs were used in the wireless sector—mobile handsets and base stations [2]. A large percentage of the remaining 32% find usage in embedded systems—cameras, sensors, audio players, and industrial and automotive systems. Baseband chips typically consist of one or more DSP cores. The largest market for DSP silicon is known as embedded solutions, generally referred to as System-on-Chip (SoC) products. Of that SoC DSP market, cell phones constitute the largest segment, with baseband modem chips being the most significant. In 2012, Qualcomm shipped 616 million Mobile Station Modem chips. An average of 2.3 DSP cores in each chip amount to a total of 1.6 billion DSPs shipped in silicon in 2013 [3].

The energy efficiency of embedded SoCs in handheld and base station space has led to adoption of DSPs in the high-performance computing (HPC) space as coprocessors that offer increased giga floating-point operations per second per milliwatt (GFLOPS/mW) to enable energy-efficient scaling in the HPC domain.

The fundamental difference between handheld devices and HPC is the power versus performance trade-off in the overall power budget of the system. Handheld devices (cell phones, tablets, etc.) being battery operated are extremely sensitive for standby power. They try to optimize leakage and dynamic power of the embedded

SoC to the fullest extent possible using hardware techniques such as automatic voltage scaling (AVS) and many hardware–software codesign techniques to reduce overall power. On the other hand, HPC kernels operate on large datasets and need the maximum performance possible in a given power budget. The design parameters are optimized differently to achieve the optimal solution to the HPC market. DSPs are broadly categorized into following three categories [4]:

- Application-specific DSP (AS-DSP): Typically customized to an application to serve high-end application performance or cost requirements, for example, DSP optimized for speech coding.
- Domain-specific DSP (DS-DSP): DS-DSPs are targeted to a wider application domain, for example, cellular modems (TI C540, TCSI Lode). They can be applied to a variety of applications within the domain and can run domain-specific algorithms efficiently.
- General purpose DSP (GP-DSP): GP-DSP is designed for markets with a volume high enough to allow specialized solutions. GP-DSPs have evolved from the classic Fast Fourier Transform (FFT)/filtering multiply-accumulate design paradigm. Examples are TI C66x, Lucent 16xx, and ADI 21. GP-DSPs are readily available, widely applicable and have a large software base.

The industry trend is to increase compute capacity by using heterogeneous processing, using dedicated processing elements to efficiently handle specific tasks, and using high level of concurrency and parallelism to get the job done faster. Memory access latency has been the bottleneck for computing as the processor advances, the processing speed is usually input-output (IO) bound that the data is not ready when core needs it. Industry trend is to improve the efficiency of data movement, by prefetching the data closer to CPU, using wider memory bandwidth to reduce the memory access latencies. The compute and memory bandwidth trends are explained in the context of two domains—mobile computing and high-performance computing.

8.2.1 Mobile Computing Trends

The mobile computing and bandwidth trend in smartphones and tables is illustrated in Figure 8.1. The 1080p video support and octa-core (8 cores) already available in smartphones today [5] illustrate the compute trend in smartphones and tablets over the last decade. A smartphone has more computing power than personal computers and for the first time, the number of tablets and smartphones crossed the number of PCs in 2010 [6].

The compute power increase in tablets and smartphones enables richer end applications that find the need and use for all the compute power available in the system. But handheld devices being battery operated need to be extremely efficient to enable a long battery life. One of the most demanding applications where semiconductors are used is in the various applications of digital video from tablet computers to home entertainment. iPad with a retina display is already at high-definition (HD) resolution (2048 × 1536 pixels), and all indications are that video is racing toward what is known as 4K resolution, also known as ultrahigh definition, 3840 × 2160 pixels,

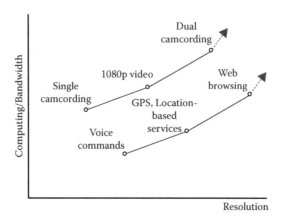

FIGURE 8.1 Mobile computing trends: Computing/bandwidth trend. (Woo, S. Samsung Analyst Day 2013, The Semiconductor Wiki Project, accessed August 2013, http://www.semiwiki.com/forum/files/S.LSI_Namsung%20Woo_Samsung%20System%20LSI%20Business-1.pdf.)

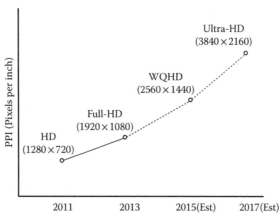

FIGURE 8.2 Mobile display trends: Illustrates memory bandwidth and compute trend. (Woo, S. Samsung Analyst Day 2013, The Semiconductor Wiki Project, accessed August 2013, http://www.semiwiki.com/forum/files/S.LSI_Namsung%20Woo_Samsung%20System%20LSI%20Business-1.pdf.)

which is roughly four times the pixels and hence four times as demanding as HD. A pictorial representation of display technology trend is shown in Figure 8.2 [7].

Camera sensors are showing a similar exponential trend as well. This is illustrated in Figure 8.3. Very sophisticated image processing algorithms enable face detection, wide dynamic range, night vision, and other advanced features for processing the captured image in sensors.

8.2.2 High-Performance Computing and DARPA Computing Challenges

The end application for HPC is typically significant data crunching for scientific research or cloud-based data crunching on large data sets. Typically, HPC is used in

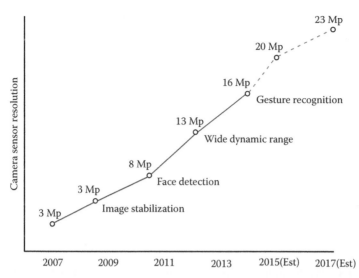

FIGURE 8.3 Mobile camera sensor trends: Illustrates memory bandwidth and algorithm trends. (Woo, S. Samsung Analyst Day 2013, The Semiconductor Wiki Project, accessed August 2013, http://www.semiwiki.com/forum/files/S.LSI_Namsung%20Woo_Samsung%20System%20LSI%20Business-1.pdf.)

oil and gas exploration, bioscience, big data mining, weather forecast, financial trading, electronic design automation, and defense. HPC systems need to be scalable, provide high computing power to meet the ever increasing processing needs with high energy efficiency for varied end applications. One of the major challenges faced in supercomputing industry is its efforts to hit exascale compute level by the end of the decade. In 1996, Intel's Accelerated Strategic Computing Initiative (ASCI) Red was the first supercomputer built under the ASCI; the supercomputing initiative of the U.S. government to achieve 1 TFLOP performance [8]. In 2008, the IBM-built Roadrunner supercomputer for Los Alamos National Laboratory in New Mexico reached the computing milestone of 1 petaflop by processing more than 1.026 quadrillion calculations per second; it ranked number 1 in the TOP500 list in 2008 as the most powerful supercomputer [9]. Moving to 1000 times capacity to ExaFLOPs is very challenging—given the fact that power density of such ExaFLOP cannot expand 1000 times. Power delivery and distribution create significant challenges. Intel, IBM, and HP, for example, are continuing on the journey for core performance on multicore, many core, as well as graphical processing unit (GPU) accelerator model, and face significant power efficiency challenges.

To reach to the ExaFLOP level of compute performance, an optimal HPC processor should have sufficient compute performance and efficient data movement on chip and cross chips as necessary. The TOP500 list of supercomputers and their performance scaling is shown in Figure 8.4 [10]. Even in defense industry, most deployed military information systems have constrained computational capability because of limited electrical power available on platforms, heat dissipation challenges, and limitations in size and weight. The end result is that many of the current intelligence, surveillance, and reconnaissance systems have sensors that collect more information than can be processed in real time resulting in potentially valuable real-time intelligence data not processed in a timely manner.

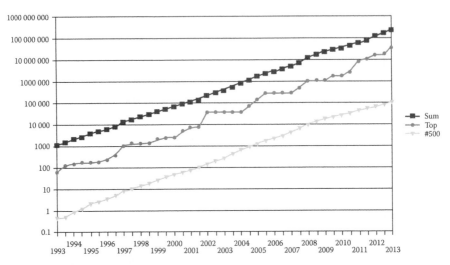

FIGURE 8.4 The logarithmic y axis shows performance in GFLOPS. The "Top" line denotes the fastest supercomputer in the world at the time. The "#500" line denotes super-computer No. 500 on the TOP500 list. The "Sum" line denotes the total combined perfor-mance of supercomputers on the TOP500 list. (TOP500, Wikipedia, accessed August 2013, http://en.wikipedia.org/wiki/TOP500)

Current embedded processing systems have power efficiencies of around 1 GFLOPS/W. Recently, Texas Instruments multicore DSP claimed to achieve 12.8 single precision GFLOPS/W [11]. Warfighters anticipate requirements of at least 75 GFLOPS/W. The goal of the Power Efficiency Revolution For Embedded Computing Technologies (PERFECT) program is to provide warfighter-required power efficiency [12]. PERFECT aims to achieve the 75 GFLOPS/W goal by taking novel approaches to processing power efficiency. These approaches include near-threshold voltage opera-tion and massive heterogeneous processing concurrency, combined with techniques to effectively use the resulting concurrency and tolerate the resulting increased rate of soft errors. To reach Defense Advanced Research Projects Agency (DARPA) PERFECT goals, it requires more than 14 times more power efficiency compared to current DSPs and 75 times more power efficiency comparing to typical HPC processors.

8.3 DSP COMPARISON WITH APPLICATION PROCESSORS AND GPU

8.3.1 WHAT MAKES A DSP A DSP? REAL-TIME PROCESSING—DSP VERSUS APPLICATION PROCESSOR

A DSP is typically characterized by a single-cycle MAC (multiply and accumulate) operation and has the ability to support hard real-time requirements. They enable a better support for real-time requirement (e.g., audio samples delivered every 100 milliseconds and the signal processing has to be completed within that window to keep up with the real-time data stream) with short pipelines, in order processing, no speculation in the processor, ability to run a high-level real-time operating system (RTOS), low latency interrupts, and so on. Further, DSPs typically also support multiple execution units with highly customized datapaths, sophisticated direct memory access (DMA), and high-bandwidth memory systems, with very efficient zero or near-zero overhead looping.

In a hard real-time system, missing a deadline would result in a total system failure. In a soft real-time system, the usefulness of the results degrades after its deadline, thereby degrading the quality of service of the system.

An application processor provides high-level OS functions such as virtualization, multiple levels of caches with MMU, speculative fetching and branching, protected memory, semaphore support, context save and restore, and threading support. An application processor such as an ARM® core in an embedded SoC typically will not be able to meet hard real-time requirements, but might be able to support soft real-time requirements.

8.3.2 VLIW AND SIMD

Very long instruction word (VLIW) architectures have multiple functional units that take advantage of vastly available instruction-level parallelism in applications. Single-instruction multiple data (SIMD) techniques operate on multiple data in a single instruction (exploiting data parallelism).

The main difference between DSP and other SIMD-capable CPUs is that the DSPs are self-contained processors with their own instruction set, whereas SIMD extensions rely on the general-purpose portions of the CPU to handle the program details, and the SIMD instructions handle the data manipulation only. DSPs also tend to include instructions to handle specific types of data, sound, or video, for instance, whereas SIMD systems cater to generic applications.

Oftentimes, a general purpose processor executes the main application and calls the SIMD processor for the compute-intensive kernels. SIMD has typically dominated HPC since Cray-1. Adding SIMD enhancements to embedded processing cores is becoming increasingly common as well. ARM® A15 cores offer SIMD instructions. The Texas Instruments C66x DSP architecture adds SIMD instructions to a VLIW architecture [11].

8.3.3 DSP VERSUS GPGPU

A GPU provides more FLOPs per chip when compared to the x86 main processor in a computer. This raw GPU performance is achieved by the following reasons:

- Each core inside the GPU is very simple, with a simple instruction set and no out-of-order, speculation, or other complex logic. Programming the GPU is more complicated, as programs are run on groups of processors and with lots of little constraints. This makes it possible to fit more cores into the same area.
- GPUs typically have very little cache memory. Hence programs have to rely on bandwidth and managing to stream data the GPU.

A GPGPU (general-purpose GPU) extends usage of GPU units to general-purpose tasks. For tasks that operate on wide data (vector processing) and data intensive (imaging, video, graphics), GPGPU programming can offer predictable algorithms that can effectively and efficiently prefetch data and stream it through the cores at a predictable rate. But GPGPUs are very bad for control code programming. SIMD processing is applied on large vectors of independent elements in parallel.

A classic single-core DSP, on the other hand, has specialized instructions in the instruction set for signal processing, support for loops in very efficient ways, and is often SIMD. Recent DSP implementations support vector processing capabilities as well. DSPs are general enough to be able to run a rudimentary OS and operate semi-independently from the main processor. In a multicore DSP cluster, each DSP operates on a different problem. So rather than one vector of a thousand elements in a video compression, each DSP might operate on independent video streams load balancing all the thousand streams across the multicore cluster. Although DSP programming is painful compared to GP processors, their programming model is much simpler compared to GPGPUs [13]. Hence, GPGPUs are very different from DSPs and are used to solve different types of problems in different ways.

8.3.4 Single Core versus Multicore

The trend to develop multicore systems is a general trend not just in DSPs but across the processor space. The thermal design power limitations and the energy efficiency requirements have forced processor developers to add multicore capabilities instead of increasing the clock frequency of the operation.

The variable parameters in the design of multicore systems include type and number of cores; the speed of operation of the cores (instances required); homogeneous or heterogeneous multicore systems; and the usual considerations of area, power, and performance.

Functions that determine the topology and architecture of a multicore platform are cache coherency and memory bandwidth. When there are caches involved in a multicore system, support for cache coherency across the entire multicore system is desired as it enables the software running on the embedded processing system to balance load and allocate tasks seamlessly between cores. In addition to the number of cores, the reduced instruction set computer (RISC) (control decision)/SIMD (data vector) processing balance for the end application determines the number of DSP or application processors required in the system. Specialized hardware accelerators typically improve specific algorithms in hardware by a factor of 10x–50x over a GPCPU or DSP. Availability of such specialized accelerators improves the overall energy efficiency of the system by powering up the accelerator logic only for the duration it is required. It also frees up available compute capability in the multicore system for other processing. High-bandwidth sophisticated DMA functions take up all the data movement tasks in the multicore platform, enabling processors to be powered up on demand for a specific data processing task. Further, the software programming model, the software tool chain availability, and support for multicore programming models, such as OpenMP, are typically taken into account in the selection of multicore architectures for a given end application.

8.4 DSP COMPUTE SCALING TREND AND FLOATING-POINT CONVERGENCE

8.4.1 GMAC SCALING TREND

GMAC is a one billion multiply-accumulate operations. The metric for measuring the performance improvement in DSPs is GMAC scaling. Figure 8.5 shows the MMAC/s (million multiply-accumulate operations per second) trend over five generations of Texas Instruments DSP architectures [14]. The MMAC improvement trend attributed to both the clock increase and parallel processing across generations shows that MMAC availability in DSP cores double every year. It can be seen that within the same architecture, technology scaling typically improves performance by a factor of two. On top of technology scaling, architectural improvements enhance performance by another factor of two. Although the MMAC compute scaling trend enables significant compute scaling, the focus has since shifted to energy efficiency.

8.4.2 FIXED VERSUS FLOATING POINT

There are many considerations for system developers when selecting DSPs for their applications. Among the key factors to consider are the computational capabilities

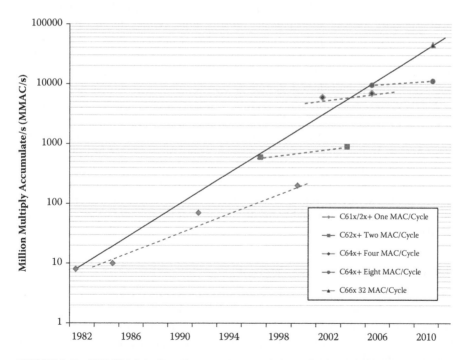

FIGURE 8.5   TI DSP MMAC scaling over the last 30 years. The dotted lines show the performance of each architecture with improvement achieved by technology scaling. The performance shift between dotted lines shows the impact of the architectural innovations between successive architectures. (R Venkatasubramanian, Texas Instruments, Inc., Dallas, TX.)

required for the end application, processor and system costs, performance attributes, and ease of development. The first and foremost factor is the fixed-point versus floating-point arithmetic support for the end application.

Underlying DSP architecture can be separated into two categories based on the numeric representation—fixed point and floating point. These designations refer to the format used to store and manipulate numeric representations of data [15].

Fixed-point systems use the bits to represent a fixed range of values, either integers or with a fixed number of integer and fractional bits. The dynamic range of values is therefore quite limited and values outside the set range must be saturated to the endpoints. Fixed-point processors usually quote their 16-bit performance as multiplies per second or MAC operations per second. Algorithms developed for fixed-point processors have to operate on a set of data that stays within the predetermined range to make the optimal use of the quoted DSP performance. Because of this, any data set that is not predictable or has a wide variation will have significant performance reduction in a fixed-point DSP.

On the other hand, floating-point representations offer a wider dynamic range by rational number representation (scientific notation), using a mantissa and an exponent for the representation. Floating-point representation was standardized by the IEEE Standard for Floating-Point Arithmetic, IEEE 754. It is a technical standard for floating-point computation established in 1985. The latest version, IEEE 754-2008 published in August 2008, extends the original IEEE 754-1985 standard and IEEE Standard for Radix-Independent Floating-Point Arithmetic, IEEE 854-1987. Single-precision floating-point format is represented in 4 bytes (32 bits) and represents a wide dynamic range of floating-point values. Double-precision floating-point format is represented by 8 bytes (64 bits) and represents an even wider dynamic range of floating-point values.

Single-precision floating-point operations where numbers are represented in 32 bits as: $(-1)^S \times M \times 2^{(N - 127)}$, where S is the sign bit, M the mantissa, and N the exponent. S is 1 bit, N represented in 8 bits and M represented with 23 bits. In this way numbers in the range $2^{-127} - 2^{128}$ can be represented with 24 bits of precision in the mantissa. By contrast, a fixed-point algorithm with 16 bits can only represent a range of 2^{16} values (the numbers 0–65535), so there is much less dynamic range inherent in the numerical representation. Hence, any data set that is not predictable or has a wide variation would naturally fit into a floating-point representation of the data set.

8.4.3 DYNAMIC RANGE AND PRECISION

The exponentiation inherent in floating-point computation assures a much larger dynamic range—the largest and smallest numbers that can be represented—which is especially important when processing extremely large data sets or data sets where the range may be unpredictable. As such, floating-point processors are ideally suited for computationally intensive applications.

It is also important to consider fixed- and floating-point formats in the context of precision—the size of the gaps between numbers. Every time a DSP generates a new number via a mathematical calculation, that number must be rounded to the nearest value that can be stored via the format in use. Rounding and/or truncating numbers during signal processing naturally yields quantization error, which is the deviation between actual analog values and quantized digital values. Since the gaps between adjacent numbers can be much larger with fixed-point processing when compared to floating-point processing, round off error can be much more pronounced. As such, floating-point processing yields much greater precision than fixed-point processing, thereby distinguishing floating-point processors as the ideal DSP when computational accuracy is a critical requirement.

The data set requirements associated with the target application typically dictate the need for fixed-point or floating-point processing.

8.4.4 CONVERGENCE OF FIXED AND FLOATING POINT

From a hardware implementation point of view, fixed-point implementations are definitely smaller and faster than floating-point implementations. However, there is a price to pay in terms of development time for certain algorithms. Typically, algorithms are developed based on computer models and used for initial system deployments. As the deployments grow in scope and usage, engineers gather real-world data and bring this back to the lab to improve the system performance by tweaking the algorithms. These new algorithms are often developed using MATLAB® or other inherently floating-point tools. The challenge then lies in translating these floating-point algorithms to fixed-point algorithms while retaining the performance of both the algorithm and of the system. Unwieldy or complex algorithms can use up a disproportionate amount of system resources thereby lowering the overall performance of the system. It is not uncommon for the process of porting code from MATLAB to a real system to take weeks or months when complicated processing is involved. If the DSP processor hardware offers floating-point support, the entire conversion from floating point to fixed point is unnecessary and would enable faster time to market. This is illustrated in Figure 8.6 [15].

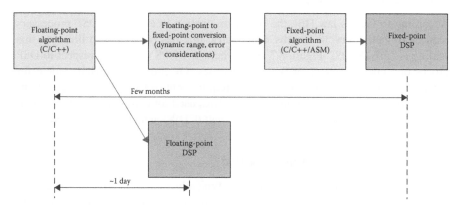

FIGURE 8.6 Effort involved in fixed-point to floating-point conversion and ease of mapping to floating-point DSP.

Across mission-critical applications, such as defense, public safety infrastructure, and avionics, floating point provides ease of development and performance lift. Not only does floating point shorten development life cycle by being able to use code directly out of MATLAB, but also floating-point implementations of many algorithms take fewer cycles to execute than fixed-point code (such as large FFT). For example, radar, navigation, and guidance systems process data that are acquired using arrays of sensors. The varying energy pattern across the many sensor elements provides the information relevant to the location and tracking of the target. This array of data must be processed as a set of linear equations to extract the desired information. Solution methods include math functions such as matrix inverse, factorization, and adaptive filtering. Image recognition, used for medical imaging such as ultrasound, as well as machine vision and industrial automation, also requires a high degree of accuracy and thus benefits from floating point. In ultrasound, signals from sound sources must be defined and processed to create output images that provide useful diagnostic information. The greater precision enables imaging systems to achieve a much higher level of recognition and definition for the user.

A well-known application area for floating-point use is in audio processing, where a high sampling rate, coupled with very tight latency requirements, can force filtering and other noise reduction algorithms toward the higher precision and larger dynamic range provided by floating point. Wide dynamic range also plays a part in robotic design. Unpredictable events can occur on an assembly line. The wide dynamic range of a floating-point DSP enables the robot control circuitry to deal with unpredictable circumstances in a predictable manner.

Texas Instruments introduced the C66x line of DSP processors in 2009, which merged fixed- and floating-point capabilities into a single processor without compromising the speed of operation [11]. Since then, almost all the major players in the DSP market space have followed suit and merged fixed-/floating-point implementations in their DSPs.

8.5   ENERGY EFFICIENCY AS THE METRIC

8.5.1   Gene's Law and Its Scaling

According to Gene's law [16], the power dissipation in embedded DSP processors reduces by half every 18 months. One metric that has been used to illustrate this scaling is "mW/MIPS." Figure 8.7 shows the mW/MIPS trend with respect to DSP processor improvements over multiple decades. Of course, the driving force for Gene's law has been all the computational requirements for various end applications as described in Section 8.2. Some of the metrics used for comparison of DSP architectures include computational density per watt (CDW) and energy per function.

8.5.2   Computational Density per Watt

In computing domain, CDW is a measure typically used to determine the energy efficiency of a particular computer architecture or system. Computational density is the measure of computational performance across range of parallelism, grouped by process technology [17]. CDW normalizes the process technology and process

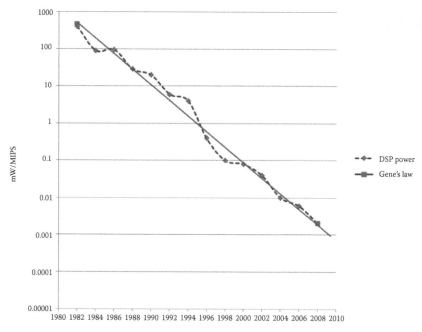

FIGURE 8.7   Gene's law. (Frantz, G., Digital signal processor trends, *IEEE Micro*, 20(6), 52–59, © 2000 IEEE.)

voltage considerations and provides a single metric that can be used to compare performance across range of parallelism.

In the context of DSP cores, the CDW metric that is typically used is the number of MMAC's supported by 1 milliwatt of power (MMAC/mW). Based on the type algorithms running in the end application, the exact computational unit is compared for CDW. Example: In HPC, the majority of the operations are double precision floating-point operations. So the number of double-precision floating-point per watt is the metric used in HPC. Other typical computational metrics include 16-bit integer per watt, 32-bit integer per watt and single-precision floating-point per watt.

Sometimes computational density is normalized to energy. In such scenarios, energy efficiency metric (MMAC/mJ) is used for comparison. Typically, all the architectural and physical design innovations in DSP design are targeted toward improving the overall power and energy efficiency of the overall system. For example, the C66x DSP core from Texas Instruments [14] showed a power efficiency improvement of 4.5× over the previous C64x+ DSP architecture as shown in Table 8.1.

8.5.3   Energy per Function

If an algorithm is predominantly using one type of function (e.g., 1024-point single-precision FFT), the exact energy for that function can be computed for diverse architectures and "Energy per function" metric can be used for comparing the architectures. Note that "Energy per function" is different from the "Energy efficiency per watt" metric. Energy efficiency per watt computes total energy consumed regardless of the end algorithm or function, whereas energy per function only looks at one

TABLE 8.1
TI DSP Power Efficiency Scaling

	C64x+ (2005)	C66x (2009)	Comments
Process and voltage	65 nm, 1.1 V	40 nm, 1 V	
Operating speed	1GHz	1.25 GHz	
Total power	960 mW	1180 mW	
16-bit fixed-point MMAC	8000	40000	
Power efficiency (MMAC/mW)	8.3	37	4.5X improvement

Source: Damodaran, R. et al., *25th International Conference on VLSI Design (VLSID)* ©2012 IEEE.

TABLE 8.2
Energy per FFT Function

Platform	Effective Time to Compute 1024 Point Complex-to-Complex Single-Precision FFT (µS)	Power (W)	Energy per FFT (µJ)
GPU: nVidia Tesla C2070	0.16	225	36.0
GPU: nVidia Tesla C1060	0.30	188	56.4
GPP: Intel Xeon Core Duo @3 GHz	1.80	95	171.0
GPP: Intel Nehalem Quad core @3.2 GHz	1.20	130	156.0
DSP: TI C6678 @1.2 GHz	0.86	10	8.6

Source: Saban N. Multicore DSP vs. GPUs, Workshop on GPU & Parallel Computing, Israel, January 2011, http://www.sagivtech.com/contentManagment/uploadedFiles/fileGallery/Multi_core_DSPs_vs_ GPUs_TI_for_distribution.pdf.

function and computes the energy for the same. For example, in HPC, FFT's are used widely. A comparison of "Energy per FFT function" between GPU, DSP, and GPP is shown in Table 8.2 [18].

8.5.4 Application Cube

Another interesting metric for DSP architecture comparison called "Application cube" was proposed by Lucent back in 1997 [19]. It is shown in Figure 8.8. The three parameters—power, performance, and cost—are defined on the axes of the cube in three dimensions. For any particular application, each DSP processor is mapped to various cubes and smaller the volume of the cube, the more suited the DSP is to the application. Ever since this metric was proposed in 1997, code size has become less important consideration, and the dimension could instead be replaced with "Memory bandwidth per function."

8.5.5 Energy Benchmarking

Benchmark suites such as BDTI [20] and EEMBC® [21] are typically used for performance and power analysis and comparison across architectures. However, benchmarking multicore architectures are significantly more complicated than benchmarking single-core devices. This is because multicore performance is not only affected by the choice of CPU, but also is heavily dependent on the memory system and the interconnect capabilities [2].

Depending on the end application, specific application kernels are used for benchmarking as shown in Table 8.3. Even though different architectures may show different results for the application kernels, finally what matters is the actual performance achieved for the system in the context of the end application.

Because of the focus on energy efficiency and power efficiency considerations in architectures, EEMBC released EnergyBench™ [21]—a benchmark suite that attempts to provide data on the amount of energy a processor consumes while running EEMBC's performance benchmarks. The EEMBC-certified Energymark™ score is a metric that normalizes the process technology and voltage and provides a metric for a processor's efficient use of power and energy.

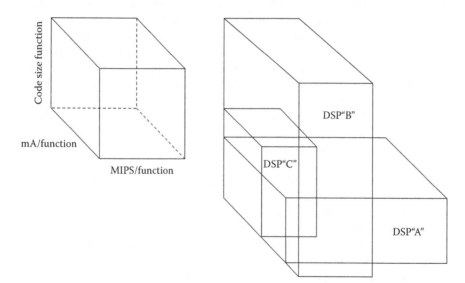

FIGURE 8.8 Application cube. (Edwin J. Tan and Wendi B. Heinzelman, DSP architectures: past, present and futures, *SIGARCH Comput. Archit. New,* 31(3), 6–19, http://doi.acm.org/10.1145/882105.882108, © 2003 IEEE.)

TABLE 8.3
Typical Benchmarks for Various Application Classes

Type of Application	Typical Algorithms
Control code	Loops, calls and returns, branch performance
Multimedia and graphics kernels	SAD, DCT/IDCT
General DSP	FFT, IFFT, IIR, matrix multiply, viterbi decoding
Application specific	Mediabench for multimedia, radar algorithms for RADAR etc.

8.6 IMPROVING ENERGY EFFICIENCY IN MEMORY AND IO INTERFACES

8.6.1 IMPROVING MEMORY BANDWIDTH AND MEMORY EFFICIENCY

On-chip embedded SRAM has been an area of active research for the last three decades. In advanced technology nodes, numerous power management techniques have been used to improve memory energy efficiency including memory pipelining to enable full bandwidth from memories and array source biasing [14]. Near-threshold operation of memories with assist circuits has been deployed to reduce the overall memory leakage power as well.

For external memory accesses, the access latency has become a major performance road block in recent years as memory performance has not kept up with the processing capacity gains from CPU frequency and massive parallelism in multicore SoCs. Next-generation high-performance memory interfaces such as Hybrid Memory Cube (HMC) and High-Bandwidth Memory Interface (HBM) are being looked at to address the memory performance bottleneck.

HMC is a computer RAM technology developed by Micron Technology that deploys 3D packaging of multiple memory dies, typically 4 or 8 memory dies per package with through-silicon vias (TSV) and microbumps [22]. It has more data banks than classic DRAM memory of the same size, and memory controller is integrated into memory package as separate logic die. According to the first public specification HMC 1.0, published in April 2013, HMC uses 16-lane or 8-lane (half-size) full-duplex differential serial links, with each lane having 10, 12.5, or 15 Gbps SerDes. The ultimate aim of HMC memory topology is to offer improved Gbps data transfer at a lower power footprint.

A standard that leverages Wide IO and TSV technologies to deliver products with memory interface ranging from 128 GB/s to 256 GB/s has been defined by the HBM initiative by JEDEC standard in March 2011 [23]. The HBM task group is defining support for up to 8-high TSV stacks of memory on a data interface that is 1024-bit wide. HBM provides very large memory bandwidth because of parallelism; it basically integrates memory into the same SoC package and thus increases processing capacity in the SoC and reduces the power per Gbps compared to HMC.

8.6.2 HIGH BANDWIDTH IOs

Consider a multicore SoC with some number of DSP cores and application processor cores (and potentially multimedia and other accelerator peripherals). For efficient processing in the cores in the SoC, enough data movement has to be enabled in and out of the SoC to feed the computational capacity offered by the cores. Figure 8.9 gives an overview of IO interfaces that can potentially be used in an embedded multicore SoC with their relative latency versus bandwidth comparison [23,24]. On one end of the spectrum is the high bandwidth applications support with 1 or 10 Gigabit Ethernet interfaces that enable data transfer over a long distance. On the other end of the spectrum, we have PCI Express and other serial interfaces that offer short links

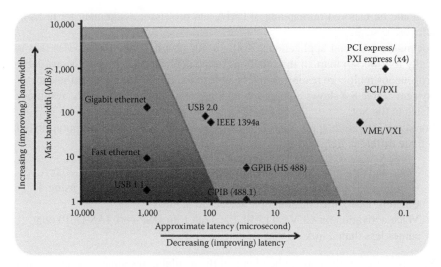

FIGURE 8.9 IO bandwidth versus latency comparison. (From National Instruments white-paper, PXI Express Specification Tutorial, 2010, http://www.ni.com/white-paper/2876/en/.)

with high data rates. Each of the categories of interfaces is evolving and continues to offer the most efficient data transfer at the least power footprint depending on the end-use case. It is important to note that getting data in and out of the multicore SoC takes up significant amount of energy and the SoC architecture has to comprehend this.

8.7 HOLISTIC APPROACH TO DSP PROCESSING: ARCHITECTURE, ALGORITHMS, AND SYSTEMS

8.7.1 ALGORITHMS MAPPING TO UNDERLYING ARCHITECTURE

AS-DSP and DS-DSP described in Section 8.2 typically would be tailored to a specific market. For example, Qualcomm's Snapdragon processor is tailored for a specific market—application processor for mobile application. The architecture is tailor made for individual functions in that application space. Similarly, a processor optimized for audio/voice domain would require a much lower power and smaller footprint DSP.

Depending on the application, loss of precision and dynamic range due to floating-point to fixed-point conversion can be avoided by choosing processor cores with native floating-point support. This results in improved time to market and better dynamic range for calculations. For every algorithm in the application space, the mapping of the algorithm to the underlying architecture is key to achieving the most energy efficient end product out of the architecture. The energy per function and the CDW metrics for the specific functions need to be compared and the architecture/algorithms optimized accordingly to enable energy efficiency improvement. Data movement within the SoC and to/from external sources consumes power and energy.

This should be factored in mapping of DMA and processor assignment for tasks in a multicore system.

For multimedia end applications, Kollig et al. [25] describe the process that is typically followed to map all the functions in the application kernel to the available heterogeneous multicore resources and be able to maximize the energy efficiency of the system. Table 8.4 shows a potential mapping of algorithm class to the resources available in the SoC.

For voice activation application in a smartphone, the power consumption linked with three different architectures is shown below:

- Voice activation on ARM® A9 (application processor) takes about 20 mA.
- Using a dedicated chip for handling always-on voice processing, using TI C55 (GP-DSP), the consumption reduces to 4.5 mA.
- Voice activation on CEVA TL410, the Teak Lite DSP core (AS-DSP) consumes less than 2 mA.

So depending on the end application and the available resources, the power and energy consumption can vary significantly. This has to be taken into account in mapping of algorithms to available resources in an SoC.

8.7.2 REDUCING IDLE AND STANDBY POWER

IDLE and Standby power is the leakage and dynamic power consumed when the processor core is not performing any active task. IDLE/Standby power management is typically achieved by either of the two implementation schemes:

1. Based on the load balancing and scheduling operations, software can use AVS implementation if it is available (on a per core granularity) to reduce the voltage of processor core that is idle.
2. Software can schedule all the data crunching activities to a given processor and enable completion of the task as soon as possible with full voltage and maximum clock frequency. Once the task is complete, the processor core can be powered down.

The decision metric to use either of the above schemes is the idle period. If the processor core is idle for a long period, powering it down would provide the most energy efficient solution. On the other hand, if the idle periods are short and happen more often, it may better qualify for automatic voltage scaling.

8.7.3 ARCHITECTURE (HARDWARE PLATFORM)

A general overview of multicore DSP platforms is provided by Karam et al. [2], Tan and Windi [26], and Oshana [27]. From an energy efficiency point of view, heterogeneous multicores offer the best algorithm mapping to the underlying architecture. Typically any end application has a control plane that requires application processing including high-level operating system tasks. The data plane is controlled by the control plane, which includes all the high-speed DMA functions and monitoring the

TABLE 8.4

Algorithm Mapping to Resources for a Multimedia SoC

Algorithm Class	Resource Mapping	Reasoning
Analog standards audio decode	Fixed-point DSP	Legacy
Digital compressed audio decode	VLIW processor	Flexibility, availability of codecs, cost of HW
Audio post processing	Fixed-point DSP	Performance, flexibility
Analog standards video decode	Function-specific HW	Performance, legacy
Digital uncompressed video decode	Function-specific HW	Performance, legacy
Digital compressed video decode: Established codecs	Function-specific HW with control processor	Performance, cost of HW
Digital compressed video decode: New codecs	VLIW processor	Flexibility, standards evolution
Picture quality: artifact repair	Function-specific HW, partly on VLIW processor	Performance, flexibility, cost of HW
Picture quality: motion adaptive picture processing	VLIW processor with HW coprocessor	Performance, flexibility, cost of HW
Content browsing and control	Industry standard CPU	Support

Source: Kollig, P. et al. *Design, Automation and Test in Europe Conference and Exhibition,* © 2009 IEEE.

data crunching operations in the DSP or GPU or hardware accelerators. High-speed IOs typically facilitate this high-speed data movement.

This heterogeneous multicore operation is illustrated in the context of the scalable multicore platform called Keystone from Texas Instruments [28]. The heterogeneous multicore platform is shown in Figure 8.10. Keystone multicore architecture supports a variable number of C66x floating-point DSP cores and ARM® Cortex A8 and A15 RISC cores. These programmable elements are supplemented by AccelerationPacs, which are configurable accelerators that offload standardized functions from the programmable cores. As the features are implemented in hardware, AccelerationPacs give significant performance improvement for the function with improved energy efficiency. Physical interconnection within the chip denoted by Teranet and Multicore Navigator offer structures for fast access to on-chip memory and to external memory and hardware-based feature to facilitate load balancing and resource sharing within and between chips. These elements are scalable, allowing the platform to tailor the SoC to the performance and power consumption needs of devices serving different markets. Finally, there are high-performance IOs which, like the AccelerationPac, vary by market.

The Keystone Multicore SoC enables three levels of compute capabilities:

1. DSP CorePac with vector processing performs the scientific, data crunching, delivers real-time low-latency analytics achieving 12.8 GFLOP/W power efficiency.
2. ARM® CorePac handling the control plane tasks including high-level OS support, task management, and protocol logic.

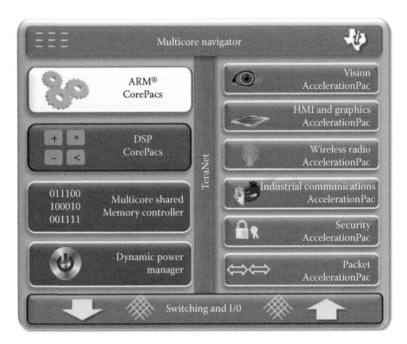

FIGURE 8.10   Keystone heterogeneous multicore platform from Texas Instruments. (From Texas Instruments Keystone Platform. http://www.ti.com/lsds/ti/dsp/keystone_arm/overview .page.)

3. Configurable and programmable AccelerationPac delivering fast-path data processing with low latency and deterministic response.

All these heterogeneous compute elements can run in parallel to achieve high capacity, power efficiency and delivers performance more than Moore's law.

A few embedded processing SoCs that belong to this platform include:

1. TI 66AK2H12: 4 Cortex-A15 processors and 8 TMS320C66x DSPs optimized for HPC, media processing, video conferencing, off-line image processing and analytics, gaming, security digital video recorders (DVR/NVR), virtual desktop infrastructure, and medical imaging.
2. TI 66AK2E05: 4 Cortex-A15 processors and 1 TMS320C66x DSP optimized for Enterprise video, IP cameras (IPNC), traffic systems (ITS), video analytics, industrial imaging, voice gateways, and portable medical devices.
3. TI TDA2X: 2 Cortex-A15 processors, 2 TMS320C66x DSP and a Vision AccelerationPac optimized for automotive vision.

Another example of a multicore platform is the CEVA-XC4000 series of multi-core processors optimized for next-generation wireless infrastructure applications [29]. The CEVA-XC4500 delivers highly powerful fixed-point and floating-point vector capabilities supplying the performance and flexibility demanded by next-generation wireless infrastructure applications. Freescale's QorIQ Qonverge is another heterogeneous multicore architecture optimized for wireless infrastructure applications [30].

8.7.4 SOFTWARE ECOSYSTEM

A seamless software ecosystem is key to any hardware multicore platform. From an energy efficiency point of view, the software stack must be able to load balance the tasks among the multicore systems and reduce idle/standby power through either AVS or powering down the idle processor cores.

The software model for wireless handsets is typically asymmetric processing, where systems are preconfigured and optimized for specific use cases (often employing more than one type of DSP core). In wireless infrastructure, the software model is to deploy symmetric multiprocessing through many similar cores. In the HPC space, multicore SoC complexity inevitably translates to software (SW) complexity.

For the sake of illustration of the software and system development capabilities, the TI software stack for DSP and ARM® in the context of the Keystone architecture is shown in Figure 8.11 [31]. The TI software development kit includes TI RTOS (real-time OS) running on DSP with real-time OS, optimized foundational SW including device drivers,

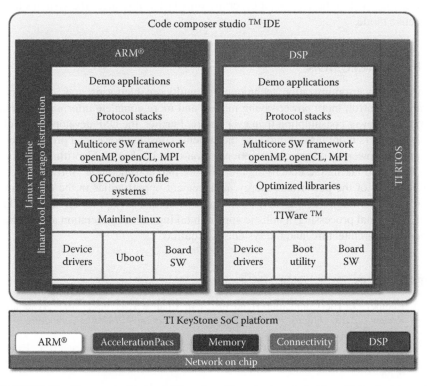

FIGURE 8.11 Keystone software platform from Texas Instruments. (From Texas Instruments Appnote. www.ti.com/lit/pdf/spry231.)

board support package, boot utility so that the user does not need to spend the time to study SoC register level details. TI RTOS also provides optimized DSP libraries, multicore SW framework with OpenMP, OpenCL, and MPI, as well as networking protocol stack to facilitate application level software development. In ARM® side, TI SW SDK is based on mainline Linux that leverages the open-source community effort to deliver reliable and high-quality SW to reduce time to market cycles. TI Eclipse-based Code composer Studio IDE enables a rich development, debug, and instrumentation environment that increases SW development productivity, optimizes SW effort, and enables best return on investment.

Power-efficient CMOS processes integrate hardware techniques that enable granularly defined low-power modes, and voltage and frequency scaling. Software APIs make these techniques readily available to the application for control through the RTOS, and test tools help the designer evaluate different implementations for power consumption. This enables such fine-grained control of low-power modes and voltage/frequency scaling on a per core granularity. Software is designed to use the DSP's internal memory wherever possible, keeping high-bandwidth memory on-chip and reserving external memory for low-speed, occasional access. Off-chip memory also serves well for booting, and can be powered down after startup. Software is typically optimized for performance to reduce the code's footprint in memory and the number of instruction fetches. Tighter code makes better use of the cache and internal instruction buffers. As it generally runs faster, it reduces the system's time in active mode.

8.8 CONCLUSION

On-chip power-optimization techniques in multicore platforms now offer more granular control, more power-saving modes, and more complete information about processor power consumption than ever before. Further, the DSP development tools give designers more insight into how the systems consume power and provide techniques for lowering power consumption via on-chip hardware. Algorithms need to be developed taking the underlying architecture into consideration. Software needs to be aware of the power and energy management opportunities in the underlying hardware and maximize the energy savings. To continue to improve on the quest for efficient signal processing, a holistic approach taking into consideration algorithms, underlying architecture, and the systems is required.

REFERENCES

1. Koomey, J. The Computing Trend that Will Change Everything. *MIT Technology Review*, 2012. http://www.technologyreview.com/news/427444/the-computing-trend -that-will-change-everything/.
2. Karam, L. J., I. AlKamal, A. Gatherer, G. A. Frantz, D. V. Anderson, and B. L. Evans. Trends in multi-core DSP platforms. *Signal Processing Magazine, IEEE* 26(6): 38–49, 2009.
3. Strass, W. Forward Concepts. eNewsletter, Oct. 2013. https://fwdconcepts.com /enewsletter-101613/.

4. Visconti, F. The Role of Digital Signal Processors (DSP) for 3G Mobile Communication Systems, *Engineering Universe For Scientific Research And Management (EUSRM),* 1(5), 2009.

5. Mick, J. Samsung Unleashes 28 nm Octacore Chip for Smartphones, Tablets, *DailyTech,* accessed August 2013, http://www.dailytech.com/Samsung+Unleashes +28+nm+Octacore+Chip+for+Smartphones+Tablets/article32023.htm.

6. Arthur, C. Tablet shipments suggest a crossing point with PCs might not be far off, *The Guardian,* accessed August 2013, http://www.theguardian.com/technology/2013 /feb/01/tablets-crossing-point-pcs.

7. Woo, S. Samsung Analyst Day 2013, *The Semiconductor Wiki Project,* accessed August 2013, http://www.semiwiki.com/forum/files/S.LSI_Namsung%20Woo_Samsung%20 System%20LSI%20Business-1.pdf.

8. ASCI Red, *Wikipedia,* accessed August 2013, http://en.wikipedia.org/wiki/ASCI_Red.

9. FLOPS, *Wikipedia,* accessed August 2013, http://en.wikipedia.org/wiki/FLOPS.

10. TOP500, *Wikipedia,* accessed August 2013, http://en.wikipedia.org/wiki/TOP500.

11. Texas Instruments Appnote SPRT619. TMS320C66x multicore DSPs for high-performance computing, accessed August 2013, http://www.ti.com/dsp/docs/dspspl ashtsp?contentId=145760.

12. DARPA PERFECT program. http://www.darpa.mil/Our_Work/MTO/Programs/Power _Efficiency_Revolution_for_Embedded_Computing_Technologies_(PERFECT).aspx.

13. General-purpose computing on graphics processing units, *Wikipedia,* accessed August 2013, http://en.wikipedia.org/wiki/General-purpose_computing_on_graphics_processing _units.

14. Damodaran, R. et al. A 1.25 GHz 0.8 W C66x DSP Core in 40nm CMOS. 25th International Conference on VLSI Design (VLSID), IEEE, pp. 286–291. Hyderabad, India, 2012.

15. Friedmann, A. TI's new TMS320C66x fixed- and floating-point DSP core conquers the "Need for Speed," Texas Instruments Whitepaper SPRY147, accessed August 2013, http://www.ti.com/lit/wp/spry147/spry147.pdf.

16. Frantz, G. Digital signal processor trends, *IEEE Micro,* 20(6), 52–59, 2000.

17. Williams, J. et al. Computational Density of Fixed and Reconfigurable Multi-Core Devices for Application Acceleration, 2008. http://rssi.ncsa.illinois.edu/proceedings /academic/Williams.pdf.

18. Saban, N. Multicore DSP vs. GPUs, Workshop on GPU & Parallel Computing, Israel, January 2011, http://www.sagivtech.com/contentManagment/uploadedFiles/fileGallery /Multi_core_DSPs_vs_GPUs_TI_for_distribution.pdf.

19. Lucent Technologies. A New Measure of DSP. Allentown, PA, 1997.

20. Berkeley Design Technology, Inc., accessed August 2013, http://www.bdti.com.

21. The Embedded Microprocessor Benchmark Consortium (EEMBC), accessed August 2013, http://www.eembc.org.

22. Hybrid Memory Cube Consortium, accessed August 2013, http://www.hybridmemorycube .org.

23. JESD235 Open Standard Specifications, JEDEC Solid State Technology Association, October 2013, http://www.jedec.org/standards-documents/docs/jesd235.

24. National Instruments whitepaper. PXI Express Specification Tutorial, 2010, http:// www.ni.com/white-paper/2876/en/.

25. Kollig, P., C. Osborne, and T. Henriksson. Heterogeneous multi-core platform for consumer multimedia applications. *Design, Automation & Test in Europe Conference & Exhibition (DATE), IEEE,* Nice, France, pp. 1254–1259, 2009.

26. Tan, E. J. and B. H. Wendi. DSP architectures: Past, present and futures. *ACM SIGARCH Computer Architecture News,* 31(3): 6–19, 2003.

27. Oshana, R. *DSP for Embedded and Real-Time Systems.* Newnes: Boston, MA, 2012.

28. Texas Instruments Keystone Platform. http://www.ti.com/lsds/ti/dsp/keystone_arm /overview.page.
29. CEVA-XC Family. http://www.ceva-dsp.com/CEVA-XC-Family.
30. Freescale QorIQ Qonverge Platform, 2013. http://www.freescale.com/webapp/sps/site /overview.jsp?code=QORIQ_QONVERGE.
31. Texas Instruments Appnote. Accelerate multicore application development with KeyStone software. www.ti.com/lit/pdf/spry231.

9 Nanoelectromechanical Relays

An Energy Efficient Alternative in Logic Design

Ramakrishnan Venkatasubramanian
and Poras T. Balsara

CONTENTS

9.1   INTRODUCTION

Complementary metal–oxide–semiconductor (CMOS) transistor scaling over the last 30 years has enabled significant integration of complex electronic circuitry providing improvements in switching speed, density, cost, and functionality of CMOS chips [1]. However, as the thermal voltage kT/q does not scale, the threshold voltage (V_T) of CMOS transistors cannot be reduced along with the lithographic dimensions [2]. The threshold voltage of CMOS transistors in the sub-40-nm regime has already been scaled to a value that balances leakage energy and dynamic energy optimally [3]. Further reduction in the threshold voltage would actually increase the amount of energy consumed per operation. Further, there is not much supply voltage scaling expected in smaller technology nodes because of the thermal limit (kT/q) [4,5].

Because of these inherent limitations in continual CMOS scaling and increasing power density of integrated circuits due to larger, faster electronic circuits [6], researchers have focused on improving energy efficiency of CMOS circuits. This has led to multicore processing, that is, recover system throughput through parallelism. The benefits of this approach are limited as well, because a minimum in total energy is reached when the active energy and leakage energy are balanced and further V_{DD} reduction will only increase the total energy per operation. The subthreshold slope of CMOS shown in Figure 9.1 is approximately 60–70 mV/decade. To achieve greater energy efficiency in logic circuits, there is always a quest to develop new devices with steeper subthreshold slopes with the ultimate goal of developing an ideal switch with zero off-state current as shown in Figure 9.1. Alternative transistor designs like the tunnel field effect transistor (FET) [7] with steeper subthreshold slope have been proposed. However, they are also fundamentally limited in energy efficiency because of nonzero off-state current.

Furthermore, CMOS logic circuit operation is susceptible to extreme environmental conditions; for example, CMOS cannot function effectively over 150°C. This poses a major limitation for usage of CMOS circuits in harsh and extreme environments such as space, defense, and operating environments with high radiation levels.

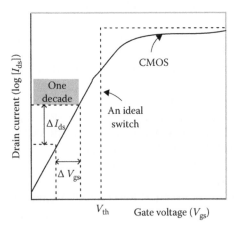

FIGURE 9.1   Subthreshold slope definition. CMOS subthreshold slope is 60–70 mV/decade. An ideal switch offers 0 mV/decade subthreshold slope.

Typically, electronic circuits used in these harsh conditions are packaged with expensive integrated circuit packages that maintain the temperature/radiation requirements and increase the cost of logic circuits used in such applications exponentially.

The current projections by the International Technology Roadmap for Semiconductors (ITRS) show that the end of the road on metal–oxide–semiconductor FET (MOSFET) scaling will arrive sometime around 2022 with an 11 nm process [8]. Process variation and increasing fabrication costs have to be addressed before CMOS can be scaled further. In 20-nm CMOS, the gate oxide is only about five atoms thick. If merely a single atom is out of place, the gate-oxide thickness varies by 20%. This process variation will only worsen with scaling [8].

The single largest hurdle to further scaling of the MOSFET is the fabrication cost. The semiconductor industry might solve the fabrication cost hurdle to enable further scaling, given its track record. However, the physical size limitations and fabrication cost restrictions will require a paradigm shift in design of electronic circuits. Many devices are being explored for cointegration with CMOS technology to solve the above issues.

9.2 CMOS REPLACEMENT DEVICES: MECHANICAL RELAYS

Numerous emerging devices that offer better energy efficiency per operation that could potentially act as CMOS replacement devices have been proposed [8–12]. This list includes a plethora of emerging devices including, but not limited to, piezoelectric devices, magnetic devices, electrostatic devices, and carbon nanotube based devices. The fundamental limitation in CMOS is the trade-off between decreasing active energy and increasing leakage energy to get optimal energy efficiency. This trade-off requirement has to be eliminated to improve the energy efficiency of any emerging device. A mechanical switch with zero OFF state leakage exhibits this behavior. Pott et al. provide a comprehensive summary of history and advances in relay-based computing over the last decade [11]. Mechanical switches exhibit abrupt switching behavior that allows V_{DD} to be decreased while maintaining relatively high ON state current. The use of mechanical switches (relays) for computing was first implemented in Zuse Z3—an electrically powered digital computer built in 1941. This computer used about 2000 relays to perform calculations for aircraft design [13]. Numerous advancements in planar processing technology over the past few decades have created a renewed interest in mechanical computing for ultralow power applications [11].

9.2.1 ELECTROSTATICALLY ACTUATED NANOELECTROMECHANICAL RELAY DEVICES

MEMS technology has seen significant advances in the last decade and the electrostatic MEMS relay is by far the most common as it requires extremely low power and relatively straightforward fabrication requirements. Improvements in silicon fabrication techniques have allowed three-dimensional structures to be realized from silicon substrates. Numerous sensor and actuator circuits have been designed using MEMS technology. MEMS devices can be free-standing or can have one or more degrees of freedom. This freedom allows them to serve many different functions. The ability to design unique structures that implement the same functionality of CMOS transistors but without the drawbacks of transistor technologies is the key advantage of MEMS devices. Scaling MEMS relays down to the nanoscale to create fast, reliable, miniature nanoelectromechanical (NEM) relays is the logical next step for the MEMS relay technology as well. Numerous MEMS devices that operate similar to an ideal switch have been reported [14,15]. NEM devices actuated electrostatically are attractive for IC applications because they are relatively easy to manufacture using conventional planar processing techniques/materials, do not consume much active power, and are scalable. The device stress and actuation potential increase with reduction in feature size. Hence scaling of relays leads to better reliability and energy efficiency [5,8,16].

NEM relays are essentially three- or four-terminal mechanical switches that are electrostatically actuated. The channel could be metal or semiconductor channel (creates a field-effect transistor when the mechanical parts come in contact). The metal channel relays are more appealing than the semiconductor channel as the latter inherently has all the limitations of CMOS with respect to energy efficiency reduction. In a metal channel relay, when the mechanical switch is turned ON, the channel creates a conducting path between source and drain. When the switch is OFF, there is no drain-source current and hence, the leakage through the device is zero. The electrostatic actuation of the mechanical switch results in a mechanical delay in the relay, which far exceeds the electrical delay of the logic implemented using the relays.

NEMS devices offer another significant advantage over CMOS. As alluded to previously, CMOS circuits are susceptible to harsh and extreme operating conditions. However, NEM relays, being mechanical switches, offer fairly reliable operation in the presence of harsh and extreme operating conditions [17,18].

9.2.2 NEM Relay Devices

9.2.2.1 Cantilever Beam and Laterally Actuated NEM Relays

A cantilever beam–based NEM relay has been reported by Chen et al. [19]. This is the simplest and most intuitive mechanical switch that is electrostatically actuated. Figure 9.2 shows a cantilever beam relay [19]. This is a four-terminal device. The cantilever beam forms the gate terminal. The base terminal is below the beam. The gate carries a metal channel and contact dimples that touch the source and drain terminals below when the beam is actuated. So, when the relay is ON, it creates a metal channel between source and drain terminals through the contact dimples. When the relay is OFF, there is no drain-source current and hence, the leakage through the device is zero. Hence, the relay behaves similar to an ideal switch.

A laterally actuated NEM relay device is reported by Chong et al. [20], in which a polysilicon beam is laterally actuated to realize the mechanical switch. This relay has been heterogeneously integrated with CMOS technology to realize a hybrid NEMS–CMOS static random access memory (SRAM) cell, which showed improved bitcell stability and lower energy dissipation.

9.2.2.2 Suspended Gate Relays

A four-terminal suspended gate NEM relay devices has been reported [5], which incorporates a movable poly-SiGe gate suspended by spring-like folded flexures above tungsten electrodes. A thin coating of titanium oxide is applied to the device to improve the reliability of the suspended gate. Relays operating at 10 V with a 90-nm nominal gap between the dimples on drain/source and the suspended gate have been fabricated and reported. Spencer et al. also discuss a scaled relay operating at 1 V with a nominal gap of 5–10 nm between the gate and drain/source [5]. Figure 9.3 shows the device construction and basic electrical/mechanical elements in the relay circuit. Figure 9.4 shows the voltage transfer characteristics of the relay. The suspended gate is electrostatically actuated whenever there is a voltage between

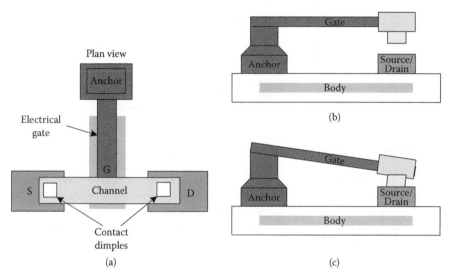

FIGURE 9.2 (a) Cantilever beam relay. (From Chen, F. et al., Integrated circuit design with NEM relays, *2008 IEEE/ACM International Conference on Computer-Aided Design [ICCAD]* © 2008 IEEE.) (b) The nonactuated relay. (c) When actuated, the contact dimples touch source and drain terminals. (Reprinted with permission from R. Venkatasubramanian, S. K. Manohar, and P. T. Balsara, NEM relay-based sequential logic circuits for low-power design, *IEEE Transactions on Nanotechnology*, 12[3], © 2013 IEEE.)

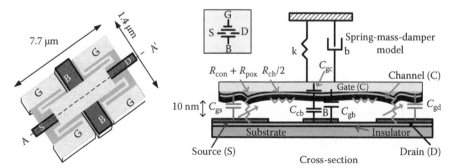

FIGURE 9.3 Suspended gate relay (From Spencer, M. et al., Demonstration of integrated micro-electro-mechanical relay circuits for VLSI applications, *IEEE Journal of Solid-State Circuits*, 46[1], © 2011 IEEE.) NEM Relay symbol shown in the inset. The cross-sectional view is along axis AA' shown in the plan view. The suspended gate is modeled as a spring-mass-damper system. (Reprinted with permission from R. Venkatasubramanian, S. K. Manohar, and P. T. Balsara, NEM relay-based sequential logic circuits for low-power design, *IEEE Transactions on Nanotechnology*, 12[3], © 2013 IEEE.)

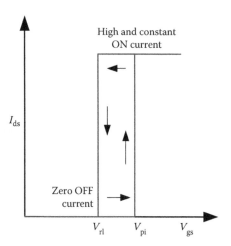

FIGURE 9.4   The voltage transfer characteristics of the relay.

the gate and base. When the electrostatic force is strong enough to overcome the spring-mass-damper system, the channel electrode touches the dimples on source and drain, thereby creating a metal channel between source and drain. When the voltage between gate and base is lowered, the electrostatic force reduces and eventually will open the relay once it crosses a release threshold voltage. Since there is no metal channel between source and drain when the relay is off, there is no leakage current.

9.3   VERILOG-A MODEL OF SUSPENDED GATE RELAY

9.3.1   DEFINITION OF NEM RELAY PARAMETERS

Pull-in voltage V_{pi} is the voltage applied between gate and base above which the electrostatic force overcomes the spring-mass-damper system and the relay turns ON. Release voltage V_{rl} is defined as the voltage below which the relay opens and the switch is OFF. The NEM relay exhibits a hysteretic property for the pull-in and release threshold voltages. The pull-in voltage (V_{pi}) is larger than the release voltage (V_{rl}). The mechanical delay involved in switching ON the relay is denoted as t_{mon} and the mechanical delay involved in switching OFF the relay is denoted as t_{moff}. The mechanical delay is an order of magnitude larger than the electrical delay of the relay (t_e). As an example, the mechanical delay of the suspended gate relay used in this work is of the order of hundreds of nanoseconds, whereas the electrical delay is of the order of tens of picoseconds.

9.3.2   ELECTROMECHANICAL SYSTEM MODEL

The electromechanical relay device is typically modeled as a spring-mass-damper system. Verilog-A model of the relay device is co-simulated with HSPICE [21] to simulate the second-order mechanical system defined in Equation 9.1 and the electrical model of the relay. This simulation comprehends all the mechanical and electrical effects involved in the electrostatic actuation of the relay in a circuit. The Verilog-A model covers the self-actuation effect but does not cover electro- or mechanical-thermal effects.

Dispersion forces such as Casimir and Van der Waals' force strongly affect the mechanical behavior of nanoscale devices. In nanoscale devices, the effects of these two forces are so significant that they must be considered even in absence or presence of electrostatic forces. Hence, it is very important to factor these dispersion forces in the Verilog-A model of the NEM relay device.

An accurate Verilog-A model of the suspended gate relay operating at 1 V with a nominal air gap of 5–10 nm developed by Venkatasubramanian et al. [22] has been used to validate the circuits described in this chapter. At any instant in time, the electrostatic force (F_{elec}) balances with the spring-mass-damper system.

As alluded to previously, in nanoscale electromechanical devices, dispersion forces such as the Casimir force and Van der Waals' force affect the pull-in stability of the relay devices significantly [23,24]. Dispersion forces decrease the pull-in deflection and actuation voltage of the relay. However, the fringing field increases the pull-in deflection while decreasing the pull-in voltage (V_{pi}) of the relay. The model also accounts for the dispersion forces.

Van der Waals' force is the dominant force of attraction when the air gap separation is below several tens of nanometers (nonretarded regime) [24]. If the air gap separation is above several tens of nanometers, the Casimir force is dominant. As the 1 V suspended gate relay has a separation well within tens of nanometers, the Casimir force is assumed to be negligible and hence Van der Waals' force alone is considered in the relay model.

Hence, the force vector equation is expressed as follows:

$$m\ddot{x} = F_{elec}(x) + F_{vdW}(x) - b\dot{x} - kx \tag{9.1}$$

where x is the displacement of the gate, m the mass of the suspended gate, k the spring constant of the gate structure, and b the damping coefficient of the motion of the gate.

The electrostatic force F_{elec} is expressed as:

$$F_{elec} = \frac{\epsilon_0 A_{ov} V_{gb}^2}{2(g_0 - x)^2} \tag{9.2}$$

where ϵ_0 is the permittivity of free space, V_{gb} the voltage between gate and base, A_{ov} the area of overlap between gate and base electrodes, and g_0 is the normal gap between electrodes when switch is OFF.

The Van der Waals' force per unit length ΔF_{vdW} is given by:

$$\Delta F_{vdW} = \frac{Aw}{6\pi(g_0 - x)^3} \tag{9.3}$$

where A is the Hamaker constant [24] (~1.6 eV for Si with no surface layer) and w the width of the suspended gate relay. The Hamaker constant, A, can be roughly defined as a material property that represents the strength of Van der Waals' interactions between macroscopic bodies. It is defined in Joules. Typical values of A varies in the range of 10^{-19}–10^{-20} J for various material bodies [25].

For two parallel and flat surfaces (as in the case of suspended gate relay), the total Van der Waals' force of the suspended gate F_{vdW} is expressed as:

$$F_{vdW} = \frac{AwL}{6\pi(g_0 - x)^3} \qquad (9.4)$$

where L is the length of the suspended gate relay. In the case of cantilever beam relays or laterally actuated relays, Equation 9.3 has to be integrated over the length of the relay, taking into account the bending effects of the beam to determine the total Van der Waals' force. The impact of Van der Waals' force on the pull-in parameters of cantilever beams is given in [26].

The voltage necessary to turn on the relay is the pull-in voltage and is derived to be as follows:

$$V_{pi} = \sqrt{\frac{8kg_0^3}{27\epsilon_0 A_{ov}}} \qquad (9.5)$$

The mechanical delay (t_m) is inversely proportional to the gate overdrive voltage $(|V_{gb}|/V_{pi})$ and the undamped angular frequency of the spring-mass-damper system $\sqrt{k/m}$ [27]. This is denoted by

$$t_m \propto \sqrt{\frac{m}{k}} \frac{V_{pi}}{|V_{gb}|} \qquad (9.6)$$

9.4   NEM RELAY–BASED DIGITAL LOGIC DESIGN

The speed of operation of a NEM relay–based digital logic circuit is limited by the mechanical delay of the switch. As the mechanical delay far exceeds the electrical delay of the switch, it has been proposed that combinatorial relay logic be implemented as complex gates to minimize the number of mechanical delays in the critical path [5,19]. It has been shown that relay logic offers up to 10× efficiency gain in low-frequency circuits (operating up to 100 MHz) by having multiple parallelized architectures to achieve a certain throughput [5,19]. To be able to build large digital logic circuits using relays, sequential and combinatorial logic circuits need to be realized, taking into account the device properties to minimize the mechanical delays in the critical paths.

9.4.1   INTRINSIC CAPACITANCE OF THE RELAY (C_{int})

In the cantilever beam relay and the suspended gate relay a true parallel plate capacitor exists between gate and base (C_{gb}). The small overlap area between gate and source/

drain also creates small parallel plate capacitances C_{gs} and C_{gd}. It is reported in [5] that C_{gb} is of the order of 1–2 fF and contributes approximately 60% of the total relay self-capacitance. Because of the near-zero leakage in the relay, charge stored in these capacitors is retained very efficiently. This intrinsic capacitance could be used in relay-based circuit design.

9.4.1.1 Relay-Based Latch Circuits

9.4.1.1.1 Bootstrap Latch

The intrinsic capacitance of the relay is used to improve the latch performance in the bootstrap latch shown in Figure 9.5a. R2 and R3 form a buffer. Relay R4 essentially bypasses the buffer when CLK is high and provides a faster Q output. The output Q is available after one mechanical delay (t_m). Note that R1 and R4 switch ON at the same time. The intrinsic capacitor is charged (in t_m time), and either of the relays R2 or R3 turn ON depending on D being "1" or "0." Note that R2 or R3, once turned ON, stay ON until the next cycle or whenever C_{int} is affected. Relay R4 electrically isolates the output when CLK = 0 and preserves the charge in C_{int}. This ensures that loading at the output Q does not affect the intrinsic capacitance C_{int}. Even though relays R2 and R3 take another t_m delay to close, this is hidden from the CLK-Q delay of the latch.

Ideally, for bootstrapping functionality, we expect output Q to replenish the charge in the intrinsic capacitance C_{int} once the input is latched. So instead of relay R4, Q could be directly connected to the C_{int} node to make it a true bootstrapping latch. But the loading on output Q might discharge the intrinsic capacitance C_{int} and the latch might not be able to hold the value stored in it.

The operation of the latch is shown in Figure 9.5c. The mechanical displacement of the suspended gate with respect to the nonactuated OFF position of the relays R1 and R2 is shown in the waveform as well. When a relay R1 is OFF, R1.$x\_pos = 0$. The CLK and D signals require a minimum pulse width of t_m in order for the latch to be functional. The minimum setup time is the electrical delay $t_e = R_{on} * C_{int}$, where R_{on} is the ON resistance of the relay and C_{int} the intrinsic capacitance. The dead time for the minimum setup is $R_{on} * C_{int} + t_m$. When D = 1, Q = 0, or D = 0, Q = 1, there is a possibility of drive fight when CLK = 1, and node d_{int} could be corrupted. To minimize this drive fight, relay R_1 should be sized stronger than R_4. If R_1 is constructed using two parallel relays, effective feedback resistance (R_{on} of R_4) would be twice that of the setup path resistance (R_{on} of R_1), thereby aiding the new value to be written into the flip–flop.

9.4.1.1.2 Keeper Latch

In the absence of an intrinsic capacitance (as in the case of the lateral actuation relay [20]), the performance of the latch can be improved using the classic back-to-back CMOS inverter topology optimized differently for NEM relays. The latch topology is shown in Figure 9.5b. Relay pairs R2–R3 and R4–R5 form buffers. The CLK-Q delay of this latch is $2t_m$ (Relay R1 to turn ON, then the buffer R2–R3 to turn ON). The mechanical delay required to turn ON the relays in the recirculating path is hidden. The minimum setup time is the electrical delay $t_e = R_{on} * C_{int}$, where R_{on} is the ON resistance of the relay and C_{int} the intrinsic

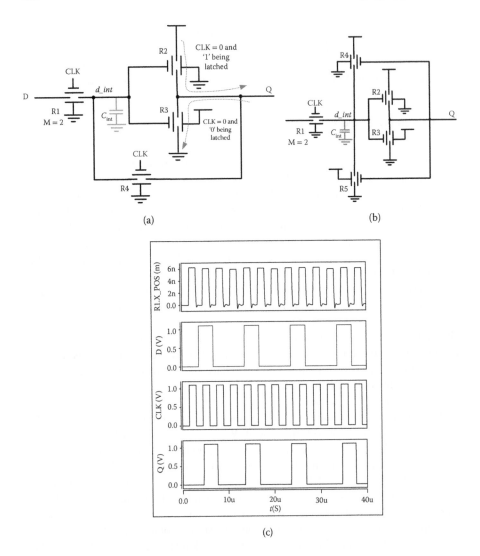

FIGURE 9.5 (a) Bootstrap latch, (b) keeper latch, and (c) bootstrap latch operation. (Reprinted with permission from R. Venkatasubramanian, S. K. Manohar, and P. T. Balsara, NEM relay-based sequential logic circuits for low-power design, *IEEE Transactions on Nanotechnology*, 12[3], © 2013 IEEE.)

capacitance. The dead time for the minimum setup is $R_{on} * C_{int} + 2 * t_m$. When D = 1, CLK = 1, and Q = 0, there is a direct path from D to ground through relays R1 and R5. Essentially, the relays will operate as a potential divider and node *d_int* will get a value of $V_{dd}/2$. If $V_{pi} > V_{dd}/2$, it will result in a setup failure and will make the latch circuit inoperable. Same condition applies for D = 0, CLK = 1, and Q = 1. To avoid the setup failure, multiple parallel relays could be added in the place of R1 (reduce the effective resistance) to ensure that node *d_*int has a swing above V_{pi}.

9.4.1.2 Relay-Based Flop Circuits: Tristate Relay Flop

A flop using only four relays and operating at a frequency of $1/2t_m$ has been proposed by Venkatasubramanian et al. [28]. The flip–flop topology is shown in Figure 9.6a. Relays R2 and R3 form a buffer. When the CLK is low, the intrinsic capacitance is

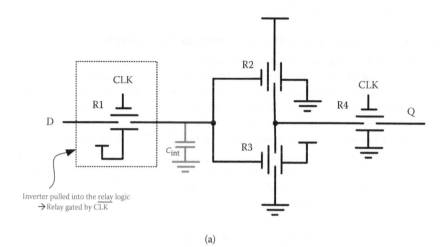

(a)

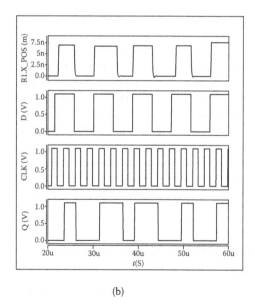

(b)

FIGURE 9.6 (a) Tristate flop, and (b) tristate flop operation. Both electrical and mechanical measurements are shown. R1.*x_pos* is the mechanical displacement of the suspended gate in the Y direction with respect to nonactuated position of the relay. (Reprinted with permission from R. Venkatasubramanian, S. K. Manohar, and P. T. Balsara, NEM relay-based sequential logic circuits for low-power design, *IEEE Transactions on Nanotechnology*, 12[3], © 2013 IEEE.)

charged and the buffer is turned ON. The latched value is available at the buffer output. When CLK turns high, the output gets the latched value after one mechanical delay (t_m). Note that the base of R1 is connected to VDD. So R1 will turn ON when CLK is low. The operation of this flop is shown in Figure 9.6b. The mechanical displacement of the suspended gate with respect to the default position is shown in the waveform for relay R1. When R1 is OFF, $R1.x\_pos = 0$.

9.4.1.3 Max Frequency of Operation and Min-Pulse Width

The clock and data signals of the sequential logic circuits should have a pulse width of at least t_m for the logic operation to succeed. Care must be taken to ensure that this min-pulse width is guaranteed on all intermediate nodes of the relay logic circuit as well. This limits the maximum frequency of operation of a relay logic circuit to at most $1/2t_m$. The setup time, hold time, and the Tclk-Q delay for each of the sequential logic circuits proposed in this work is listed in Table 9.1. The electrical delay in each sequential circuit is denoted as t_e ($= R_{on} * C_{int}$). This work proposes a flop (tristate flop) and a latch (bootstrap latch) that can work at a frequency of $1/2t_m$—which is theoretically the fastest sequential circuit possible for a NEM relay logic circuit for a given nominal overdrive voltage $|V_{gb}|/V_{pi}$.

9.4.1.4 Performance, Energy, and Area Comparison

The performance and area (relay count) comparison between the various sequential logic topologies is shown in Table 9.1. The electrical delay in each circuit is denoted as t_e ($= R_{on} * C_{int}$). Setup time reported is the drop-dead setup time and CLK-Q delay reported is for infinite setup time—in line with the definition of setup and CLK-Q delay for CMOS circuits. For latch circuits, setup time is measured when the latch is getting disabled. In flip–flop circuits, setup time is measured at the rising edge of the clock. The tristate flop gives the maximum performance for the smallest area penalty. Among latches, the bootstrap latch gives the maximum performance as the tristate flop.

The Verilog-A model is co-simulated with HSPICE to prove the functional correctness of the sequential logic circuits and to obtain the energy per transition. The co-simulation comprehends all the mechanical and electrical effects involved in the electrostatic actuation of the relay in a circuit. The relay operates at 1 V with V_{pi} = 0.6 V, V_{rl} = 0.55 V, t_{mon} = 130 nanoseconds. The average current I_{avg}, energy per

---

TABLE 9.1

Performance Metrics of NEM Relay Sequential Circuits

Type	No. of Relays (count)	Setup Time	Hold Time	CLK-Q Delay	Comment
Bootstrapping Latch	5	t_e	0	t_m Latch	Fast
Keeper latch	6	t_e	0	$2t_m$	
Tristate Flop	4	t_e	0	t_m	Fastest flop For a given $V_{overdrive}$

transition for 0 to 1 transition ($E_{0 \to 1}$), and 1 to 0 transition ($E_{1 \to 0}$) are also reported for each of the sequential cells in Table 9.2.

9.4.1.5 Improving Performance of NEM Relay Circuits Using Charge Boosting

As the speed of operation of NEM relay logic circuits is primarily limited by the mechanical delay, multiple circuit techniques are explored to improve the speed of operation of NEM relay logic circuits.

The suspended gate relay illustrated in this chapter has V_{pi} as approximately 0.6 V and the nominal operating voltage is 1 V. So the nominal gate overdrive voltage ($|V_{gb}|/V_{pi}$) = 1.66x. The mechanical delay (t_m) is inversely proportional to the gate overdrive voltage ($|V_{gb}|/V_{pi}$) as shown in Equation 9.6. This is illustrated in Figure 9.7. t_{moff} does not vary as much with overdrive voltage for the suspended gate device used in this work. Charge boosting enables faster relay operation at the expense of the device reliability. So, as long as the device can handle an overdrive voltage reliably, this concept can be used to improve the mechanical delay of the relay. Metal engineering techniques can be used to improve the reliability of operation for overdriven relays. However, this might increase the relay fabrication cost.

A localized charge-boosting scheme to overdrive V_{gb} to improve the performance of the mechanical switch was first proposed by Venkatasubramanian et al. [29]. The

---

TABLE 9.2

NEM Relay Flops/Latches: Power/Energy Comparison

	I_{avg}	$E_{0 \to 1}$	$E_{1 \to 0}$
Sequential cell	(A)	(J)	(J)
Bootstrap latch	1.19e-11	1.18e-16	2.63e-11
Tristate flop	1.03e-15	2.43e-15	1.69e-16

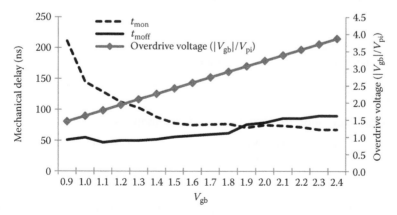

FIGURE 9.7 Mechanical delay (t_{mon}, t_{moff}) versus overdrive voltage. For an overdrive of 2x, t_{mon} reduces by approximately 2x. t_{moff} does not vary as much with overdrive voltage for the suspended gate model referred to in Section 9.3.

parallel plate capacitance between the gate and base of the relay is significant enough and hence is used to realize the storage capacitance. The feasibility of localized charge boosting is tested out by integrating the charge-boosting logic into the tristate flop.

An integrated charge-boosting flip–flop using a cross-coupled voltage doubler is proposed by Venkatasubramanian et al. [28]. The doubler could be integrated with a single flip–flop or a group of flip–flops (vector flops) thereby minimizing the area overhead incurred for the doubler circuit.

9.4.1.5.1 NEM Voltage Doubler and Integrated Charge-Boosting Flip–Flop

It is important to recollect that the relay turns on whenever $|V_{gb}| > V_{pi}$. The operation of the relay for various gate/base voltage combinations is shown in Table 9.3.

A two-phase two relay cross-coupled voltage doubler shown in Figure 9.8a is used to generate differential signals X, $\overline{X}$ which swing between $(V_{dd}, 2V_{dd})$ [30]. A constant local supply of $V_{out} = 2V_{dd}$ is generated as well. The operation of the doubler is shown in Figure 9.8b. Nonoverlapping two-phase clock is typically generated using break-before-make circuit [31].

The operation of the doubler is controlled by nonoverlapping clocks φ_1 and φ_2 and a set of delayed nonoverlapping clocks φ_{1d} and φ_{2d}. Relays R_{c1} and R_{c2} are configured in such a way that the parallel plate capacitor between the gate and the base terminals act as the fly capacitor (C_{FLY}) for the doubler operation. Depending on the loading on the doubler circuit, multiple parallel relays are required to realize the total doubler capacitance. Relays R_1 and R_2 are turned ON on alternating phases of the clock and the relays R_{c1} and R_{c2} are charged to V_{dd}, respectively. Early clocks φ_1 and φ_2 set up the latch initially and thereby decouple the charging and discharging operations of the doubler. This also eliminates the short circuit current in the doubler altogether. The outputs X and $\overline{X}$ swing between V_{dd} and $2V_{dd}$. Relays R_3 and R_4 along with multiple parallel relays R_{c3} to realize the variable storage capacitor (C_{ST}) generate a constant local $2V_{dd}$ supply. This is shown in Figure 9.9b.

The NEM voltage doubler concept is extended to develop an integrated charge-boosting flip–flop and is described in detail by Venkatasubramanian et al. [29]. Basically, the integrated charge-boosting flip–flop that deploys localized charge boosting should provide the following:

- Rail-to-rail output swing of $[0, 2V_{dd}]$
- Reduce mechanical delay of operation to $t_m/2$

TABLE 9.3
NEM Relay Operation for Various V_g/V_b Combinations

V_g	V_b	Relay State	Comments
0	0	OFF	
0	$V_{dd}, 2V_{dd}$	ON	
V_{dd}	V_{dd}	OFF	
V_{dd}	$0, 2V_{dd}$	ON	Special case of $V_g = V_{dd}$ and $V_b = 2V_{dd}$ has to be handled
$2V_{dd}$	$2V_{dd}$	OFF	
$2V_{dd}$	$0, V_{dd}$	ON	Special case of $V_g = 2V_{dd}$ and $V_b = V_{dd}$ has to be handled

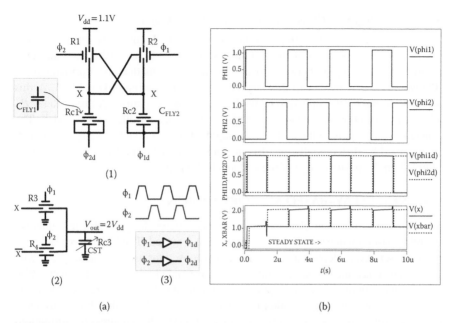

FIGURE 9.8 (a) NEM voltage doubler that uses relays for storage capacitance, and (b) NEM voltage doubler operation. (Reprinted with permission from R. Venkatasubramanian, S. K. Manohar, and P. T. Balsara, NEM relay-based sequential logic circuits for low-power design, *IEEE Transactions on Nanotechnology*, 12[3], © 2013 IEEE.)

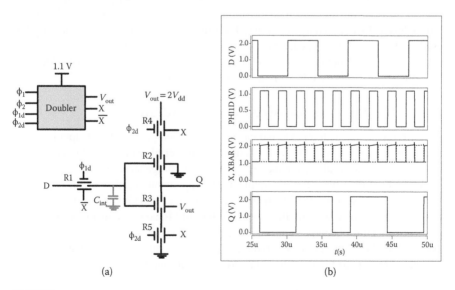

FIGURE 9.9 (a) Integrated charge-boosting flip–flop. (b) Operation. (Reprinted with permission from R. Venkatasubramanian, S. K. Manohar, and P. T. Balsara, NEM relay-based sequential logic circuits for low-power design, *IEEE Transactions on Nanotechnology*, 12[3], © 2013 IEEE.)

9.4.5.1.2   Integrated Charge-Boosting Flip–Flop Design

The integrated charge-boosting flip–flop is shown in Figure 9.9a and the timing diagram for the operation of the flip–flop is shown in Figure 9.9b. Note that D is assumed to be swinging between $[0,2V_{dd}]$ as the previous flop (or) a primary input (PI) would be driving it with this larger swing. As the swing of D is $[0, 2V_{dd}]$, this overdrive voltage gets applied to relays R_2 and R_3 and they turn ON faster. Relay R_1, which samples D on the negative edge of the clock, is controlled by φ_{1d} and $\overline{X}$. The constant local supply voltage of $2V_{dd}$ is supplied by the doubler. Relays R_4 and R_5 are sampled by differentially overdriven clock φ_{2d} and X. The latched value of D is sampled on the positive edge of the system clock (φ_{1d}).

9.5   NEM RELAY–BASED MEMORY DESIGN

Memories are important elements in electronic circuit design and a few interesting relay-based memory architectures have been reported. The bistable seesaw relay device described by Jeon et al. [32] has been proposed for use as a memory cell with one seesaw relay and one access transistor. A suspended relay–based memory architecture has been proposed by Gupta [33]. A multivalued memory architecture using relays has been proposed by Stalter [34]. Three new parallel readout bitcell architectures are proposed by Venkatasubramanian et al. [35], which address the limitations of Gupta [33] and simultaneously improve the memory access time.

9.5.1   THREE-RELAY RANDOM ACCESS MEMORY (3RRAM)

A three-relay RAM (RRAM) architecture proposed by Gupta [33] has three-relay devices in each bitcell. The memory architecture is shown in Figure 9.10a. Each bitline has a global precharge circuitry. The intrinsic capacitance of the relay (C_{int}) is used to store the charge in each bitcell. The fundamental limitation of this architecture is that the readout is serial in nature. Data stored in word "0" is accessible only if all the other words (words 1...n) are enabled, and so on.

9.5.2   TWO-RELAY RANDOM ACCESS MEMORY (2RRAM)

A two-relay bitcell with an external capacitor has been proposed by Gupta [33] and is shown in Figure 9.10b. This architecture has the serial readout limitation as well. In addition, an external capacitance is assumed for charge storage.

As both of the above bitcell architectures implement serial readout, they have higher read access time (RAT). Further, as each bitcell read involves all the bitcells in the memory array column, this poses a reliability issue in the memory operation. The entire column becomes nonfunctional even if one of the bits in that column is not reliable. Also, serial readout architecture is not conducive for low-voltage operation because of their higher voltage headroom requirements.

9.5.3   P4RRAM: PARALLEL 4R RAM BITCELL

P4RRAM—Parallel 4R RAM bitcell is based on the tristate flop topology (Figure 9.6). The bitcell architecture is shown in Figure 9.11a.

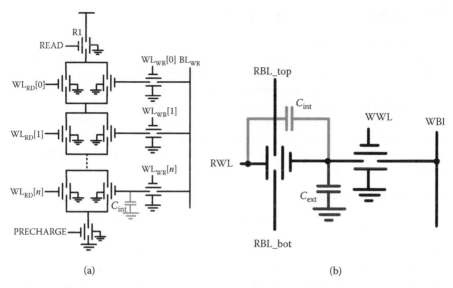

FIGURE 9.10 (a) 3RRAM memory bitcell and memory architecture, and (b) 2RRAM memory bitcell.

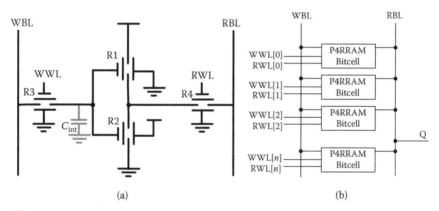

FIGURE 9.11 (a) P4RRAM Bitcell. (b) P4RRAM Array Column.

Tristate flop operation is explained in Section 9.4.1.2. Relays R1 and R2 form a buffer. When the CLK is low, the intrinsic capacitance is charged and the buffer is turned ON. The latched value is available at the buffer output. When CLK turns high, the output gets the latched value after one mechanical delay (t_m). This concept is used in the topology of the P4RRAM bitcell. As it is a parallel readout architecture, the read bitline (RBL) can be shared across all the bitcells in the column and can be tristated when there is no read operation. No precharge or predischarge of RBL is necessary.

This memory bitcell is very reliable as it stores both "1" and "0" and drives the output directly from the supply. Also, the RAT is one mechanical delay (t_m) for both the read and write operations. The memory array column for this bitcell is shown in Figure 9.11b.

As long as the RC delay required to discharge the RBL capacitance is less than one mechanical delay, additional sense amplifier circuitry is not required to improve the RAT.

9.5.4 P3RRAM: Parallel 3R RAM Bitcell

P3RRAM bitcell (Parallel 3R RAM) is a variant of the 4R bitcell described in Section 9.5.3, in which zero storage capability has been removed to minimize the bitcell area. The bitcell architecture is shown in Figure 9.12a. Hence, the memory RBL has to be predischarged before every read operation. This results in additional delay for the RAT. The memory array column for this bitcell is shown in Figure 9.12b.

The basic read operation of this bitcell is shown in Figure 9.13a. Bit[0] is storing a "0" and Bit[1] is storing a "1." The predischarge cycle before every read operation ensures that the RBL is discharged. When Bit[1] is read out by asserting RWL[1], the RBL gets asserted to "1."

9.5.5 P3RRAM-SNR

This bitcell topology proposes a notion of storage node refresh (SNR) where the bitcell state is replenished through the write bitline every time a memory write operation is performed. The bitcell architecture is shown in Figure 9.13b. An external capacitance C_1 is added to each bitcell and it holds the state of the bitcell and facilitates faster readout. The circuit topology ensures that the internal capacitance C_1 is replenished and the bitcell state is preserved during any memory write operation. Charge replenishment enables a reliable memory operation and also improves the memory RAT.

9.5.6 Comparison of the NEM Relay Bitcell Architectures

The low- and high-noise margins (NM_L and NM_H, respectively) for a relay circuit can be defined by examining the DC transfer characteristic of a relay-based buffer [33]. The DC characteristics of a buffer are shown in Figure 9.14.

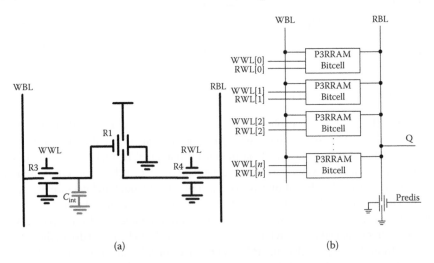

(a)   (b)

FIGURE 9.12   (a) P3RRAM Bitcell. (b) P3RRAM Array Column.

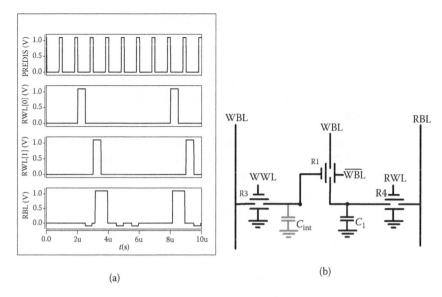

(a)

(b)

FIGURE 9.13 (a) P3RRAM Bitcell operation. (b) P3RRAM-SNR Bitcell.

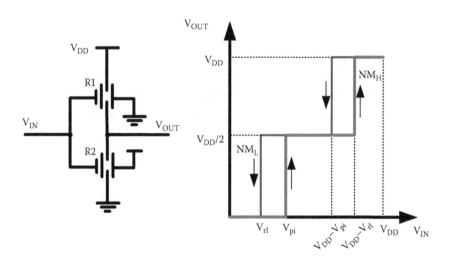

FIGURE 9.14 Relay buffer (left) and its DC transfer characteristics (right).

The write margin, read margin, and hold margin of a memory bitcell are the key metrics that define the stability and noise tolerance characteristics of the bitcell. For each of the bitcell architectures proposed, the margin metrics are shown in Table 9.4.

The relay parameters used in Table 9.4 are as follows:

- t_m: the mechanical switching delay
- R_{on}: ON resistance of the relay
- C_{int}: the intrinsic capacitance of the relay
- C_1: the external capacitance added in P3RRAM-SNR
- C_{RBL}: the capacitance of the RBL

The electrical delay in each of the bitcell circuits is a function of the relay ON resistance R_{on} and the capacitance involved in charge storage. The advantages and disadvantages of each of the bitcell architectures are shown in Table 9.5.

1. *Stability metrics*: Write margin is the maximum noise that the bitcell can handle during a write operation without corrupting the value being stored. Read margin is the maximum noise that the bitcell can handle that will not corrupt the value stored in the memory during a read operation. Hold margin is a measure of the data retention property of the bitcell and is defined as the noise voltage above which the bitcell will fail to hold the value, thereby corrupting the memory location.
2. *Performance metrics*: RAT, write access time (WAT), and cycle time (T_{cc}) are the typical performance care-abouts for a memory bitcell. The comparison of various performance metrics and stability metrics is shown in Table 9.4.

TABLE 9.4

Performance/Stability Metrics for the Bitcell Architectures

	P4RRAM	P3RRAM	P3RRAM–SNR
Write access time (WAT)	$t_m + R_{on} * C_{int}$	$t_m + R_{on} * C_{int}$	$t_m + R_{on} * C_{int}$
Read access time (RAT)	$t_m + 2 * R_{on} * C_{RBL}$	$t_m + 2 * R_{on} * C_{RBL}$	Reading "1": $t_m + 2 * R_{on} * C_{RBL}$ Reading "0": 0 (RBL is predischarged)
Memory cycle time (T_{cc})	$max\{RAT, WAT\}$	$max\{RAT, WAT\}$ + predischarge cycle	$max\{RAT, WAT\}$ + predischarge cycle
Write margin	Storage of "1": $V_{DD} - V_{rl}$ Storage of "0": V_{rl}	Storage of "1": $V_{DD} - V_{rl}$ No "0" storage	Dependent on C_1 charge Storage of "1": For full C_1 charge, $V_{DD} - V_{rl}$ No "0" storage
Read margin	∞	∞	∞
Hold margin	Holding "1": $V_{DD} - V_{pi}$ Holding "0": V_{pi}	Holding "1": $V_{DD} - V_{pi}$	Holding "1": $V_{DD} - V_{pi}$

TABLE 9.5

Summary of NEM Relay–Based Memory Bitcell Architectures

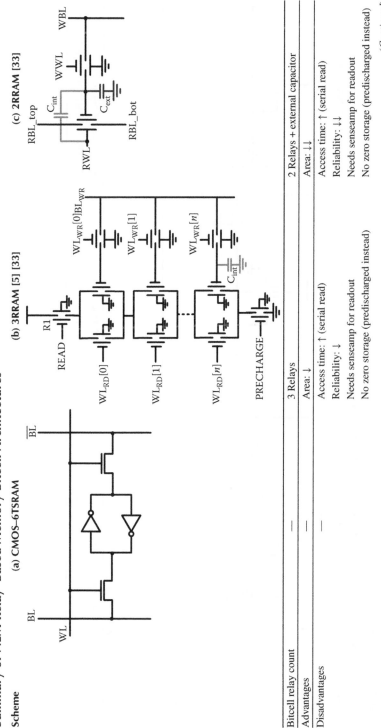

Scheme	(a) CMOS–6TSRAM	(b) 3RRAM [5] [33]	(c) 2RRAM [33]
Bitcell relay count	—	3 Relays	2 Relays + external capacitor
Advantages	—	Area: ↓	Area: ↓↓
Disadvantages	—	Access time: ↑ (serial read)	Access time: ↑ (serial read)
		Reliability: ↓	Reliability: ↓↓
		Needs senseamp for readout	Needs senseamp for readout
		No zero storage (predischarged instead)	No zero storage (predischarged instead)

(Continued)

TABLE 9.5 (CONTINUED)
Summary of NEM Relay–Based Memory Bitcell Architectures

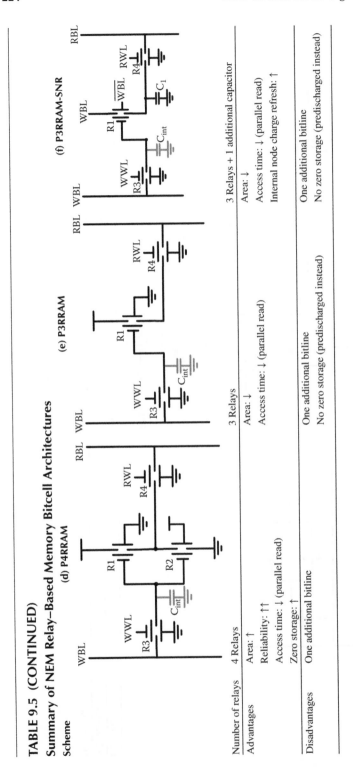

Scheme	(d) P4RRAM	(e) P3RRAM	(f) P3RRAM-SNR
Number of relays	4 Relays	3 Relays	3 Relays + 1 additional capacitor
Advantages	Area: ↑ Reliability: ↑↑ Access time: ↓ (parallel read) Zero storage: ↑	Area: ↓ Access time: ↓ (parallel read)	Area: ↓ Access time: ↓ (parallel read) Internal node charge refresh: ↑
Disadvantages	One additional bitline	One additional bitline No zero storage (predischarged instead)	One additional bitline No zero storage (predischarged instead)

In summary, the P3RRAM bitcell offers a good compromise between area, speed, and reliability among all the NEM-relay bitcell architectures. The P3RRAM-SNR bitcell provides faster RAT with an additional external capacitor.

9.6 CONCLUSION

Over the past few decades, CMOS technology scaling has enabled constant reduction of energy-per-operation in integrated circuits. But the threshold voltage of the transistors has already been scaled to its limit and any further reduction in the threshold voltage would actually increase the amount of energy consumed per operation. So the need for new devices with steeper subthreshold slope such as NEM relay is necessary to enable further reduction of energy-per-operation in integrated circuits. To be able to build large logic circuits using NEM relays, the key building blocks such as combinatorial logic, sequential logic, SRAMs, and IOs need to be studied in detail. This chapter gives an overview of two of the key building blocks of NEM relay logic—sequential latch and flip–flops have been shown to have a delay of $1/2t_m$, which is theoretically the fastest frequency possible for a given overdrive voltage.

According to ITRS-2009 Emerging Research Devices report [36], "Approaches to minimize switching energy in which mechanical energy is stored and released to assist the transition between logic states merit further development (in NEM relays)." Several innovative circuits related to relay-based circuit architecture have been reviewed in this chapter. Conceivably, these circuits could form a relay logic–based cell library that could be used to build large-scale electronic circuits using NEM relays or NEMS–CMOS heterogeneous integrated circuits.

REFERENCES

1. The International Technology Roadmap for Semiconductors (ITRS), 2011. http://public .itrs.net.
2. N. S. Kim, T. Austin, D. Baauw, T. Mudge, K. Flautner, J. S. Hu, M. J. Irwin, M. Kandemir, and V. Narayanan. Leakage current: Moore's law meets static power. *Computer*, 36(12): 68–75, 2003.
3. D. Markovic, V. Stojanovic, B. Nikolic, M. A. Horowitz, and R. W. Brodersen. Methods for true energy-performance optimization. *IEEE Journal of Solid-State Circuits*, 39(8): 1282–1293, 2004.
4. H. Kam. *MOSFET Replacement Devices for Energy-Efficient Digital Integrated Circuits*. PhD thesis, University of California, Berkeley, CA, 2010.
5. M. Spencer, F. Chen, C. C. Wang, R. Nathanael, H. Fariborzi, A. Gupta, Hei Kam et al. Demonstration of integrated micro-electro-mechanical relay circuits for VLSI applications. *IEEE Journal of Solid-State Circuits*, 46(1): 308–320, 2011.
6. B. Nikolic. Design in the power-limited scaling regime. *IEEE Transactions on Electron Devices*, 55(1): 71–83, 2008.
7. W. M. Reddick and G. A. J. Amaratunga. Silicon surface tunnel transistor. *Applied Physics Letters*, 67(4): 494–496, 1995.
8. M. Haselman and S. Hauck. The future of integrated circuits: A survey of nanoelectronics. *Proceedings of the IEEE*, 98(1): 11–38, 2010.
9. Advanced CMOS and Emerging Technologies—Sematech, 2012. http://www.sematech .org/meetings/archives/symposia/9105/Jammy_Raj.pdf.

10. R. Jammy. Materials, Process and Integration Options for Emerging Technologies. Technical report, SEMATECH, 2009.

11. V. Pott, H. Kam, R. Nathanael, J. Jeon, E. Alon, and T-J. K. Liu. Mechanical computing redux: Relays for integrated circuit applications. *Proceedings of the IEEE*, 98(12): 2076–2094, 2010.

12. The Economist. New transistors: Mechanical advantage, July 15, 2011, p. 82, 2011. www.economist.com/node/18956106.

13. T.-J. K. Liu, E. Alon, V. Stojanovic, and D. Markovic. The relay reborn. *IEEE Spectrum*, 49(4): 40–43, 2012.

14. P. M. Zavracky, S. Majumder, and N. E. McGruer. Micromechanical switches fabricated using nickel surface micromachining. *Journal of Microelectromechanical Systems*, 6(1): 3–9, 1997.

15. D. A. Czaplewski, G. A. Patrizi, G. M. Kraus, J. R. Wendt, C. D. Nordquist, S. L. Wolfley, M. S. Baker, and M. P. de Boer. A nanomechanical switch for integration with CMOS logic. *Journal of Micromechanics and Microengineering*, 19(8): 085003, 2009.

16. H. Kam, V. Pott, R. Nathanael, J. Jeon, E. Alon, and T. -J. K. Liu. Design and reliability of a micro-relay technology for zero-standby-power digital logic applications. In *2009 IEEE International Electron Devices Meeting (IEDM)*, pp. 1–4, Baltimore, MD, 2009.

17. X. Wang, S. Narasimhan, S. Paul, and S. Bhunia. NEMTronics: Symbiotic integration of nanoelectronic and nanomechanical devices for energy-efficient adaptive computing. In *2011 IEEE/ACM International Symposium on Nanoscale Architectures (NANOARCH)*, pp. 210–217, San Diego, CA, 2011.

18. H. F. Dadgour. Evolutionary and Disruptive Approaches for Designing Next-Generation Ultra Energy-Efficient Electronics. PhD thesis, University of California, Santa Barbara, CA, 2010.

19. F. Chen, H. Kam, D. Markovic, T.-J. K. Liu, V. Stojanovic, and E. Alon. Integrated circuit design with NEM relays. In *2008 IEEE/ACM International Conference on Computer-Aided Design (ICCAD)*, pp. 750–757, San Jose, CA, 2008.

20. S. Chong, K. Akarvardar, R. Parsa, J. –B. Yoon, R. T. Howe, S. Mitra, and H.-S. P. Wong. Nanoelectromechanical (NEM) relays integrated with CMOS SRAM for improved stability and low leakage. In *IEEE/ACM International Conference on Computer-Aided Design—Digest of Technical Papers (ICCAD) 2009*, pp. 478–484, San Jose, CA, 2009.

21. HSPICE. User Guide: Simulation and Analysis, 2010.

22. R. Venkatasubramanian. Energy efficient circuit design using Nanoelectromechanical relays. PhD thesis, Department of Electrical Engineering, University of Texas, Dallas, TX, 2012.

23. R. Soroush, A. Koochi, A. S. Kazemi, A. Noghrehabadi, H. Haddadpour, and M. Abadyan. Investigating the effect of Casimir and van der Waals attractions on the electrostatic pull-in instability of nano-actuators. *Physica Scripta*, 82(4): 045801, 2010.

24. R. Maboudian and R. T. Howe. Critical Review: Adhesion in surface micromechanical structures. *Journal of Vacuum Science Technology B: Microelectronics and Nanometer Structures*, 15(1): 1–20, 1997.

25. J. N. Israelachvili. *Intermolecular and Surface Forces*, Revised 3rd Edition. Academic Press, Elsevier Science, USA, 2011.

26. A. Ramezani, A. Alasty, and J. Akbari. Influence of van der Waals force on the pull-in parameters of cantilever type nanoscale electrostatic actuators. *Microsystem Technologies*, 12: 1153–1161, 2006.

27. W-C. Chuang, H-L. Lee, P-Z. Chang, and Y-C. Hu. Review on the modeling of electrostatic MEMS. *Sensors*, 10(6): 6149–6171, 2010.

28. R. Venkatasubramanian, S. K. Manohar, and P. T. Balsara. NEM relay-based sequential logic circuits for low-power design. *IEEE Transactions on Nanotechnology*, 12(3): 386–398, 2013.

29. R. Venkatasubramanian, S. K. Manohar, and P. T. Balsara. Improving performance of NEM relay logic circuits using integrated charge-boosting flip flop. In *2011 IEEE/ACM International Symposium on Nanoscale Architectures (NANOARCH)*, pp. 37–44, San Diego, CA, 2011.
30. F. Pan and T. Samaddar. *Charge Pump Circuit Design*. McGraw-Hill electronic engineering series. McGraw-Hill, USA, 2006.
31. R. Venkatasubramanian, S. K. Manohar, and P. T. Balsara. Ultra low power high efficiency charge pump design using NEM relays. In *2011 IEEE 54th International Midwest Symposium on Circuits and Systems (MWSCAS)*, pp. 1–4, Seoul, South Korea, 2011.
32. J. Jeon, V. Pott, H. Kam, R. Nathanael, E. Alon, and T-J. K. Liu. Seesaw relay logic and memory circuits. *Journal of Microelectromechanical Systems*, 19(4): 1012–1014, 2010.
33. A. Gupta. NEM Relay Memory Design. Master's thesis, University of California, Berkeley, CA, 2009.
34. D. T. Stalter. Digital Logic and Multi-Valued Memory Using NEMS Switches. Master's thesis, Case Western Reserve University, Cleveland, OH, 2010.
35. R. Venkatasubramanian, S. K. Manohar, V. Paduvalli, and P. T. Balsara. NEM relay based memory architectures for low power design. In *12th IEEE Conference on Nanotechnology (IEEE-NANO), 2012*, pp. 1–5, Birmingham, England, 2012.
36. Emerging Research Devices. *International Technology Roadmap for Semiconductors,* 2009 edition. http://www.itrs.net/links/2009ITRS/2009Chapters_2009Tables/2009_ERD .doc, 2009.

10 High-Performance and Customizable Bioinformatic and Biomedical Very-Large-Scale-Integration Architectures

Yao Xin, Benben Liu, Ray C.C. Cheung, and Chao Wang

CONTENTS

Very-large-scale-integrated circuits (VLSIs) and field-programmable gate arrays (FPGAs) are definitely competitive in computation-intensive applications, benefiting from their inherent parallelism, customizable property, and low power consumption. They are mostly employed to accelerate various algorithms since their introduction, by which large design space can be obtained to explore different levels of parallelization.

In this chapter, three key objectives are focused on: (1) The computing power and potential of parallel VLSI architectures in applications of bioinformatics and biomedical engineering are revealed and discussed. (2) According to intrinsic bottleneck of algorithms, parallel computing can be generalized into data-intensive and compute-intensive problems. Two successful bioinformatic and biomedical VLSI architectures are presented, which are representatively data-intensive and compute-intensive, respectively. The first one is a parameterizable VLSI architecture for geometric biclustering (GBC), which is a data mining technique allowing simultaneous clustering of the rows and columns of a matrix. The architecture is also scalable with the size of column data and the resources provided on FPGA. Compared with CPU implementation, it achieves considerable acceleration as well as low energy consumption. The second one is a hardware architecture for the stochastic state point process filter (SSPPF), which is an effective tool for coefficients tracking in neural spiking activity research. This architecture is scalable in terms of both matrix size and degree of parallelism. Experimental result shows its superior performance comparing to the software implementation, while maintaining the numerical precision. (3) Design techniques for different parallelism levels are classified and discussed. Hybrid design techniques are employed in the architectures to achieve multilevel parallelism exploration. This also provides a macroscopic view into high-performance architecture design.

10.1 EMERGING BIOINFORMATICS AND BIOMEDICAL ENGINEERING

10.1.1 Bioinformatics and Biomedical Algorithms

Nowadays, bioinformatics and biomedical engineering have become indispensable parts of modern life. The bioinformatics integrates biological information and newly updated computational technologies, and the target is to undertake two major tasks [1]. The first is to retrieve and organize data to allow researchers to readily access and manage existing information. In genetics and genomics research, for instance, the sequencing technique provides a way to determine the sequence of nucleotide bases in a DNA molecule. The famous comprehensive database GenBank contains nucleotide sequences that are publicly available for more than 260,000 named organisms [2]. The National Center for Biotechnology Information builds and maintains the GenBank with data obtained from worldwide individual laboratories, or from large-scale sequencing projects. The second task is to develop methodologies and tools to facilitate the analysis of biological data. This purpose is oftentimes achieved by applying well-developed algorithms or computational software/hardware, which could be in computer science or applied math, into the analysis process. One example is the Burrows–Wheeler transform [3], an algorithm employed in data compression, which has been successfully applied to DNA sequence alignment [4] such as the well-known tools Burrows–Wheeler Aligner [5] and Bowtie 2 [6].

Similarly, biomedical engineering is also a multidisciplinary field, utilizing engineering principles (such as electrical, chemical, optical, mechanical) to understand, modify, or control biological systems for the purpose of health care. Taking the neural coding problem for example, new techniques such as high-density multielectrode array recordings and multiphoton imaging techniques [7] make it possible to record from hundreds of neurons, rather than the previous repeated single neuron stimulating. Neurons can be characterized as electrical devices and can be directly analyzed as information processing devices using signal processing theory [8]. In addition, the neural coding is a fundamentally statistical problem since neural responses to stimuli are stochastic. Therefore, many classical statistical methods, which are widely utilized in communication and signal processing area, have been applied to build models to describe neural behavior [9].

Being extraordinary extensive fields, the major research of bioinformatics covers sequence analysis, gene expression analysis, computational evolutionary biology, and so on. On the other hand, biomedical engineers focus on a broader domain [10]. Some representative subdisciplines include biomechanics, prosthetic devices and artificial organs, neural engineering, and genetic engineering. In this chapter, we mainly focus on the specific problems of bio-sequence analysis and neural coding applications.

Analysis for bio-sequence such as DNA, RNA, and protein sequence is one of the most promising and challenging tasks [11]. Specifically, it refers to analyze the sequences to reveal the functions, structures, or mutations. This hidden gene information in the nucleotide sequence is in charge of encoding proteins for organs. Such analysis enables us to explore similarities between different species and to discover fatal diseases that oftentimes represent in forms of DNA base mutations. The technique has been widely applied into personalized medicine, biological research, and forensics. Some major types of these analyses consist of (1) DNA/RNA sequence alignment, which is meant to identify similarities between two or multiple sequences or to map short sequences to a reference genome database; (2) biclustering, originally for data mining application, which is an important method to analyze large-scale biological sequence set [12]; (3) sequence assembly, which is to construct the original

DNA sequence with small fragments after sequencing process; and (4) phylogenetic analysis of DNA or protein sequences, which is an important method to find the evolutionary history of organisms from bacteria to humans [13].

The study of neural coding refers to searching and characterizing the correlations between neural action potentials or spikes and external-world sensory stimuli, motor actions, or mental states. The coding involves encoding and decoding corresponding to two opposite mapping directions [9]. The encoding is to predict the spike response of neurons to various types of stimuli. Conversely, the decoding problem concerns the estimation of the stimulus that gave rise to certain neuronal pattern. The neural coding algorithms need to characterize the models of neural spiking activity, which is a stochastic point process. Therefore, statistical theory, probability methods, and stochastic point processes are always applied in this field.

10.1.2   HIGH-PERFORMANCE COMPUTING REQUIREMENTS

In general, the major critical areas of requirement for the use of high-performance computing (HPC) in bioinformatics and biomedical research include (1) organizing, managing, mining, and analyzing the large volume of biological data; and (2) modeling and simulation, particularly in multiscale problems (modeling from the neuron, genome to the organism) [14]. The following sections will illustrate these requirements specifically.

Massively parallel next-generation sequencing techniques together with the need of gene expression analysis have resulted in a huge explosion in the amount of digital biological data. For around 20 years recently, the sizes of the genomic banks have experienced an exponential growth. Figure 10.1 is a picture of GenBank Statistics [15], which illustrates the growth of GenBank from its inception to 2013. The number of bases in GenBank has doubled approximately every 18 months. The size of the famous protein database—UniProtKB/TrEMBL—has increased by about 380 times from 1996 to 2013 (Release 1, November 1996). It retains 39,870,577 sequence entries consisting of 12,710,398,609 amino acids according to the latest release statistics (Release 2013_07) [16].

Furthermore, the improvement of sequencing machines and other progress in biotechnologies would further promote the future growth of biological databases. Thus, the essential problem is that the analysis of resulting data is unable to keep up with the producing pace. Thus, the processing of this avalanche of data to transform them into biological understanding is a significantly challenging task. Nowadays, most bioinformatics analysis tools are running on general purpose CPU platforms that are no longer scalable to the ever growing rate of publicly available data. Current approaches to this problem using conventional computers may take tens to hundreds of thousands of CPU hours to complete the alignment requirements for a single human genome. Hence, there is a need for faster computing platforms to cope with the database growth and people begin to resort to HPC to accelerate the processing and reduce bulk of computing time.

Except the requirement from biological data, in other applications, such as the neural modeling, the complexity of neural models continues to increase when realistic conditions and attributes, to emulate models for more accurate assessment of

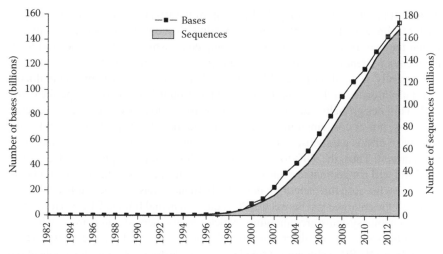

FIGURE 10.1   The graph depicts the number of bases and the number of sequence records in each release of GenBank, beginning with Release 3 in 1982. (Data from National Center for Biotechnology Information, accessed September 15, 2013, http://www.ncbi.nlm.nih.gov /genbank/statistics.)

the actual model, are incorporated, such as larger populations, varied ionic conductances, more detailed morphologies, and so on [17,18]. Neural modelers often increase the computational requirements by orders of magnitude, which is often constrained by the limited performance of conventional personal computers. Therefore, HPC platforms and tools are desperately needed to facilitate such research.

HPC is defined as a set of hardware and software techniques developed to build computer systems capable of rapidly performing large amounts of computation. These techniques generally rely on harnessing the computing power of large numbers of processors working in parallel, either in tightly coupled shared-memory multiprocessors or loosely coupled clusters of PCs. Emerging HPC platforms such as FPGAs [4], Cell Broadband Engine [19], and graphics processing units [20] have been explored in biology-related research. In famous HPC centers such as San Diego Supercomputer Center [21] and Pittsburgh Supercomputing Center [22], bioscience applications occupy a large portion of computation missions.

10.2   HIGH-PERFORMANCE VERY-LARGE-SCALE INTEGRATION

10.2.1   Need for Very-Large-Scale Integration in High-Performance Computing

According to the so-called *Moore's Law*, the performance of microprocessors has doubled every 18 months or so during the past four decades [23]. Although the conventional CPUs have experienced a significant improvement on computing performance, many limitations or bottlenecks still remain preventing them to successfully keep up with the demands from HPC applications. Simply increasing in clock rates and instruction-level parallelism could hardly be delivered further, because transistors are reaching the limits of miniaturization under current semiconductor technology. Besides, the memory bandwidth and disk access speeds could not match the ever-increasing CPU speeds completely, while the techniques to reduce the memory latency bottleneck are at the expense of extra circuitry complexity.

Currently, general-purpose CPU vendors change the strategy to rely on multicore and multithreaded architecture, to continue the proportional scaling of performance [24]. This design helps manage the thermal dissipation, because when frequency increases, power consumption escalates to impractical levels [25]. However, to take advantage of this parallelism, sophisticated hardware and developing software tools are required. Thus, the shift to multicore multithreaded paradigms introduces overhead and will not guarantee a linear enhancement in speed with increased number of processors because the memory bandwidth bottleneck has not been fully addressed.

To fill the enlarged gap between HPC requirement and limited CPU performance, heterogeneous systems are emerging, which are augmented hardware accelerators as coprocessors, as an alternative to CPU-only systems [25]. Very-large-scale-integration (VLSI) technology provides several custom solutions for application-oriented accelerators such as application-specific integrated circuit and FPGA. The conventional CPUs require decoding instructions and transferring data from cache or internal memory, which occupy a large portion of the overall computing process [26]. In contrast, application-specific VLSI could customize the pipeline, instruction flow, and data structure in correspondence with certain applications. Nevertheless, the hardware coprocessor faces a challenge that restraints its adaptability: the payback of designing a circuit for one application is low due to the long development time and high research and development costs [23].

The FPGA is an attractive solution to enlarge the diversity of applications. An FPGA consists of a large array of configurable logic blocks, block random-access memory (RAM), digital signal processing blocks, and input/output blocks [25]. The reconfigurability makes it suitable to a variety of applications. Containing a massive amount of programmable logic, FPGAs are able to provide different levels of parallelism: multiple processing elements (PEs) can be created in one single chip and work at the same time; dataflows between different operators can be controlled and streamed clock by clock; efficient pipeline architectures of variable depth are supported; on-chip memory reduces the bandwidth bottleneck for memory access. Besides, the FPGA is also a competitive candidate in embedded applications because of its power efficiency. All these properties have guaranteed the unshakable position of FPGAs in HPC community.

10.2.2 High-Performance Multilevel Parallel Designs

To outperform general-purpose CPUs, reconfigurable computers require techniques to design architectures that are efficient enough and achieve maximal parallelism. It means in each algorithmic step, the number of clock cycles is small while the computation carried out is as much as possible. The clock frequency is typically in the range of 2 to 3 GHz for high-performance processors. However, the hardware architectures always operate between 100 and 300 MHz for state-of-art FPGAs. Therefore, reconfigurable computers have to overcome a 10 times performance deficit to compete with CPUs.

There are several levels of parallelism to explore [27,28], which are enumerated as follows:

- Task-level parallelism: Conventionally exploited by cluster computing, which is achieved by making each processor execute a respective thread on the same or different data set. It concerns the number of coarse-grained tasks that can be executed in parallel for FPGAs.
- Instruction-level parallelism: Only a limited number of instructions can be executed on conventional processors. With much deeper pipelining controlled cycle by cycle, FPGAs can therefore support a variety of simultaneously executing *in-flight* instructions.
- Bit-level parallelism: This parallelism is achieved by increasing processor word size. In hardware, customizable bandwidth of data and instruction can be set according to specific requirement.
- Data-level parallelism: *Divide and Conquer* method, which refers to spreading the data onto different processing nodes that all carry out the same instructions. The FPGAs can take advantage of the fine-grained architecture for parallel execution. Multiple PEs could be constructed to execute on a large number of data sets simultaneously. Therefore, the equivalent performance of plenty of conventional CPUs can be achieved in a single FPGA chip.

By exploiting efficient mapping of all these levels of parallelism to the fabric, an FPGA operating at 200 MHz can outperform a 3 GHz processor by orders-of-magnitude or more, with greatly reduced power consumption [29]. Other definitions of types of parallelism such as kernel-level, problem-level, loop-level, expression-level parallelism, and so on [30,31] exist. They are specific to certain applications and can be covered by the four basic parallel types, thus are not elaborated here.

Other than parallelism level, parallelism is also classified in the term of granularity, which means the ratio of computation in relation to the amount of communication. Based on that, fine-grained, coarse-grained, and embarrassing parallelism is derived according to the order of decreasing in frequency of data communication among processors. The finer the granularity, the greater the potential for parallelism and hence speed-up would be achieved, but the higher the overheads of synchronization and communication.

The VLSI/FPGA parallel architectures are customizable and application specific. Different applications require specific computations and data structures, thus various unique parallel designs should be employed to meet these requirements. Therefore, to design a specific hardware architecture, the types of applications need to be summarized and classified. A technical report from UC Berkeley [32] had summarized 13 basic classes of problems that are explored by high-performance parallel architectures. They referred each class as a dwarf, which is a set of algorithms classified by similarity in computation and data movement, or memory access modes. Such classification is far different from other classification methods by subjects. From the perspective of parallel computing, however, this high-level abstraction allows reasoning about the dwarves' behavior across a wide range of applications. Subjects, algorithms, together with hardware platforms evolve with time, but the underlying computation and data flow patterns remain still into the future.

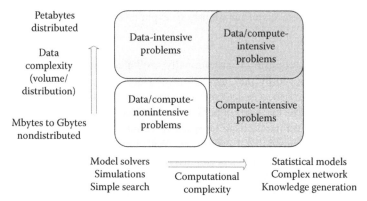

FIGURE 10.2 Classification of the application space between data-intensive and compute-intensive problems. (From Gorton, I. et al., *Computer*, 41(4), 30–32, 2008. With permission.)

From a macroscopic perspective, parallel computing applications can also be generally classified as data intensive and compute intensive [33]. Data-intensive application refers to processing and managing massive amounts of data volumes by parallel computing, and associated data analysis cycles are significantly reduced so that decisions can be made fast and timely [34]. Most of the execution time is spent on input/output (I/O) or memory transferring, so data-intensive applications are data-path oriented. The data volumes could be in size of terabytes or petabytes collected from experiments or simulations. Processing requirements usually scale near linearly according to data size and are normally amenable to straightforward parallelization methods. In contrast, compute-intensive computing devotes most of execution time to complex computation with smaller volumes of data required. Processing requirements normally scale superlinearly with data size [33]. Figure 10.2 shows a diagram to illustrate the application space between these problems [33]. Generally, there is always an overlap between these two types of computing.

In the following sections of this chapter, two high-performance parallel architectures in bioinformatics and biomedical engineering have been selected because they are representative in parallel computing: one is a data-intensive application and the other is a compute-intensive application. They also belong to two different dwarves in the Berkeley report [32]: the former comes from the Graph Traversal dwarf, and the second comes from the Dense Linear Algebra dwarf. Parallel techniques for each problem differ greatly. The techniques for data-intensive problem often involve decomposing or partitioning the big data into multiple segments and distributing to PEs. Each PE independently stores a segment and is responsible for computing its value. In this case, nonlocal communication should be minimized. This partition method is consistent with the definition of data-level parallelism. The more efficient the data be subdivided and distributed, the finer granularity can be explored in parallel processing. On the other side, compute-intensive processing usually involves task-level or bit-level parallelism. Because the data flow in compute-intensive problems is typically treated as a whole, parallelization should be performed within an application or an algorithm flow. The overall process can be resolved into separated distinct tasks and executed in parallel on appropriate platforms. If data dependency between successive steps exists and serial processing is unavoidable, bit-level or instruction-level parallelism would be exploited efficiently.

Indeed, more than one parallel technique is employed in each example. Exploiting parallelism at a single level is not an optimal solution and it would be wiser to use hybrid parallelism at all design levels. The hybrid design method is illustrated in detail in the following sections.

10.3 SCALABLE VERY-LARGE-SCALE-INTEGRATION ARCHITECTURE FOR DATA-INTENSIVE EXAMPLE

The data-intensive applications are becoming central to discovery-based science, especially in bioinformatics [35]. High-performance VLSI architectures are thus explored widely. In this section, a scalable GBC accelerator for DNA microarray data analysis is introduced. The bio-data analysis is a proper data-intensive application because the DNA database presents an exponential growth rate as aforementioned. The algorithm in GBC problem is not that complex. However, the large data volume involved makes the architecture design challenging. In this example, both data-level and task-level parallelism are explored efficiently.

10.3.1 Introduction to Geometric Biclustering

In the past decade, genomic data analysis has become one of the most important topics in biomedical research. Biologists correlate the gene information with diseases. For example, a cancer class can be discovered based on tissue classification [36]. Genes that have similar expression patterns may have common biological functions. There can be tens of thousands of genes involved in a microarray experiment, so efficient data analysis methods are indispensable for finding similar gene expression patterns. Figure 10.3 describes the flow of the gene analysis from expressing the genes to DNA microarray for pattern analysis.

Biclustering is an important method to analyze large biological and finical datasets [37,38], which is a process that partitions the data samples based on a number of similar criteria so that the existing patterns in the data can be analyzed. The biclustering process can be formulated in multidimensional data space for analysis [37].

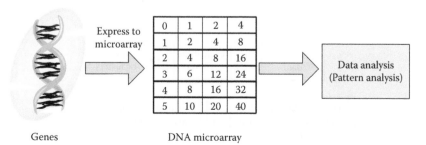

FIGURE 10.3   The flow of expressing a gene to a DNA microarray for data analysis.

GBC algorithm [39] identifies the linearities of microarray in a high-dimensional space of biclustering algorithm. It provides higher accuracy and reduces the computational complexity for coherent pattern detection compared to other methods. A hypergraph partitioning method [40] is proposed to further optimize the performance of the GBC algorithm. This method reduces the size of matrix in each operation to speed up the GBC algorithm using a software partition tool called hMetis. However, it is still time-consuming to search for the patterns of large-scale gene microarray data. A 100 by 100 matrix requires 2.8 minutes to identify the biclusters using the hypergraph partitioning method processed by the MATLAB program that runs on a machine with an Intel Core i7-920 2.66 GHz processor and DDR3 3 GB memory. But a normal gene matrix can have more than 10,000 rows or columns. Therefore, it is essential to accelerate this process [20].

10.3.2 GEOMETRIC BICLUSTERING ALGORITHM

10.3.2.1 Geometric Biclustering Work Flow

Figure 10.4 shows the overall flow of the GBC algorithm. The microarray data is converted to two-dimensional column-pair space sub-biclusters by using the Hough transform (HT) and additive and multiplicative pattern plot (AMPP) [39]. The HT detects linear data points and forms sub-biclusters in column-pair space. AMPP classifies the collinear points in sub-biclusters for different patterns. The column-pair biclusters are combined to maximal biclusters, and the number of genes in the merged biclusters is reduced in this process. In the evaluation of the maximal biclusters, the biclusters are filtered out if the number of conditions is fewer than the given parameter. Finally, the valid maximal biclusters are the target search patterns.

Assume there is an m by n data matrix. In column-pair space, each column forms a column pair with all other columns. There are $(n)(n-1)/2$ column pairs in total. Figure 10.5 illustrates an example of transforming a 5 by 4 data matrix (M_C) to a 5 by 6 column-pair matrix (M_S) in the GBC work flow. The target is finding an additive coherent pattern in the data matrix, where the value of the next column is equal to the value of the current column plus a constant ($Ci = Cj + $ constant). The input matrix M_C is transformed to M_S by using HT and AMPP for additive coherent value. For example, column pair $S_{1,2}$ in M_S represents the comparison of C1 and C2 in M_C. The value "1" in $S_{1,2}$ indicates there is an additive coherent relationship of that position in C1 and C2. All the column pairs in M_S are combined in the combination process.

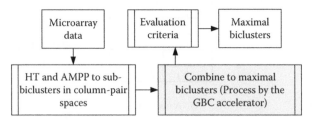

FIGURE 10.4   The work flow of the geometric biclustering algorithm.

C1	C2	C3	C4
10	11	12	13
9	10	11	12
8	20	9	22
7	20	0	0
6	20	8	9

5 × 4 matrix (Mc)

HT and AMPP to
column-pair matrix
for additive coherent values
(Ci = Cj + constant),
constant = 1

$S_{1,2}$	$S_{1,3}$	$S_{1,4}$	$S_{2,3}$	$S_{2,4}$	$S_{3,4}$
1	0	0	1	0	1
1	0	0	1	0	1
0	1	0	0	0	0
0	0	0	0	0	0
0	0	0	0	0	1

5 × 6 column-pair matrix (Ms)

FIGURE 10.5 Example of transformation from data matrix to column-pair matrix. The set bit (logic '1') in M_S represents two points at the two columns in M_C have coherent relationship.

10.3.2.2 Geometric Biclustering Combination Process

The combination algorithm is shown in Algorithm 1. Assuming that M_C is an m by n gene data matrix. The transformed column-pair matrix M_S contains $R = (n)$ $(n - 1)/2$ column pairs in total. The column pairs are denoted as $[S_{1,2}, S_{1,3} \dots S_{i,j}]$, where $1 \leq i,j \leq n$. The column is represented in a bit vector that only contains logic "0" or "1." The threshold value TH1 is used to determine whether the combined column pair is valid or not. This threshold controls the sensitivity of the matching. Higher threshold value allows matching of less similar patterns. First, the column pairs that contain the number of "1" less than TH1 are eliminated to reduce the number of combines. Next, all the remaining column pairs are combined. The combined column pairs are included in M_S' if they are valid. The combination process is finished after all column pairs are visited. Figure 10.6 illustrates an example of the combination of the column-pair matrix in Figure 10.5. In each step of the process, two column pairs (S1, S2) are selected with common column in M_C to combine, and the column pair has at least one column different to other one. For example, $S_{1,2}$ and $S_{1,3}$ are common in column C1 in M_C and different in column C2 and column C3, then they can be combined. The visited S1 will not be selected again. S2 is combined with S1 by using arithmetic AND logic to $S_{combine}$. If the number of "1" in $S_{combine}$ is greater than or equal to TH1, it is included in M_{temp}. After visiting all S2, M_{temp} is joined with M_S'. Then, the process selects next S1 for combination.

Algorithm 1: Column-pair combination
```
1 Input: column-pair matrix Ms, Si,j ∈ Ms, where 1 ≤ i,j ≤ n
2 Input: TH1, a threshold to accept combined column-pairs
3 Output: Combined matrix M's
  1://Remove the column-pair with total "1" less than TH
  2: for all Si,j ∈ Ms do
  3: if Total "1" of Si,j ≥ TH1 then
  4: M's ← Si,j
  5: end if
  6: end for
  7://Compare each column pair in M's
```

```
 8: for all S1 ∈ M'_S  do
 9: for all S2 ∈ M'_S , column in S2 ⊄ S1 and column in S1 ⊄ S2
    do
10: if At least one column in S1 and S2 is in common then
11: if At least one column in S1 and S2 is different
    then
12:   S_combine = S1&S2
13:   if Total "1" of S_combine ≥ TH1 then
14:     M_temp ← S_combine
15:   end if
16: end if
17: end if
18: end for
19: M'_S = M_S' ∪ M_temp
20: end for
```

After visiting all S1 in M'_S, evaluation criteria selects the maximal biclusters from M'_S. The elements in M'_S with a maximum number are chosen as the final solution. In the example of a column-pair matrix in Figure 10.5, $S_{1,2,3,4}$ is obtained as the maximal column pairs with column value [11000]. This represents the maximal bicluster containing columns [C1, C2, C3, C4] in the original matrix M_C, the position with

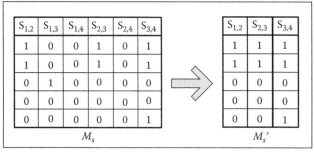

(a)

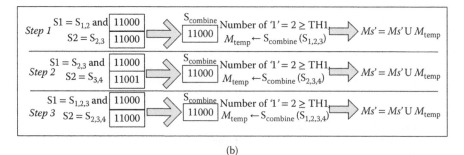

(b)

FIGURE 10.6 Example of combining a column-pair matrix, assuming TH1 = 2. (Ms). (a) Remove the column-pairs with total '1' less than TH1 (Ms'), and (b) combine each column-pair in Ms'.

value "1" is selected. The final additive coherent matrix after masking M_C with the bit matrix [39] can be obtained. This is the most computational intensive part of the GBC algorithm, which consumes over 80% of the computation time. Therefore, the aim is to accelerate this process.

10.3.2.3 Parallelism Exploration for the Geometric Biclustering Combination Process

The combination process is the most time-consuming part in the GBC algorithm. It involves combining two column pairs each time. It is obvious that this process can be speeded up by combining several two column pairs at the same time. In the example in Figure 10.6, no data dependency is found for these two column pairs: $S_{1,2}$, $S_{1,3}$ and $S_{1,2}$, $S_{1,4}$, the selected input is independent to the output at the same processing cycle. Given this, the column pairs can be divided and processed by multiple PEs running side by side in parallel. Data-level parallelism is thus easily achieved. This parallelism can also be treated as task level because each column-pair combing process can be seen as an independent task. For example, *Step 1* and *Step 2* in Figure 10.6 can be processed in the same cycle. $S_{1,2,3}$ is dependent on $S_{1,2}$, $S_{1,3}$, and $S_{2,3}$, therefore the process of $S_{1,2,3}$ must be after combining all $S_{1,2}$, $S_{1,3}$, and $S_{2,3}$. The parallel ability of FPGA can facilitate this process.

Data-intensive problem often face the problem of I/O bandwidth [41] as the conventional CPUs. The gap between off-chip memory and processing units can significantly diminish the parallel advantage, due to the fact that huge data volume could not be stored cost-effectively on chip. Fortunately, the reconfigurability of FPGAs enables us to configure as many pins as possible to be I/O ports. In this design, only a 64-bit wide interface is dedicated to communicate with external memory because of reality restrictions.

10.3.3 ARCHITECTURE DESIGN

10.3.3.1 Architecture

A scalable GBC accelerator to facilitate the parallelism of the GBC algorithm is proposed. Figure 10.7 shows the architecture of GBC accelerator; the column pairs in matrix M_S are stored in external memories. There are four main components in the FPGA design to process the Algorithm 1; they include (1) Memory controller, (2) Clock management, (3) Data controller, and (4) N PEs.

The memory controller handles the read/write signals and sequence of data transferring to and from the external memory. It is the interface between external memory and data controller. The number of read port ($Port_R$) and write port ($Port_W$) depends on the external memory.

The clock management generates different clocks required for memories (memory clock), PEs (PE clock), and internal logics (system clock).

The data controller contains four units to process the combination process. First, S1, S2 selection unit selects the column pairs S1, S2 as in Algorithm 1. Second, PE management unit manages the available PEs to process the combination, distributes the selected column pairs to PEs, and collects the combined result $S_{combine}$.

Third, data R/W unit asserts the read/write signals and handles the column pair's data reading/writing for the memory controller. Fourth, state machine controls different states of the accelerator such as initialization and the finish of process.

The accelerator contains at most N PEs processing at the same time. The architecture of each PE is described in Figure 10.8, which consists of an independent set of FIFOs to store the two input column pairs (S1, S2) and the combined column pairs ($S_{combine}$). This distributed local memory makes the parallel process more efficient by keeping the balance of data transfer rate between external memory and PEs. The combine logic in the PE executes a bitwise *AND* operation for the two input column pairs. The count "1" logic is a shift register, a k-bit decoder (decode to $\log_2 k$-bit) to check the number of "1" in *AND_RESULT* register, and an adder to sum up all the "1" values. Meanwhile, the *AND_RESULT* is stored in *Result_FIFO* for writing $S_{combine}$ to memory. The threshold comparer checks if the combination is valid. The combination is valid if the total number of logic "1," which is stored in register

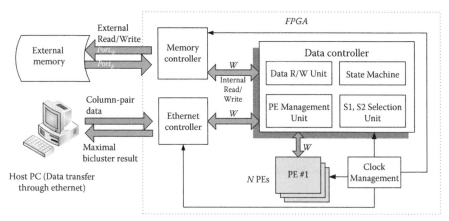

FIGURE 10.7 The architecture of geometric biclustering accelerator. W is the data width, N is the maximum number of processing elements, $Port_R$ is the number of memory read port, $Port_W$ is the number of memory write port.

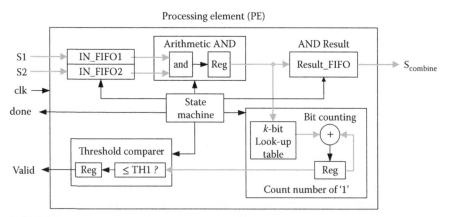

FIGURE 10.8 The architecture of a processing element.

NUMBER_ONE, is greater than or equal to the threshold TH1. The state machine counts the total number of bytes combined by using a counter and asserts signal *done* when the combination of two column pairs is finished.

The data width *W* in the PE is the number of bits that can be executed in each operation. The PE can execute *W* bits of *AND* and check *W/k* bits of "1" at the same time. The number of clock cycles to execute *m* bit data in each PE is:

$$\mathrm{PE}_{cycle} = \frac{m}{W}\left(1 + \frac{W}{k}\right) \qquad (10.1)$$

10.3.3.2 Data Structure

A custom data structure for the GBC accelerator to provide high efficient combination process is designed. Figure 10.9 shows the data structure of the column pairs in matrix M_S stored in the external memory. The bit "1" in S_ID indicates the columns in matrix M_C that column pair represents. For example, the column pair $S_{1,2}$ represents column C1 and column C2 in M_C. The S_ID of $S_{1,2}$ is [0000 … 011]. The width of S_ID is *Q* bits. After storing the S_ID, the bit-vector data of each column pair is stored in the memory. This representation method facilitates the selection of S1 and S2, which reduces data transfer. In the combination Algorithm 1, the accelerator selects two column pairs with at least one common column and one different column in M_C. The column-pair bit vectors are combined only if the conditions are matched. In the example of Figure 10.9, it first reads the S_ID of $S_{1,2}$ (S_ID1) and $S_{1,3}$ (S_ID2) for checking the conditions. The conditions are checked by using simple AND and XOR gates. The column-pair bit vectors are combined only if both conditions are matched. This saves the time to read the data if the conditions are not matched. If the combined column pair $S_{combine}$ ($S_{1,2,3}$) is valid, the S_ID of $S_{1,2,3}$ is equal to S_ID1 *OR* S_ID2, which is [0000 … 111]. The new S_ID and $S_{combine}$ are then written to the free memory space.

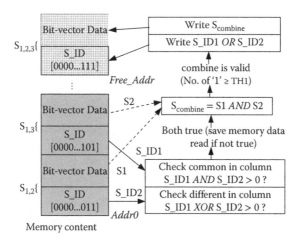

FIGURE 10.9 The data structure of the column pairs in memory.

10.3.4 EXPERIMENTAL RESULTS

In this section, the GBC accelerator on an FPGA is implemented to study the performance by a set of genetic benchmarks. The impact of the PE is studied. A GBC software program is also realized and compared to the GBC accelerator.

10.3.4.1 Genetic Benchmarks

The GBC algorithm is mainly employed for the analysis of large-scale gene microarray data. The gene data is taken as benchmarks to evaluate the performance of the GBC accelerator implemented on FPGA. Six benchmarks of various column-pair matrix size are chosen. The benchmarks are real biological data, which can be downloaded from the National Cancer Institute [42], the Department of Information Technology and Electrical Engineering of ETH Zurich [43] and the National Center for Biotechnology Information [44]. The details of the benchmarks are shown in Table 10.1.

10.3.4.2 Optimization of Geometric Biclustering Accelerator

Internal Structure of PE: Table 10.2 shows the information of the resources used in the GBC accelerator, which is implemented on Xilinx ML605 development platform [45] (using XC6VLX240T-1 FPGA). The number of PEs, data-width, and frequency are scalable depending on the size of the FPGA. Each PE uses three block RAMs for FIFOs. If the architecture and number of PEs are optimized, a higher performance accelerator can be implemented. The impact of the k-bit decoder on the frequency of PEs is shown in Table 10.3. The k-bit decoder counts the number of "1" from k-bit data. The output of the decoder is the number of "1", which is $\log_2 k$-bits data. From Equation 10.1, when the bit width of the decoder is increased, the number of clock cycles to count the number of "1" decreases. However, this causes the frequency of the PE decreases as shown in Table 10.3. It is because the decoder becomes more complex when k increases. The implementation tool Xilinx Synthesis Tool (Xilinx,

---

TABLE 10.1
Information of the Six Genetic Benchmarks

Benchmarks	Size of M_C ($m \times n$)	Size of M_S ($m \times R$)	Description
Breast	100×78	$2,888 \times 3,003$	Breast cancer microarray data
RNAexp	$6,000 \times 59$	$6,000 \times 1,711$	NCI-60 cell line mRNA expression data set
Yeast	$2,993 \times 173$	$2,993 \times 14,878$	Yeast DNA microarray
Cancer_Carcinoma	$7,457 \times 36$	$7,457 \times 630$	Carcinoma cancer data
Cancer_Filter	$6,953 \times 117$	$6,953 \times 6,786$	Filtered gene expression data set of multiple human organs
Cancer_All	$16,063 \times 117$	$16,063 \times 6,786$	Gene expression data set of multiple human organs

TABLE 10.2

Information of the Geometric Biclustering Accelerator and Reference C Software Implemented

GBC Accelerator	
Number of PE and data width	$N = 130$, $W = 64$ bit
Slice usage of one PE	209
RAMB36E1 usage of one PE	3
Total slices usage in FPGA	30,144 (80% of total slices)
Total block RAMs usage in FPGA	403 (97% of total block RAMs)
System/memory frequency	200 MHz/400 MHz
External memory	DDR3 512 MB
Number of memory read/write port	$\text{Port}_R = 1$, $\text{Port}_W = 1$
Machine Running Software Reference Design	
CPU	Intel Core 2 Duo 2.8 GHz
Memory	DDR2 2 GB RAM

TABLE 10.3

Optimization of Processing Elements (16-Bit Decoder Cannot Be Synthesized)

k-bit decoder	1	2	4	8	16
PE frequency (MHz)	306	263	269	269	x
Clock cycles (64-bit data)	64	32	16	8	4
Time (64-bit data) (millisecond)	20.9	12.2	5.9	3.0	x

San Jose, CA) [46] cannot synthesize the 16-bit decoder because the decoder has 2^{16} input combinations, which are not efficiently implemented. As a result, 8-bit decoder is the most time efficient to compute 64-bit data.

Number of PEs: It is expected that using more PEs can reduce the processing time. Power consumption is a major issue in modern technology. However, it is obvious that more PEs cause higher power consumption, as shown in Figure 10.10. The maximum number of PEs that can be implemented is 130. Therefore, the number of PEs against power and speed of the accelerator should be studied. The power consumption of the accelerator core, which excludes the memory access, is measured by Xilinx XPower Estimator [47]. The average time for combinations of benchmarks is also analyzed. In Figure 10.10, although the power consumed in more PEs is higher, the time to process the GBC algorithm is much lower than less PEs. The energy (power × time) consumed by 130 PEs is the lowest. Therefore, there should be 130 PEs implemented in the accelerator.

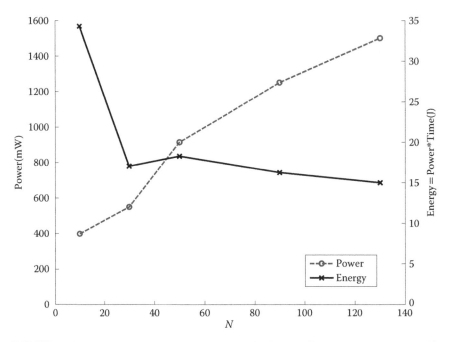

FIGURE 10.10   Power and energy consumption of using a different number of processing elements (maximum 130 processing elements can be implemented).

10.3.4.3   Comparison to Software Reference Design

The hypergraph partitioning method is used to speed up the GBC algorithm in Wang et al. [40]. The time for combining the synthetic benchmark of 100 by 100 matrix is about 2.8 minutes running on the MATLAB environment. The GBC accelerator uses 125 milliseconds to combine this synthetic benchmark, which is over 1350 times faster than Wang et al. [40]. However, it is not a fair comparison because the hypergraph partitioning method is running on an inefficient MATLAB platform. Therefore, the software realization for the GBC algorithm using the highly efficient C language is benchmarked. The algorithm of the GBC software is the same as the GBC accelerator based on Algorithm 1. The software reference design is running on a machine described in Table 10.2. Figure 10.11 shows the comparison between software and accelerator using 130 PEs. On average, the GBC accelerator speeds up the process by four times. In addition, a power meter is utilized to measure the power consumption including memory access power. The software consumes 12.5 W power on average while the accelerator only consumes 7.5 W power when processing. The power reduction is 40%.

10.4   CUSTOMIZABLE VERY-LARGE-SCALE-INTEGRATION ARCHITECTURE FOR COMPUTE-INTENSIVE EXAMPLE

In this section, a FPGA-based customizable SSPPF for neural spiking activity is present. The SSPPF is a kind of adaptive filter that has been applied to the coefficients estimation for neuron models, such as the generalized Laguerre–Volterra model (GLVM).

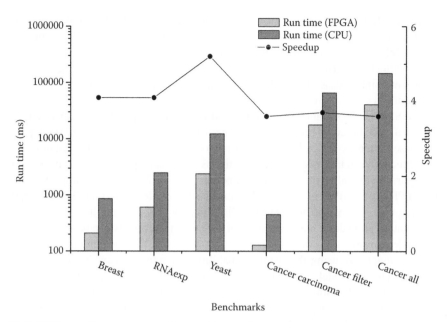

FIGURE 10.11 Work flow of the geometric biclustering algorithm.

The computation for the SSPPF involves complex matrix/vector calculation. The size of matrix/vector (corresponding to the number of model coefficients) increases exponentially along with the growth of model's inputs, making the calculation inefficient. This problem is typically compute intensive, which involves calculation on multiple types of data block. Because the data flow throughout the computation cannot be split or divided readily due to the high data dependency in each step, task-level parallelism is not suitable for this case. Nevertheless, hybrid parallel design is adopted, integrating both bit-level and instruction-level parallelism techniques.

10.4.1 Introduction to Generalized Laguerre–Volterra Model and Stochastic State Point Process Filter

As a burgeoning topic in biomedical engineering research, cognitive neural prosthesis requires a silicon-based prosthetic device that can be implanted into the mammalian brain. This device should be capable of performing bidirectional communications between the intact brain regions and bypass the degenerated region [48]. A well-functioning mathematical model has to be established beforehand for effective processing of neural signals. The GLVM, proposed by Song et al. [49], is a data-driven model. It is first applied to predict mammalian hippocampal CA1 neuronal spiking activity based on detected CA3 spike trains, so that the expected neuroprosthetic function can be achieved.

Within the generalized Laguerre–Volterra (GLV) algorithm, the most computation-intensive stage in the overall calculation flow lies in the estimation of the Laguerre coefficients, which has to be done beforehand using the recorded input/output data. Previous silicon-based implementations of the GLVM [50,51], based on the simplified single-input and single-output model, are unable to represent real situation, because a model output is normally affected by multiple inputs of spiking activity. The multi-input, multi-output GLVM was implemented later on an FPGA-based platform [52]. However, the adopted coefficients tracking method is the steepest decent point process filter (SDPPF), which is simple in mathematical expressions but sacrifices certain levels of accuracy, thus being less effective.

The SSPPF is a suitable choice for the aforementioned problem [5]. Derived by Eden et al. [53] based on Bayes rule Chapman–Kolmogorov paradigm and point process observation models, it has been verified to be effective for tracking dynamics of neural receptive fields under various conditions. Chan et al. [54] successfully applied the algorithm to realize the estimation function of the GLVM. A major improvement of SSPPF over SDPPF is the introduction of adaptive learning rate. The adopted learning rate is constant in Li et al. [52]. Although this can simplify the iterative computation procedure, brain activities of the behaving animals can be time variant and subject to stochastic variations. To be more realistic, the learning rate should be updated adaptively using the firing probability computed previously [52] together with the detected model output at present. Up to now, the SSPPF algorithm has been only implemented in software and executed on a desktop setup, resulting in certain limitation in calculation process. Because of the complex computational burden and a potential demand of embedded platform with low power consumption in the future, there is a need of high-performance hardware architecture for this algorithm.

10.4.2 STOCHASTIC STATE POINT PROCESS FILTER ALGORITHM

The SSPPF is proposed to adaptively estimate model parameters for point process observations when analyzing neural dynamics [53]. The model of neural firing is defined as a time-varying parameter vector $C(k)$, which is a linear evolution process with Gaussian errors. The $C(k)$ and its covariance matrix $R(k)$ are updated by following recursive equations:

$$R(k)^{-1} = [R(k-1)+Q]^{-1} + \left[\left(\frac{\partial \log P(k)}{\partial C(k)}\right)^T P(k)\left(\frac{\partial \log P(k)}{\partial C(k)}\right)\right.$$
$$\left. -(y(k)-P(k))\frac{\partial^2 \log P(k)}{\partial C(k)\partial C(k)^T}\right] \tag{10.2}$$

$$C(k) = C(k-1) + R(k)\left[\left(\frac{\partial \log P(k)}{\partial C(k)}\right)^T (y(k)-P(k))\right] \tag{10.3}$$

where $P(k)$, $y(k)$, and k are the firing probability of spike firing, spike observed, and the discrete time steps, respectively. Applying Equations 10.2 and 10.3 to the GLVM [54], the two equations can be rewritten into the following forms:

$$R(k) = [(R(k-1)+Q)^{-1} + k_1 M(k)^T M(k)]^{-1} \tag{10.4}$$

$$C(k) = C(k-1) + k_2 R(k) M(k)^T \tag{10.5}$$

where the vector $M(k)$ is the convolution product between the model inputs/output and the Laguerre basis functions. The k_1 and k_2 can be derived as follows:

$$k_1 = \alpha^2 P(k) + \beta[y(k) - P(k)] \tag{10.6}$$

$$k_2 = \alpha[y(k) - P(k)] \tag{10.7}$$

$$\alpha = \frac{1}{\sqrt{2\pi P(k)}} \exp[-w(k)^2] \tag{10.8}$$

$$\beta = \frac{w(k)}{\sqrt{2\pi P(k)}} \exp[-w(k)^2] + \frac{1}{2\pi P(k)^2} \exp[-2w(k)^2] \tag{10.9}$$

The GLVM functions for the calculation of $M(k)$, $y(k)$, $P(k)$, and $w(k)$ were introduced in Chan et al. [54], hence they are not elaborated here. In practice, if the number of parameters in $C(k)$ to estimate is N, $R(k)$ and Q would be an N by N matrix, and M would be a 1 by N vector, while C would be an N by 1 vector.

The SSPPF algorithm is reformed into two major parts. The first part is a collection of a large amount of matrix computation, the most time-consuming and principal aspect of the overall computation. The second part involves intensive arithmetic operations that are difficult for the hardware to implement. Therefore, our architecture of SSPPF is designed to lay a special focus on the calculation stages expressed by Equations 10.4 and 10.5, which has more potential for parallelization. A desktop computer is responsible for the second part and $M(k)$ update, communicating with FPGA in real time.

10.4.3 ARCHITECTURE DESIGN

10.4.3.1 Parallelism Exploration for the Stochastic State Point Process Filter Algorithm

First, the parallel design techniques should be considered comprehensively to realize an efficient architecture. The overall computation flow is decomposed into seven steps, which are presented in Table 10.4. It is straightforward to see that the computation flow is difficult to be separated into subtasks that can be performed in parallel, as data dependency exists between consecutive steps. These tasks need to be carried out in a sequential way. In addition, the data involved in SSSPF is either matrix or vector; data partitioning and assembling method is not that appropriate when certain operations like matrix inversion are performed, which would make the overall complexity high. Therefore, neither task-level nor data-level parallelism is optimized to utilize.

TABLE 10.4

Steps to Finish One Iteration of Stochastic State Point Process Filter (N: Vector/Matrix Size)

Step Number	Operation
1	$r_1(k) = R(k-1) + Q$
2	$r_2(k) = r_1(k)^{-1}$
3	$r_3(k) = k_1 \times M(k)^T \times M(k)$
4	$r_4(k) = r_2(k) + r_3(k)$
5	$R(k) = r_4(k)^{-1}$
6	$r_5(k) = k_2 \times R(k) \times M(k)^T$
7	$C(k) = C(k-1) + r_5(k)$

Exploiting bit-level and instruction-level parallelism makes sense for the architecture of SSPPF. Bit-level exploration implies increasing the data width, by concatenating multiple operands to form a vector and processing at one time. This method is also referred to as vector processing [55]. The data organizational format in SSPPF algorithm is suitable to be handled in a vector manner, and the intrinsic properties of VLSI/FPGA facilitate this technique. First, the on-chip memory blocks (block RAMs) are customizable in data width, because multiple block RAMs can be cascaded to cache wide words. By this, a vector of data in one memory address is stored and accessed, not requiring extra instructions, which reduces power consumption. Second, vector processing demands certain data alignment to meet specific requirements of the underlying algorithm. FPGAs provide the design freedom to create a custom alignment unit, which would further improve performance and power consumption [56]. Finally, the reconfigurability of an FPGA allows us to explore multiple degrees of vectorization and choose the optimal one to meet the area and throughput constraints [55].

Instruction-level parallel design needs to execute multiple instructions in parallel. Pipelining is an efficient architecture mechanism that exploits instruction-level parallelism at run time [57]. It can be viewed as a form of parallelism where a process is divided into a set of stages, all of the stages run concurrently, and the output of each stage acts as the input to the next one. The goal is to hide the latency of operations in parallel, increase system throughput by overlapping (in time) the execution of the stages, and maximize the instruction efficiency [28,57].

10.4.3.2 Architecture for Stochastic State Point Process Filter

The parallel VLSI architecture for SSPPF is constructed taking into consideration the discussed techniques. The data representation adopted is single precision floating-point format in the IEEE-754 standard, so each data is represented by 32 bits. Unlike most hardware designs [52,58], the matrix size can be dynamically changed to be arbitrary in each updating iteration in this architecture, without preconfiguration. The size of matrix/vector is only limited by on-chip memory resource. The architecture is also scalable in degree of parallelism, because the computing units are capable of concatenating to increase the degree of vector processing.

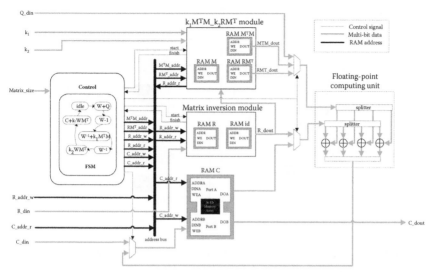

FIGURE 10.12 Overall architecture of stochastic state point process filter.

Figure 10.12 shows the overall architecture for SSPPF. The top layer module implements two submodules according to principal steps in SSPPF calculation: the computation of $k_1M(k)^TM(k)$ and $k_2R(k)M(k)^T$ is conducted in k_1 $M^TM\_k_2RM^T$ module; the inversion of $R(k)$ is performed by matrix inversion module. Vectors and matrices involved are stored in block RAMs some of which are arranged to true dual-port mode to enable concurrent reading and writing.

At the top and matrix inversion module, vector processing method is employed: a vector of four floating-point data constitutes the basic operand, while a computing unit consists of four floating-point operators. The intrinsic FPGA parallelism can be further explored by combining more computing units in parallel, with data-width for operation and storage increased accordingly. Operations are performed in horizontal sweep fashion, which is to process multiple elements of a row in parallel until the entire row is scanned.

The matrix inversion module is shown in Figure 10.13, which is based on the Gauss–Jordan elimination algorithm with partial pivoting. Given an N by N square matrix A, the inversion process begins by building an augmented matrix with the size N by $2N$: $[AI]$. The identity matrix I is on the right side of A. Row operations are then performed iteratively to transform $[AI]$ into $[IA^{-1}]$. Two sets of operators and RAMs are needed for original and identity matrix simultaneously. The calculations in two time-consuming phases—normalization and elimination—are fully pipelined. Elimination module is in charge of these two calculations as shown in Figure 10.14. The pivot row normalization step is realized by multiplying the pivot's reciprocal calculated in advance, by each row element. Two block RAMs are set to synchronize multiple operand flows in the computing units: one to cache a normalized pivot row and another to cache all elements in the column where the pivot is located. The control logic continuously generates address signals to various RAMs for reading and writing. The true dual-port RAMs facilitate the pipelining operation

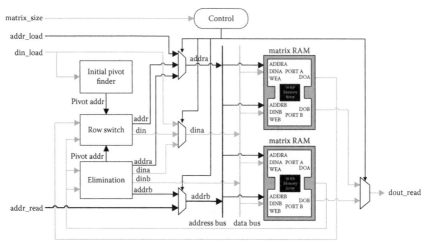

FIGURE 10.13 Matrix inversion module. Maximal parallelism is explored in this module by vector processing and pipelining.

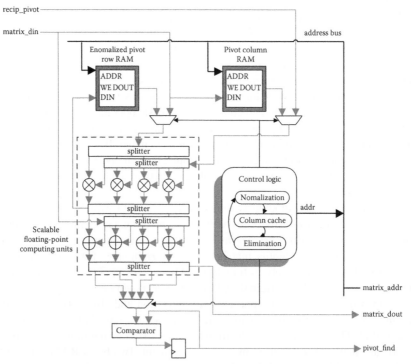

FIGURE 10.14 Elimination module is the focus of computation, because the elimination phase is most time-consuming throughout the inversion process. Indeed, two sets of computing units are implemented for original and identity matrix, but only one set is presented in this figure.

through concurrent operations: reading vectors of data from one port and receiving results in another port.

To achieve efficient pivot location, a comparator is set as the last node in the computing pipeline, directly receiving data flow from adder outputs. Pivot search is thus performed with only one extra clock in general. The found pivot row address is output for next-round row interchange when the elimination process is done. In this module, both bit-level and instruction-level parallelism are fully achieved through deep pipelining and vector processing.

In the $k_1M^TM\_k_2RM^T$ module, the memory of vector $M(k)$ together with floating-point operators is shared by two distinct calculations, as shown in Figure 10.15. Only a minimum number of operators are implemented in this module, because the two involved computations are not bottlenecks affecting the overall performance, which only consume a small portion of the total calculation time.

The approximated number of clock cycles in each round of $C(k)$ update is estimated in Equation 10.10. The D_p denotes the parallel degree (it is four in the current design), and N denotes matrix/vector size. In this estimation, both the latency of operator intellectual property cores and the control signal overheads are overlooked; each would be fairly trivial when the size N increases.

$$(2N^2 + 28N)\left\lceil \frac{N}{D_P} \right\rceil + 4N^2 + N \qquad (10.10)$$

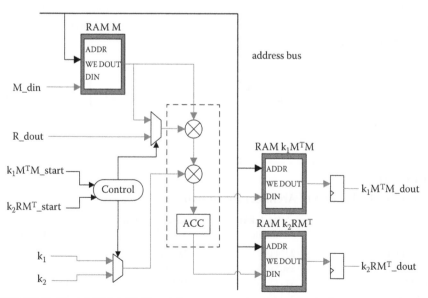

FIGURE 10.15 $k_1M^TM\_k_2RM^T$ module, where the calculation of $k_1M(k)^TM(k)$ and $k_2R(k)M(k)^T$ are performed.

10.4.4 EXPERIMENTAL RESULTS

The architecture is synthesized, placed, and routed on a Xilinx Virtex-6 FPGA (XC6VLX240T-1). The resources occupied by the whole architecture and the main modules are summarized in Table 10.5. The upper limit of matrix/vector size N is set to 256 under consideration of available on-chip memory. The resource usage is rather small except for block RAMs.

The functionality of the SSPPF architecture needs to be verified, so two sets of synthetic experimental data are selected as the initial input into the hardware. The size of vector $C(k)$ is set to be 49 and 139 in experiments, which accounts for the coefficients to estimate in a multi-input single-output model, and under two specific conditions: (1) only second-order self-kernel is applied, (2) both self-kernel and cross kernel in second order are applied. All kernels (besides the 0th) have five inputs, and the number of Laguerre basis functions L is set at 3 [52,59]. Meanwhile, the Laguerre coefficients estimation in MATLAB description of SSPPF algorithm is realized with a double precision floating-point format as a reference. The estimated vector $C(k)$ of one iteration calculated by two platforms is shown in Figure 10.16. The results seem to be the same, and error analysis needs to be carried out to identify the achievable accuracy.

In error analysis, a comparison is made between the results achieved by FPGA design and MATLAB realization in double precision. Initial input coefficient vector and covariance matrix are randomly generated. Absolute maximum value of k_1 and k_2 under worst conditions is chosen to evaluate possible maximum error. There are 100 independent experiments performed to update $C(k)$ and calculate the average mean error (ME) and mean squared error (MSE) of FPGA implementation results in each SSPPF iteration. The ME and MSE are computed with Equations 10.11 and 10.12 (the i is the element index in $C(k)$), respectively (Figure 10.17). Although both values increase as size N grows because of the error cumulation, the MSE for $N = 250$ remains under 2×10^{-8}.

$$\text{ME} = \frac{1}{100} \sum_{k=1}^{100} \left[\frac{1}{N} \sum_{i=1}^{N} |C_{\text{FPGA}}(k)(i) - C_{\text{MATLAB}}(k)(i)| \right] \tag{10.11}$$

TABLE 10.5
Resource Utilization of Architecture Design

	$k_1 M^T M\_k_2 R M^T$ Module	Matrix Inversion Module	Overall Design
Slice LUTs	2344 (1%)	12,068 (8%)	16,667 (11%)
Slice registers	2905 (1%)	14,139 (4%)	19,035 (6%)
RAMB36E1	57 (13%)	118 (28%)	176 (42%)
RAMB18E1	2 (1%)	1 (1%)	3 (1%)
Maximum frequency	261.575 MHz	220.653 MHz	208.247 MHz

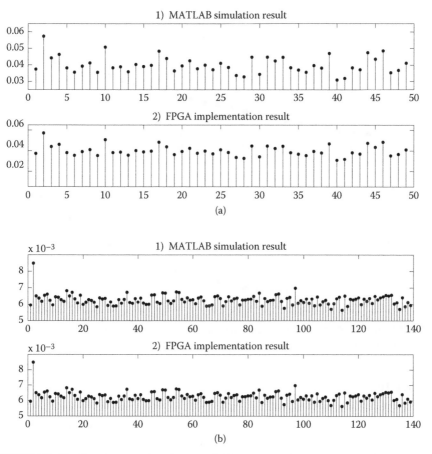

FIGURE 10.16 Result comparison between field-programmable gate array implementation and MATLAB simulation. (a) The upper part is the result for N = 49, and (b) the lower part is the result for N =139.

$$\text{MSE} = \frac{1}{100}\sum_{k=1}^{100}\left[\frac{1}{N}\sum_{i=1}^{N}(C_{FPGA}(k)(i) - C_{\text{MATLAB}}(k)(i))^2\right] \tag{10.12}$$

In a performance evaluation for execution speed, the hardware architecture is compared with software running on a commercial CPU platform. The software implementation of SSPPF algorithm in C language has been benchmarked. Configuration of the software platform is given in Table 10.6. Different datasets with the number of coefficients ranging from 50 to 250 are executed on two platforms. The FPGA architecture is driven by its possible maximum frequency clock in the evaluation.

The bar graph in Figure 10.18 illustrates the execution-time comparison between two platforms for various matrix sizes. The FPGA-based hardware platform can achieve up to seven times speedup compared to the software implementation in

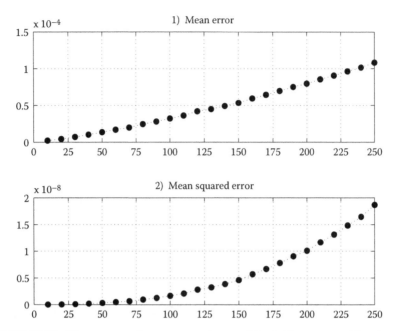

FIGURE 10.17 Mean error and mean squared error of the field-programmable gate array results with different matrix size N.

TABLE 10.6
Software Execution Platform

CPU	Intel *Core*(TM) i5-250M @ 2.50GHz
RAM	4GB
Compiler	Visual Studio 2010

calculation efficiency. Considering the degree of vector processing in this design is only four, the speedup would be much more significant by an exploration of higher degree of parallelism in the future work.

10.5 CONCLUSIONS

The VLSI technology plays an essential role in HPC field, and is widely applied in modern bioinformatics and biomedical engineering. There has been a transfer of research focus from conventional processors to parallel hardware architecture, to explore more efficient computation.

In this chapter, we have discussed the immense potential of VLSI architecture in bioinformatic and biomedical applications. Two grouping methods of the parallel computing problems are provided from different perspectives, together with

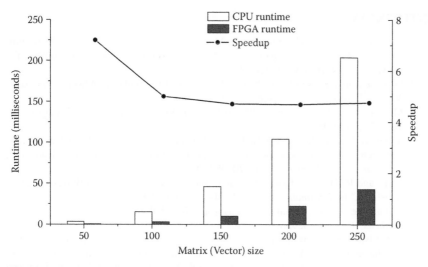

FIGURE 10.18 Execution time comparison between hardware and software platform, with different matrix size N.

several design techniques. Two high-performance VLSI architectures in applications of bioinformatics and biomedical engineering are presented. These two applications belong to data-intensive and compute-intensive problems, which are representatives in parallel architecture design. Multiple parallel levels and design techniques are explored efficiently.

The first design is a scalable FPGA-based accelerator for a GBC algorithm. By exploring hybrid parallelism, it has reached a considerable acceleration as well as low energy consumption, when compared with a software implementation on a CPU platform. The second design is a hardware architecture for the SSPPF, which is an adaptive filter to estimate neuron model coefficients in neural spiking activity. Bit-level and instruction-level parallel techniques are employed. This architecture is scalable in terms of both matrix size and degree of parallelism. Experimental results show that it is possible to outperform software implementation by making full use of the intrinsic parallel property of VLSIs.

In the future, more efficient VLSI architectures should be explored. In current GBC accelerator, the performance is restricted by the limited memory bandwidth, because data-intensive applications demand a high throughput. More powerful platforms equipped with large-capacity chip and high-bandwidth external memory would be utilized to break the bottleneck of PE number and data transfer throughput. The current SSPPF architecture only adopts a parallelism degree of four, so there is still a large space to explore for a deeper bit-level parallelism. Besides, more effective design techniques in different parallelism levels should also be considered, to reduce power consumption and enhance the computation efficiency.

REFERENCES

1. N. M. Luscombe, D. Greenbaum, and M. Gerstein, "What is bioinformatics? An introduction and overview," Department of Molecular Biophysics and Biochemistry, Yale University, New Haven, CT, Tech. Rep., 2001.
2. D. A. Benson, I. Karsch-Mizrachi, D. J. Lipman, J. Ostell, and D. L. Wheeler, "Genbank," *Nucleic Acids Research,* 36(suppl 1): D25–D30, 2008.
3. M. Burrows and D. J. Wheeler, "A block-sorting lossless data compression algorithm," Tech. Rep. 124, 1994.
4. Y. Xin, B. Liu, B. Min, W X. Y Li, R. C. C. Cheung, A. S. Fong, and T. F. Chan, "Parallel architecture for DNA sequence inexact matching with Burrows-Wheeler Transform," *Microelectronics Journal,* 44(8): 670–682, 2013.
5. H. Li and R. Durbin, "Fast and accurate short read alignment with Burrows-Wheeler transform," *Bioinformatics,* 25(14): 1754–1760, July 2009.
6. B. Langmead and S. L. Salzberg, "Fast gapped-read alignment with Bowtie 2," *Nature Methods,* 9(4): 357–359, March 2012.
7. G. Christine and K. Arthur, "Imaging calcium in neurons," *Neuron,* 73(5): 862–885, 2012.
8. C. Eliasmith and C. H. Anderson, *Neural Engineering (Computational Neuroscience Series): Computational, Representation, and Dynamics in Neurobiological Systems.* Cambridge, MA: MIT Press, 2002.
9. L. Paninski, J. Pillow, and J. Lewi, "Statistical models for neural encoding, decoding, and optimal stimulus design," in *Computational Neuroscience: Progress in Brain Research,* ed. P. Cisek, T. Drew, and J. F. Kalaska. Amsterdam, the Netherlands: Elsevier, 2006.
10. J. Enderle, J. Bronzino, and S. Blanchard, *Introduction to Biomedical Engineering,* ser. Academic Press Series in Biomedical Engineering. Burlington, MA: Elsevier Academic Press, 2005.
11. L. Hasan, Z. Al-Ars, and S. Vassiliadis, "Hardware acceleration of sequence alignment algorithms-an overview," *Design Technology of Integrated Systems in Nanoscale Era,* pp. 92–97. *DTIS International Conference on,* Rabat, Morocco, September 2007.
12. X. Gan, A. Liew, and H. Yan, "Discovering biclusters in gene expression data based on high-dimensional linear geometries," *BMC Bioinformatics,* 9(1): 1–15, 2008.
13. M. Nei, "Phylogenetic analysis in molecular evolutionary genetics," *Annual Review of Genetics,* 30(1), 371–403, 1996.
14. C. A. Stewart, R. Z. Roskies, and S. Subramaniam, "Opportunities for biomedical research and the NIH through high performance computing and data management," Indiana University, Tech. Rep., 2003.
15. National Center for Biological Information, "Genbank Statistics." Available at http://www.ncbi.nlm.nih.gov/genbank/statistics, accessed September 15, 2013.
16. "UniProtKB/TrEMBL Protein Database Release Statistics." Available at http://www.ebi.ac.uk/uniprot/TrEMBLstats, accessed September 15, 2013.
17. R. K. Weinstein and R. H. Lee, "Architectures for high-performance FPGA implementations of neural models," *Journal of Neural Engineering,* 3(1): 21, 2006.
18. E. Graas, E. Brown, and R. Lee, "An FPGA-based approach to high-speed simulation of conductance-based neuron models," *Neuroinformatics,* 2(4): 417–435, 2004.
19. M. Gschwind, H. P. Hofstee, B. Flachs, M. Hopkin, Y. Watanabe, and T. Yamazaki, "Synergistic processing in cell's multicore architecture," *Micro, IEEE,* 26(2): 10–24, 2006.
20. B. Liu, C. W. Yu, D. Z. Wang, R. C. C. Cheung, and H. Yan, "Design exploration of geometric biclustering for microarray data analysis in data mining," *Parallel and Distributed Systems, IEEE Transactions on,* PP(99): 1, 2013.
21. "San Diego Supercomputer Center." Available at http://www.sdsc.edu/, accessed September 15, 2013.

22. "Pittsburgh Supercomputing Center." Available at http://www.psc.edu/, accessed September 15, 2013.
23. "Accelerating high-performance computing with FPGAs," Altera, Tech. Rep., 10, 2007.
24. H. Esmaeilzadeh, E. Blem, R. St. Amant, K. Sankaralingam, and D. Burger, "Dark silicon and the end of multicore scaling," *Micro, IEEE*, 32(3): 122–134, 2012.
25. "High performance computing using FPGAs," Xilinx, Tech. Rep., 9, 2010.
26. B. Schmidt, *Bioinformatics: High Performance Parallel Computer Architectures*. Boca Raton: FL: CRC Press, 2010.
27. B. Mackin and N. Woods. "FPGA Acceleration in HPC: A Case Study in Financial Analytics." *XtremeData Inc. White Paper*, 2006.
28. M. Walton, O. Ahmed, G. Grewal, and S. Areibi, "An empirical investigation on system and statement level parallelism strategies for accelerating scatter search using Handel-C and Impulse-C," *VLSI Des.*, 2012: 5:5–5:5, January 2012.
29. An Xtremedata, "Scalable FPGA-Based Computing Accelerates C Applications." *The Free Library*, 2006.
30. F. I. Cheema, Zain-ul-Abdin, and B. Svensson, "A design methodology for resource to performance tradeoff adjustment in FPGAs," in *Proceedings of the 7th FPGAworld Conference*, ser. FPGAworld '10, 14–19. New York: ACM, 2010.
31. S. Scott and F. Alessandro, "Where's the beef? Why FPGAs are so fast," Microsoft Research, Tech. Rep., September 2008.
32. K. Asanovic, R. Bodik, B. C. Catanzaro, J. J. Gebis, P. Husbands, K. Keutzer, D. A. Patterson, W L. Plishker, J. Shalf, S. W. Williams, and K. A. Yelick, "The landscape of parallel computing research: A view from Berkeley", EECS Department, University of California, Berkeley, Tech. Rep., December 2006.
33. I. Gorton, P. Greenfield, A. Szalay, and R. Williams, "Data-intensive computing in the 21st century," *Computer*, 41(4): 30–32, 2008.
34. B. Furht and A. Escalante, *Handbook of Cloud Computing*. New York: Springer, 2010.
35. T. Bunker and S. Swanson, "Latency-optimized networks for clustering FPGAs," in *Field-Programmable Custom Computing Machines (FCCM)*, pp. 129–136, *2013 IEEE 21st Annual International Symposium*, Seattle, WA, 2013.
36. T. R. Golub, D. K. Slonim, P. Tamayo, C. Huard, M. Gaasenbeek, J. P. Mesirov, H. Coller et al., "Molecular classification of cancer: Class discovery and class prediction by gene expression monitoring," *Science*, 286(5439): 531–537, 1999.
37. X. Gan, A. W. C. Liew, and H. Yan, "Discovering biclusters in gene expression data based on high-dimensional linear geometries," *BMC Bioinformatics*, 9(1): 209, 2008.
38. S. Liu, Y Chen, M. Yang, and R. Ding, "Bicluster Algorithm and used in market analysis," in *Second International Workshop on Knowledge Discovery and Data Mining*, pp. 504–507, Moscow, 2009.
39. H. Zhao, A. W C. Liew, X. Xie, and H. Yan, "A new geometric biclustering algorithm based on the Hough transform for analysis of large-scale microarray data," *Journal of Theoretical Biology*, 251(3): 264–274, 2008.
40. D. Z. Wang and H. Yan, "Geometric biclustering analysis of DNA microarray data based on hypergraph partitioning," in *IDASB Workshop on BIBM2010*, pp. 246–251, Hong Kong, People's Republic of China, 2010.
41. M. Gokhale, J. Cohen, A. Yoo, W. Miller, A. Jacob, C. Ulmer, and R. Pearce, "Hardware technologies for high-performance data-intensive computing," *Computer*, 41(4): 60–68, 2008.
42. Cellminer Database. "NCI-60 Analysis Tools." Available at http://discover.nci.nih.gov /cellminer, accessed September 15, 2013.
43. Department of Information Technology and Electrical Engineering of ETH Zurich. Available at http://www.tik.ee.ethz.ch/~sop/bimax/SupplementaryMaterial/Datasets /BiologicalValidation, accessed September 15, 2013.

44. National Center for Biotechnology Information. Available at http://www.ncbi.nlm.nih.gov/guide, accessed September 15, 2013.

45. Xilinx, Inc., "Getting Started with the Xilinx Virtex-6 FPGA ML605 Evaluation Kit, UG533 (v1.5)," 2011. Available at http://www.xilinx.com/support/documentation/boards_and_kits/ug533.pdf, accessed September 15, 2013.

46. Xilinx, Inc., "XST User Guide for Virtex-6 and Spartan-6 Devices, UG687 (v12.3)," 2010. Available at http://www.xilinx.com/support/documentation/swmanuals/xilinx124/xstv6s6.pdf, accessed September 15, 2013.

47. Xilinx, Inc., "XPower Estimator User Guide," 2011. Available at http://www.xilinx.com/support/documentation/userguides/ug440.pdf, accessed September 15, 2013.

48. T. W. Berger, D. Song, R. H. M. Chan, and V. Z. Marmarelis, "The neurobiological basis of cognition: Identification by multi-input, multioutput nonlinear dynamic modeling," *Proceedings of the IEEE*, 98(3), 356–374, March 2010.

49. D. Song, R. H. M. Chan, V. Z. Marmarelis, R. E. Hampson, S. A. Deadwyler, and T. W. Berger, "Nonlinear dynamic modeling of spike train transformations for hippocampal-cortical prostheses," *IEEE Transactions on Biomedical Engineering*, 54(6): 1053–1066, June 2007.

50. T. W. Berger, A. Ahuja, S. H. Courellis, S. A. Deadwyler, G. Erinjippurath, G. A. Gerhardt, G. Gholmieh et al., "Restoring lost cognitive function," *Engineering in Medicine and Biology Magazine, IEEE*, 24(5): 30–14, September–October 2005.

51. M. C. Hsiao, C. H. Chan, V. Srinivasan, A. Ahuja, G. Erinjippurath, T. Zanos, G. Gholmieh et al., "VLSI implementation of a nonlinear neuronal model: A 'Neural Prosthesis' to restore Hippocampal Trisynaptic Dynamics," in *Engineering in Medicine and Biology Society, 2006. EMBS '06. 28th Annual International Conference of the IEEE*, pp. 4396–4399, New York, NY, August 30, 2006–September 3, 2006.

52. W. X. Y. Li, R. H. M. Chan, W Zhang, R. C. C. Cheung, D. Song, and T. W. Berger, "High-performance and scalable system architecture for the real-time estimation of generalized Laguerre-Volterra MIMO model from neural population spiking activity," *IEEE Journal on Emerging and Selected Topics in Circuits and Systems*, 1(4): 489–501, December 2011.

53. U. T. Eden, L. M. Frank, R. Barbieri, V. Solo, and E. N. Brown, "Dynamic analysis of neural encoding by point process adaptive filtering," *Neural Computation*, 16(5): 971–998, May 2004.

54. R. H. M. Chan, D. Song, and T. W. Berger, "Nonstationary modeling of neural population dynamics," in *Engineering in Medicine and Biology Society, 2009. EMBC 2009. Annual International Conference of the IEEE*, pp. 4559–4562, September 2009.

55. X. Chen and V. Akella, "Exploiting data-level parallelism for energy-efficient implementation of LDPC decoders and DCT on an FPGA," *ACM Trans. Reconfigurable Technology Systems*, 4(4): 37:1–37:17, December 2011.

56. J. Oliver and V. Akella, "Improving DSP performance with a small amount of field programmable logic," in *Field Programmable Logic and Application*, ser. Lecture Notes in Computer Science, P. Cheung and G. Constantinides, Eds. New York: Springer Berlin Heidelberg, 2003, 2778: 520–532.

57. Z. Guo, W. Najjar, F. Vahid, and K. Vissers, "A quantitative analysis of the speedup factors of FPGAs over processors," in *Proceedings of the 2004 ACM/SIGDA 12th international symposium on Field programmable gate arrays*, ser. FPGA '04. pp. 162–170, New York: ACM, 2004.

58. X. Zhu, R. Jiang, Y. Chen, S. Hu, and D. Wang, "FPGA implementation of Kalman filter for neural ensemble decoding of rat's motor cortex," *Neurocomputing*, 74(17): 2906–2913, 2011.

59. D. Song, R. H. M. Chan, V. Z. Marmarelis, R. E. Hampson, S. A. Deadwyler, and T. W. Berger, "Nonlinear modeling of neural population dynamics for hippocampal prostheses," *Neural Networks*, 22: 1340–1351, 2009.

11 Basics, Applications, and Design of Reversible Circuits

Robert Wille

CONTENTS

11.1 INTRODUCTION

Great progress has been made in the development of computing machines. Over the last 20–30 years, a design flow was developed that allows for the specification and implementation of circuits and systems that are composed of millions of components. To handle the steadily increasing complexity, several abstractions are thereby applied during the design process—represented, for example, by the specification level, the *electronic system level* (ESL), the *register transfer level* (RTL), and the gate level. Here, the design of circuits and systems based on conventional computation paradigms has been focused.

However, conventional technologies are going to face serious challenges in the near future. The ongoing miniaturization (according to *Moore's Law*), one of the main drivers of the last accomplishments in circuit and system design, eventually will come to a halt. At least when feature sizes of single transistors approach the atomic scale, no further improvement can be expected. Besides that, power dissipation represents a major obstacle. In the last decades, the amount of power dissipated in the form of heat to the surrounding environment of a chip increased by orders of magnitude. As excessive heat may decrease the reliability of a chip (or even destroys it), power dissipation is one of the major barriers to progress the development of smaller and faster computer chips. As a consequence, researchers and engineers are intensely considering alternative technologies.

Reversible computation marks a promising new computing paradigm where all operations are performed in an invertible manner. That is, in contrast to the conventional paradigm, all computations can be reverted (i.e., the inputs can be obtained from the outputs and vice versa). A simple standard operation such as the logical AND already illustrates that reversibility is not guaranteed in conventional systems. Indeed, it is possible to obtain the inputs of an AND gate if the output is set to 1 (then both inputs must be set to 1 as well). But, it is not possible to determine the input values if the AND outputs 0. In contrast, reversible computation allows bijective operations only, that is, n-input n-output functions that map each possible input vector to a unique output vector.

In this chapter, promising applications of this alternative computation paradigm are briefly reviewed. More precisely, we show how reversible circuits help in the design of quantum circuits—an emerging technology that makes it possible to solve many practically relevant problems much faster than conventional solutions. Besides that, possible benefits in the domain of low-power design are discussed, for example, a recent achievement in the design of low-power encoders where reversible circuits have successfully been applied. Afterwards, differences in the design of this kind of circuits are discussed and the current state of the art of the respective design methods is outlined. However, first, the most important basics on the reversible computation paradigm are provided.

11.2 BASICS ON REVERSIBLE CIRCUITS

Logic computations can be defined as a function over Boolean variables. More precisely: A *Boolean function* is a mapping $f : \mathbb{B}^n \to \mathbb{B}$ with $n \in \mathbb{N}$. A function f is defined over its *primary input* variables $X = \{x_1, x_2, \ldots, x_n\}$ and hence is also denoted by $f(x_1, x_2, \ldots, x_n)$. The precise mapping is described in terms of Boolean expressions that are formed over the variables from X and operations such as $\wedge$ (*AND*), $\vee$ (*OR*), or $\div$ (*NOT*).

A *multi-output Boolean function* is a mapping $f : \mathbb{B}^n \to \mathbb{B}^m$ with $n, m \in \mathbb{N}$. More precisely, it is a system of Boolean functions $f_i (x_1, x_2, \ldots, x_n)$ with $1 \leq i \leq m$. The respective functions f_i are also denoted as *primary outputs*.

Multi-output functions are also denoted as n-input, m-output functions or $n \times m$ functions, respectively. Reversible functions are a subset of multi-output functions. More precisely, a multi-output function $f : \mathbb{B}^n \to \mathbb{B}^m$ is *reversible* if:

- Its number of inputs is equal to the number of outputs (i.e., $n = m$).
- It maps each input pattern to a unique output pattern.

In other words, each reversible function is a bijection that performs a permutation of the set of input patterns. A function that is not reversible is termed *irreversible*.

TABLE 11.1 Boolean Functions

a. Irreversible (Adder)					b. Irreversible						c. Reversible					
x_1	x_2	x_3	f_1	f_2		x_1	x_2	x_3	f_1	f_2	f_3	x_1	x_2	x_3	f_1	f_2	f_3
0	0	0	0	0		0	0	0	0	0	0	0	0	0	0	0	0
0	0	1	0	1		0	0	1	0	0	0	0	0	1	0	1	0
0	1	0	0	1		0	1	0	0	1	0	0	1	0	1	0	0
0	1	1	1	0		0	1	1	0	1	1	0	1	1	1	0	1
1	0	0	0	1		1	0	0	1	0	0	1	0	0	0	0	1
1	0	1	1	0		1	0	1	1	0	1	1	0	1	0	1	1
1	1	0	1	0		1	1	0	1	1	1	1	1	0	1	1	0
1	1	1	1	1		1	1	1	1	1	0	1	1	1	1	1	1

Table 11.1a shows the truth table of a three-input, two-output function representing a 1-bit adder. This function is irreversible, as $n \neq m$. Also, the function in Table 11.1b is irreversible. Here, the number n of inputs indeed is equal to the number m of outputs, but there is no unique input–output mapping (e.g., both inputs 000 and 001 map to the output 000). In contrast, the 3×3 function shown in Table 11.1c is reversible, as each input pattern maps to a unique output pattern.

Reversible functions are realized by reversible circuits. A *reversible circuit G* is a cascade of reversible gates, where fanout and feedback are not directly allowed [1]. A reversible gate has the form g $(C;T)$, where $C = \{x_{i1},\ldots,x_{ik}\} \subset X$ is the set of *control lines* and $T = \{x_{j1},\ldots,x_{jl}\} \subset X$ with $C \cap T =$; is the set of target lines. C may be empty. Control lines and unconnected lines always pass through the gate unaltered. In contrast, the gate operation is applied on the target iff all control lines are assigned to 1. The most frequently used reversible gate is the *Toffoli gate* [2], which inverts the value of the (single) target line in this case.

Figure 11.1a shows a Toffoli gate drawn in standard notation, that is, control lines are denoted by •, whereas the target line is denoted by ⊕. A circuit composed of several Toffoli gates is depicted in Figure 11.1b. This circuit maps, for example, the input 111 to the output 110 and vice versa.

11.3 QUANTUM COMPUTATION

11.3.1 GENERAL BACKGROUND

Quantum computation [1] is a promising application of reversible logic. Here, information is represented in terms of *qubits* instead of bits. In contrast to Boolean logic, qubits do not only allow the representation of Boolean 0s and Boolean 1s but also

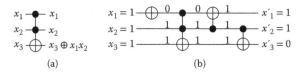

FIGURE 11.1 (a) Toffoli gate and (b) Toffoli circuit.

the superposition of both. In other words, a qubit is a two-level quantum system, described by a two-dimensional complex Hilbert space. The two orthogonal quantum states

$$| 0 \rangle \equiv \begin{pmatrix} 1 \\ 0 \end{pmatrix} \text{ and } | 1 \rangle \equiv \begin{pmatrix} 0 \\ 1 \end{pmatrix} \tag{11.1}$$

are used to represent the Boolean values 0 and 1. Any state of a qubit may be written as $|\Psi\rangle = \alpha\, |0\rangle + \beta\, |1\rangle$, where α and β are complex numbers with $|\alpha|^2 + |\beta|^2 = 1$. The quantum state of a single qubit is denoted by the vector

$$\begin{pmatrix} a \\ \beta \end{pmatrix} \tag{11.2}$$

The state of a quantum system with $n > 1$ qubits is given by an element of the tensor product of the respective state spaces and can be represented as a normalized vector of length 2^n, called the state vector. The state vector is changed through multiplication of appropriate $2^n \times 2^n$ unitary matrices. Thus, each quantum computation is inherently reversible, but manipulates qubits rather than pure logic values.

Using quantum computation and qubits in superposition, functions can be evaluated with different possible input assignments in parallel. Unfortunately, it is not possible to obtain the current state of a qubit. Instead, if a qubit is measured, either 0 or 1 is returned depending on a respective probability, that is, 0 is returned 0 with probability of $|\alpha|^2$ and 1 is returned with probability of $|\beta|^2$. After the measurement, the state of the qubit is destroyed.

Nevertheless, using these quantum mechanical phenomena, quantum computation allows for breaching complexity bounds which are valid for computing devices based on conventional mechanics. The Grover search [3] and the factorization algorithm by Shor [4] rank among the most famous examples for quantum algorithms that solve problems in time complexities, which cannot be achieved using conventional computing. The first algorithm addresses thereby the search of an item in an unsorted database with k items in time $O(\sqrt{k})$, whereas conventional methods cannot be performed using less than linear time. Shor's algorithm performs prime factorization in polynomial time, that is, the algorithm is exponentially faster than its best known conventional counterpart. First physical realizations of quantum circuits have been presented in the work of Vandersypen et al. [5].

11.3.2 Mapping Reversible Circuits to Quantum Circuits

Reversible circuits are of interest in the domain of quantum computation as all quantum operations inherently are reversible. As most of the known quantum algorithms include a large Boolean component (e.g., the database in Grover's search algorithm and the modulo exponentiation in Shor's algorithm), the design of these components is often conducted by a two-stage approach: (1) realizing the desired functionality as a reversible circuit, and (2) mapping the resulting circuit to a functionally equivalent quantum circuit.

Different libraries of quantum gates have thereby been introduced for this purpose. In the literature, the *NCV gate library* [6] is frequently applied. This library is composed of the following universal set of quantum gates:

- *NOT* gate $T(\emptyset, t)$: A single qubit t is inverted which is described by the unitary matrix $\begin{pmatrix} 0 & 1 \\ 1 & 0 \end{pmatrix}$.
- *ControlledNOT* (CNOT) gate $T(\{c\},t)$: The target qubit t is inverted if the control qubit c is 1.
- *ControlledV* gate $V(\{c\},t)$: The operation described by the unitary matrix

$$V = \frac{1+i}{2}\begin{pmatrix} 1 & -i \\ -i & 1 \end{pmatrix}$$ is performed on the target qubit t if the control qubit c is 1.

- *ControlledV⁺* gate $V^+(\{c\},t)$: The operation described by the unitary matrix

$$V^+ = \frac{1-i}{2}\begin{pmatrix} 1 & i \\ i & 1 \end{pmatrix}$$ is performed on the target qubit t if the control qubit c is 1.

The V^+ gate performs the inverse operation of the V gate as $V^+ = V^{-1}$.

The V and V^+ gates often are referred to as *controlled square-root-of-NOT* gates as two adjacent identical V, or V^+, gates are equivalent to a CNOT gate.

If circuits with Boolean inputs use NCV gates only, the value of each qubit at each stage of the circuit is restricted to one of $\{0,v_0,1,v_1\}$ where $v_0 = \frac{1+i}{2}\begin{pmatrix} 1 \\ -i \end{pmatrix}$ and $v_1 = \frac{1+i}{2}\begin{pmatrix} -i \\ 1 \end{pmatrix}$. The *NOT, V,* and V^+ operations over these four-valued logic are given by:

x	$NOT(x)$	$V(x)$	$V^+(x)$
0	1	v_0	v_1
v_0	v_1	1	0
1	0	v_1	v_0
v_1	v_0	0	1

As shown, *NOT* is a complement operation, V is a cycle, and V^+ is the inverse cycle.

An alternative library has recently been introduced by Sasanian et al. [7]. Here, four-valued quantum gates are considered and summarized to the *NCV-4 gate library*. The NCV-4 gate library is composed of (1) the three unitary gates (i.e., gates without a control line) performing the NOT-, V-, and V⁺-operation as well as (2) respective controlled versions of these gates. In contrast to the NCV-library,

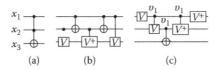

FIGURE 11.2 Mapping a Toffoli gate to respective quantum gates. (a) A Toffoli gate T ($\{x_1, x_2\}, x_3$), (b) to a functionally equivalent quantum circuits composed of NCV gates, and (c) NCV-4 gates.

the controlled gates perform the respective operation not in case the control line is 1, but in case the control line is set to the highest logic value (v_1 in this case).

Quantum circuits composed of gates from both libraries can easily be derived from reversible circuits. For this purpose, each reversible gate simply has to be mapped to a corresponding cascade of quantum gates. As an example, Figure 11.2 shows the mapping of (a) a Toffoli gate T ($\{x_1, x_2\}, x_3$), (b) to a functionally equivalent quantum circuits composed of NCV gates, and (c) NCV-4 gates, respectively. Improved mappings for other configurations of reversible gates or entire reversible circuits are available in the works of Barenco et al. [6], Miller et al. [8], and Soeken et al. [9].

11.4 LOW-POWER DESIGN

11.4.1 GENERAL BACKGROUND

Pioneering work by Landauer [10] showed that, regardless of the underlying technology, losing information during computation causes power dissipation. More precisely, for each *lost* bit of information, at least $k \cdot T \cdot \log$ (2) joules are dissipated (where k is the Boltzmann constant and T is the temperature). Since today's computing devices are usually built of elementary *gates* such as AND, OR, NAND, and so on, they are subject to this principle and, hence, dissipate this amount of power in each computational step.

Although the theoretical lower bound on power dissipation still does not constitute a significant fraction of the power consumption of current devices, it nonetheless poses an obstacle for the future. Figure 11.3 illustrates the development of the power consumption of an elementary computational step in recent and expected future CMOS generations (based on values from the work of Zeitzoff and Chung [11]). The figure shows that today's technology is still a factor of 1000 away from the Landauer limit and that the expected CMOS development will reduce this to a factor of 100 within the next 10 years. However, a simple extrapolation also shows that the trend cannot continue with the current family of static CMOS gates as no amount of technological refinement can overcome the Landauer barrier. Moreover, the Landauer limit is only a *lower* bound on the dissipation. Gershenfeld has shown that the actual power dissipation corresponds to the amount of power used to represent a signal [12], that is, Landauer's barrier is closer than immediately implied by the extrapolation from Figure 11.3.

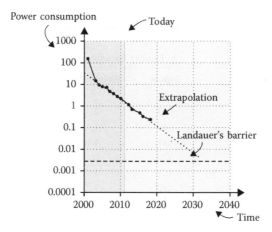

FIGURE 11.3   Power consumption Q in different CMOS generations.

As reversible circuits bijectively transforms data at each computation step, the above-mentioned information loss and its resulting power dissipation does not occur. Because of this, reversible circuits manifest themselves as the only way to break this fundamental limit. In fact, it has been proven that to enable computations with no power consumption at all, the underlying realization must follow reversible computation principles [13]. These fundamental results together with recent experimental evaluations [14] motivate researchers in investigating this direction further.

11.4.2   REVERSIBLE CIRCUITS IN THE DESIGN OF LOW-POWER ENCODERS

A precise application of reversible circuits within the domain of low-power design can be found in the design of encoders for on-chip interconnects. With the rise of very deep submicron and nanometric technologies, these interconnects are increasingly affecting the overall energy consumption in today's chips [15]. An established method to address this includes the application of coding strategies to modify the communication of the interconnects and, thus, to reduce the energy consumption [16,17].

The idea behind this is briefly illustrated by means of the *probability based mapping* (pbm) [16]. Here, it is assumed that the power consumption of the interconnect is related to the number of 1s transmitted through it. Hence, the pbm scheme tries to minimize the transmission of patterns with a large Hamming weight: Instead of transmitting the original data, each pattern is encoded depending on its probability.

As an example, Table 11.2a shows a set of data inputs with their corresponding probability of occurrence. Based on that, an encoder should map the most frequently occurring data input (i.e., 100) to a bit string with the lowest Hamming weight (i.e., 000). Then, the second-most frequently occurring data inputs should be mapped to a bit string with the second-lowest Hamming weight and so on. That is, a coding is desired, which leads to patterns with Hamming weights as shown in Table 11.2b. A precise coding satisfying this property is given in Table 11.2c.

TABLE 11.2   Illustrating the Objective of an Encoder

a. Pattern Probability		b. Desired Encoding		c. Possible Encoding	
Inputs	Probability (%)	Inputs	Weight (H)	Inputs	Encoding
000	8	000	2	000	101
001	8	001	2	001	011
010	10	010	1	010	010
011	10	011	1	011	001
100	40	100	0	100	000
101	10	101	1	101	100
110	8	110	2	110	110
111	6	111	3	111	111

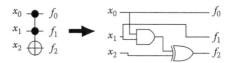

FIGURE 11.4   Mapping Toffoli gate to conventional circuit.

Unfortunately, determining the best possible or even just one *good* coding is a nontrivial task. Already for 3-bit data inputs as considered in Table 11.2a, eight different codings are possible. In the general case with m-bit data inputs, this number significantly increases to $\prod_{i=0}^{m}\binom{m}{i}$.

However, as encoders realize reversible one-to-one mappings, the application of synthesis approaches for reversible logic is a reasonable choice. In fact, the underlying reversibility of existing design methods for reversible circuits (see Section 11.5) enables to determine the desired codings in a more elegant and powerful way than their conventional counterparts. As the resulting reversible circuits represent a logic description, they can also easily be mapped to a conventional gate library. As an example, Toffoli gates with two control lines can be mapped to a netlist composed of one AND-gate and one XOR-gate as shown in Figure 11.4. From such a netlist, the established optimization and technology mapping steps can be performed.

11.5   DESIGN OF REVERSIBLE CIRCUITS

The basics and the applications for reversible circuits as introduced in Sections 11.2, 11.3, and 11.4 already represent the main models and motivation for research in the domain of reversible circuit design. Nevertheless, reversible circuits have not been intensively studied by researchers before the year 2000. The main reason for that may lie in the fact that applications of reversible logic have been seen as "dreams of the future." But, this changed as first physical realizations of reversible circuits emerged, that is, in terms of quantum circuits [5] or in terms of reversible CMOS realizations with certain low-power properties [18]. Therewith, proofs of concept were available motivating reversible circuits as a promising research area.

As a consequence, in the last years, researchers started to develop new methods for the design of this kind of circuits. In the following, the current state of the art is briefly reviewed.

11.5.1 Considering the Number of Circuit Lines

To synthesize compact reversible circuits, the number of used circuit lines is crucial—in particular for applications in the domain of quantum computation (see Section 11.3). Here, each circuit line corresponds to a qubit—a limited resource. Besides that, a large number of lines may decrease the reliability of the resulting system. For this, the number of circuit lines (or qubits) has to be kept as small as possible. However, often additional circuit lines are inevitable.

The following observations are thereby applied: If the function $f : B^n \rightarrow B^n$ to be synthesized is reversible, then obviously only n circuit lines are needed. In contrast, if irreversible functions are synthesized, additional lines might be required.

As an example, consider again the adder function from Table 11.1a. The adder obviously is irreversible, because (1) the number of inputs differs from the number of outputs, and (2) there is no unique input–output mapping. Even adding an additional output to the function (leading to the same number of inputs and outputs) would not make the function reversible. Then, without loss of generality, the first four lines of the truth table can be embedded with respect to reversibility as shown in the right-most column of Table 11.3a. However, as $f_1 = 0$ and $f_2 = 1$ already appeared two times (marked bold), no unique embedding for the fifth line is possible any longer. The same also holds for the lines shown in italic. Hence, additional outputs are needed. This may lead to more circuit lines.

In general, at least $\lceil \log_2(\mu) \rceil$ additional outputs (also called *garbage outputs*) are required to make an irreversible function reversible [19], whereby μ is the maximal number of times an output pattern is repeated in the truth table. Thus, to realize an

---

TABLE 11.3 Embeddings of Boolean Functions

a. Incomplete Embedding							b. Complete Embedding							
x_1	x_2	x_3	f_1	f_2			**0**	x_1	x_2	x_3	f_1	f_2	–	–
0	0	0	0	0	0		0	**0**	**0**	**0**	**0**	**0**	0	0
0	0	1	**0**	**1**	0		0	**0**	**0**	1	**0**	1	1	1
0	1	0	**0**	**1**	1		0	**0**	1	**0**	**0**	1	1	0
0	1	1	*1*	*0*	*0*		0	**0**	1	1	1	**0**	0	1
1	0	0	0	1	?		0	1	**0**	**0**	**0**	1	0	0
1	0	1	*1*	*0*	*1*		0	1	**0**	1	1	**0**	1	1
1	1	0	*1*	*0*	*?*		0	1	1	**0**	1	**0**	1	0
1	1	1	1	1	1		0	1	1	1	1	1	0	1
							1	0	0	0	1	0	0	0

irreversible function $f : B^n \rightarrow B^m$, at least $m + \lceil \log_2(\mu) \rceil$ circuit lines are required. The resulting circuit then would have n primary inputs, $m + \lceil \log_2(\mu) \rceil - n$ constant inputs, m primary outputs, and $\lceil \log_2(\mu) \rceil$ garbage outputs. The values of the constant inputs and the garbage outputs can thereby be chosen arbitrarily as long as they ensure a unique input/output mapping. Note that if the number of primary inputs n is larger than $m + \lceil \log_2(\mu) \rceil$, of course at least n circuit lines are needed to realize the function.

In case of the adder function, $\lceil \log_2(3) \rceil = 2$ additional outputs are required, because at most three output patterns are repeated, that is, $\mu = 3$. Thus, besides the three primary inputs and two primary outputs, a reversible realization of an adder must consist of at least one constant input and two garbage outputs. A possible assignment to the newly added values is depicted in Table 11.3b (the original adder function is highlighted in bold).

For small functions to be synthesized (i.e., for functions given in terms of a truth table), keeping the number of circuit lines small is easily possible [19]. In fact, most of the respective approaches [20–22] generate circuits with a minimal number of circuit lines. However, for larger functions (i.e., for functions that cannot be represented in terms of a truth table any longer), keeping the number of circuit lines minimal is significantly harder. In fact, most of the existing approaches [23–25] often lead to circuits that are way beyond the optimum of the number of lines [26].

11.5.2 SYNTHESIS

Synthesis is one of the most important steps while building complex circuits. Considering the conventional design flow, synthesis is carried out in several individual steps such as high-level synthesis, logic synthesis, mapping, and routing. To synthesize reversible logic, adjustments and extensions are needed. For example, throughout the whole flow, the restrictions caused by the reversibility (no fanout and feedback) and a completely new gate library must be considered.

Existing synthesis approaches addressing these issues can be categorized as follows:

- Boolean synthesis approaches for small functions: These approaches can handle small Boolean functions, for example, provided in terms of permutations [20,27], truth tables [21,28,29], positive-polarity Reed-Muller expansion [22], or Reed-Muller spectra [30]. Besides, exact approaches exist that do not only generate a reversible circuit for a given function, but in addition ensure that the resulting circuit is composed of a minimal number of gates [31].
- The scalability of all these approaches is thereby limited. Usually, only circuits for functions containing not more than 30 variables can be obtained with them. Hence, if larger functions should be synthesized, more compact function descriptions and, accordingly, other synthesis approaches have to be considered.

- Boolean synthesis approaches for larger functions: Synthesis methods for larger functions make use of more compact function representations. In particular, approaches based on exclusive sum of products [23] and approaches based on binary decision diagrams [24] fall into this category. Here, the structure of the function representation is mapped to respective reversible subcircuits. By cascading the resulting subcircuits, the overall function is realized.
- Synthesis approach based on a programming language: Although the previous approaches rely on Boolean descriptions for the synthesis of reversible circuits, recently a programming language has been proposed for this purpose [25]. This language, called *SyReC*, enables the design of more complex reversible systems.

11.5.3 OPTIMIZATION

After synthesis, the resulting circuits often are of high cost. In particular, dedicated technology-specific constraints are not considered by synthesis approaches. To address this, optimization methods have been introduced. In particular, the reduction of the quantum cost of given circuits has been considered [28,32–34]. Also, reducing the number of circuit lines [35] or optimization of further, more technology-dependent cost metrics [36] was the subject of research activities.

11.5.4 VERIFICATION AND DEBUGGING

To ensure that the respective results (e.g., obtained by optimization) still represent the desired functionality, verification is applied. For this purpose, first verification approaches have been introduced [37–40]. Moreover, even automatic debugging methods aimed at supporting the designer to detect the error in case of a failed verification are already available [41].

11.6 CONCLUSIONS

In this chapter, the basics, promising applications, and the current state of the art of the design of reversible circuits have been briefly reviewed. Although circuits based on the conventional computation paradigm will reach fundamental limits in the future, reversible circuits find useful applications in emerging technologies such as quantum computing, and have benefits to be exploited in the domain of low-power design.

However, design of reversible circuits is just at the beginning. Considering the existing design methods, many of the proposed approaches have limitations with respect to the size of the function to be synthesized or the circuit to be considered. Furthermore, circuits obtained by current synthesis approaches often are of significant costs. Besides that, no integrated design flow for complex reversible circuits exists so far. Hence, further research in this promising area is required.

REFERENCES

1. M. Nielsen and I. Chuang. *Quantum Computation and Quantum Information.* Cambridge University Press: NYC, 2000.

2. T. Toffoli. Reversible computing. In *Automata, Languages and Programming*, W. de Bakker and J. van Leeuwen (Eds.), p. 632. Technical Memo MIT/LCS/TM-151. Springer: Berlin, Heidelberg, 1980.

3. L. K. Grover. A fast quantum mechanical algorithm for database search. In *Proceedings of Theory of Computing*, pp. 212–219. Philadelphia, PA, 1996.

4. P. W. Shor. Algorithms for quantum computation: Discrete logarithms and factoring. In *Proceedings of Foundations of Computer Science*, pp. 124–134. Santa Fe, NM, 1994.

5. L. M. K. Vandersypen, M. Steffen, G. Breyta, C. S. Yannoni, M. H. Sherwood, and I. L. Chuang. Experimental realization of Shor's quantum factoring algorithm using nuclear magnetic resonance. *Nature*, 414: 883–887, 2001.

6. A. Barenco, C. H. Bennett, R. Cleve, D. P. DiVinchenzo, N. Margolus, P. Shor, T. Sleator, J. A. Smolin, and H. Weinfurter. Elementary gates for quantum computation. *The American Physical Society*, 52: 3457–3467, 1995.

7. Z. Sasanian, R. Wille, D. M. Miller, and R. Drechsler. Realizing reversible circuits using a new class of quantum gates. In *Design Automation Conference*, pages 36–41. San Francisco, CA, 2012.

8. D. M. Miller, R. Wille, and Z. Sasanian. Elementary quantum gate realizations for multiple-control Toffoli gates. In *International Symposium on Multi-Valued Logic*, pp. 288–293. Tuusula, Finland, 2011.

9. M. Soeken, Z. Sasanian, R. Wille, D. M. Miller, and R. Drechsler. Optimizing the mapping of reversible circuits to four-valued quantum gate circuits. In *International Symposium on Multi-Valued Logic*, pp. 173–178. Victoria, BC, 2012.

10. R. Landauer. Irreversibility and heat generation in the computing process. *IBM Journal of Research and Development*, 5: 261–269, 1961.

11. P. Zeitzoff and J. Chung. A perspective from the 2003 ITRS. *IEEE Circuits and Systems Magazine*, 21: 4–15, 2005.

12. N. Gershenfeld. Signal entropy and the thermodynamics of computation. *IBM Systems Journal*, 35(3–4): 577–586, 1996.

13. C. H. Bennett. Logical reversibility of computation. *IBM Journal of Research and Development*, 17(6): 525–532, 1973.

14. A. Berut, A. Arakelyan, A. Petrosyan, S. Ciliberto, R. Dillenschneider, and E. Lutz. Experimental verification of Landauer's principle linking information and thermodynamics. *Nature*, 483: 187–189, 2012.

15. S. Pasricha and N. Dutt. *On-Chip Communication Architectures: Systems on Chip.* Morgan Kaufman, Burlington, MA, 2008.

16. S. Ramprasad, N. R. Shanbhag, and I. N. Hajj. A coding framework for low-power address and data busses. *IEEE Transactions on Very Large Scale Integration (VLSI) Systems*, 7(2): 212–220, 1999.

17. L. Benini, A. Macii, E. Macii, M. Poncino, and R. Scarsi. Architectures and synthesis algorithms for power-efficient bus interfaces. *IEEE Transactions on Computer-Aided Design of Integrated Circuits and Systems*, 19: 969–980, 2000.

18. B. Desoete and A. De Vos. A reversible carry-look-ahead adder using control gates. *Integration, the VLSI Journal*, 33(1–2): 89–104, 2002.

19. D. Maslov and G. W. Dueck. Reversible cascades with minimal garbage. *IEEE Transactions on Computer-Aided Design*, 23(11): 1497–1509, 2004.

20. V. V. Shende, A. K. Prasad, I. L. Markov, and J. P. Hayes. Synthesis of reversible logic circuits. *IEEE Transactions on Computer-Aided Design*, 22(6): 710–722, 2003.

21. D. M. Miller, D. Maslov, and G. W. Dueck. A transformation based algorithm for reversible logic synthesis. In *Design Automation Conference*, pp. 318–323. Anaheim, CA, 2003.
22. P. Gupta, A. Agrawal, and N. K. Jha. An algorithm for synthesis of reversible logic circuits. *IEEE Transactions on Computer-Aided Design*, 25(11): 2317–2330, 2006.
23. K. Fazel, M. A. Thornton, and J. E. Rice. ESOP-based Toffoli gate cascade generation. In *IEEE Pacific Rim Conference on Communications, Computers and Signal Processing (PacRim 2007)*, pp. 206–209. Victoria, BC, 2007.
24. R. Wille and R. Drechsler. BDD-based synthesis of reversible logic for large functions. In *Design Automation Conference*, pp. 270–275. San Francisco, CA, 2009.
25. R. Wille, S. Offermann, and R. Drechsler. SyReC: A programming language for synthesis of reversible circuits. In *Forum on Specification and Design Languages*, pp. 184–189. Southampton, UK, 2010.
26. R. Wille, O. Keszöcze, and R. Drechsler. Determining the minimal number of lines for large reversible circuits. In *Design, Automation and Test in Europe*, pp. 1204–1207. Grenoble, France, 2011.
27. M. Saeedi, M. S. Zamani, M. Sedighi, and Z. Sasanian. Synthesis of reversible circuit using cycle-based approach. *ACM Journal of Emerging Technologies Computing Systems*, 6(4), 2010.
28. D. Maslov, G. W. Dueck, and D. M. Miller. Toffoli network synthesis with templates. *IEEE Transactions on Computer-Aided Design*, 24(6): 807–817, 2005.
29. R. Wille, D. Grosse, G. W. Dueck, and R. Drechsler. Reversible logic synthesis with output permutation. In *VLSI Design*, pp. 189–194. New Delhi, India, 2009.
30. D. Maslov, G. W. Dueck, and D. M. Miller. Techniques for the synthesis of reversible Toffoli networks. *ACM Transactions on Design Automation of Electronic Systems*, 12(4), 2007.
31. D. Grosse, R. Wille, G. W. Dueck, and R. Drechsler. Exact multiple control Toffoli network synthesis with SAT techniques. *IEEE Transactions on Computer-Aided Design*, 28(5): 703–715, 2009.
32. K. Iwama, Y. Kambayashi, and S. Yamashita. Transformation rules for designing CNOT-based quantum circuits. In *Design Automation Conference*, pp. 419–424. New Orleans, LA, 2002.
33. J. Zhong and J. C. Muzio. Using crosspoint faults in simplifying Toffoli networks. In *IEEE North-East Workshop on Circuits and Systems*, pp. 129–132. Gatineau, Quebec, Canada, 2006.
34. D. M. Miller, R. Wille, and R. Drechsler. Reducing reversible circuit cost by adding lines. In *International Symposium on Multi-Valued Logic*, pp. 217–222. Barcelona, Spain, 2010.
35. R. Wille, M. Soeken, and R. Drechsler. Reducing the number of lines in reversible circuits. In *Design Automation Conference*, pp. 647–652. Anaheim, CA, 2010.
36. M. Saeedi, R. Wille, and R. Drechsler. Synthesis of quantum circuits for linear nearest neighbor architectures. *Quantum Information Processing*, 10(3): 355–377, 2011.
37. G. F. Viamontes, I. L. Markov, and J. P. Hayes. Checking equivalence of quantum circuits and states. In *International Conference on Computer-Aided Design*, pp. 69–74. San Jose, CA, 2007.
38. S. Gay, R. Nagarajan, and N. Papanikolaou. QMC: A model checker for quantum systems. In *Computer Aided Verification*, pp. 543–547. Springer, Berlin, Germany, 2008.
39. S.-A. Wang, C.-Y. Lu, I-M. Tsai, and S.-Y. Kuo. An XQDD-based verification method for quantum circuits. *IEICE Transactions*, 91-A (2): 584–594, 2008.
40. R. Wille, D. Grosse, D. M. Miller, and R. Drechsler. Equivalence checking of reversible circuits. In *International Symposium on Multi-Valued Logic*, pp. 324–330. Naha, Okinawa, Japan, 2009.
41. R. Wille, D. Grosse, S. Frehse, G. W. Dueck, and R. Drechsler. Debugging of Toffoli networks. In *Design, Automation and Test in Europe*, pp. 1284–1289. Nice, France, 2009.

12 Three-Dimensional Spintronics

Dorothée Petit, Rhodri Mansell, Amalio Fernández-Pacheco, JiHyun Lee, and Russell P. Cowburn

CONTENTS

12.1 INTRODUCTION

Spintronics is an emerging technology in which the spin of the electron is used to form a new generation of radiation-hard, nonvolatile, low-energy devices. To date, most interest has focused on memory applications such as hard drive read heads or on future replacements for dynamic random-access memory and battery-backed static random-access memory. In this chapter, we discuss the possibility of using spintronics as a route to three-dimensional integrated circuits. Spintronics is an interesting candidate for three-dimensional architectures because high levels of functionality can be achieved from simple, thin magnetic films without the need for any top contacts or gate electrodes. Furthermore, thin magnetic layers can be easily coupled to each other simply by inserting a carefully selected nonmagnetic material between them. Finally, spintronic devices can usually be designed to minimize thermal dissipation from the magnetic element itself, meaning that power dissipation from the center of a three-dimensional ensemble of elements would not be excessive. These features open up the possibility of stacking highly functional but structurally simple units on top of each other, allowing data to be held, moved, and processed in a vertical direction as well as in the conventional lateral directions.

In this chapter, we present a numerical study of such a three-dimensional spintronic device. In this system, binary data is coded into chiral phase shifts in the antiparallel (AP) arrangement of antiferromagnetically coupled multilayers (MLs) with well-defined anisotropy. These are localized, stable, and can be controllably and synchronously propagated in the vertical direction of the ML by using an external rotating magnetic field.

Spintronics is a new technology in which the spin of the electron is used to create devices.[1] Because the electron spin is most commonly associated with ferromagnetic materials, spintronic devices are usually of a hybrid construction in which ferromagnetic materials such as nickel, iron, and cobalt are integrated into semiconductors. Because ferromagnetism persists even in the absence of any power supply, spintronic devices are usually nonvolatile and as such make excellent memories. Because many of the physical effects that are exploited in spintronics are interfacial effects, magnetic materials are usually found in spintronic devices in thin or ultrathin form usually between 0.6 and 10 nm thick. The nonlinearity of the magnetic hysteresis loops of thin film magnetic materials leads to a wide range of interesting functionalities that stretch beyond simple binary memories, including logic functions[2–4] and microwave sources.[5] Spintronics is a particularly interesting candidate when it comes to thinking about three-dimensional devices. To date, most three-dimensional complementary metal–oxide–semiconductor (CMOS) devices have been implemented at a packaging level by stacking multiple dies on top of each other. Although the metal interconnect network on a very large-scale integrated device is intrinsically three-dimensional, most CMOS still retains only a single layer of active transistors at the bottom of the devices. Because high levels of functionality can be achieved in a single magnetic layer only 0.6 nm thick, one can begin to imagine three-dimensional devices in which many active layers, each less than a nanometer thick, are fabricated on top of each other using standard ML growth methods. One of the challenges in achieving an enormously high density three-dimensional spintronic device is finding a scheme for coding and transporting digital information between layers. Physical mechanisms such as dipolar interactions[6,7] and Ruderman–Kittel–Kasuya–Yosida (RKKY) coupling[8] allow neighboring layers of magnetic material to couple to each other from an energetic point of view. The challenge is finding ways of converting that simple physical coupling into a functional information carrying channel. This relates closely to the physics of complex magnetic structures and noncollinear spin textures, which can exist in coupled magnetic system. Understanding these is the route to finding ways of using nearest neighbor interactions to represent and move information.

More fundamentally, complex magnetic structures are the subject of intense research in a variety of fields of condensed matter, from designing magnetically controlled ferroelectric materials[9] to chiral magnetic kink soliton lattices[10] and skyrmionic matter,[11–16] to cite only a few examples. In these systems, the antisymmetric Dzyaloshinsky–Moriya[17,18] exchange interaction is a key ingredient to the existence of a noncollinear spin state. Nevertheless, chiral noncollinear structures can arise in systems with symmetric interactions. Magnetic bubbles in thin films[19] or disks[20] with perpendicular anisotropy are such an example. Frustration induced by the competition between different antiferromagnetic (AF) interactions can also lead to the formation of helical magnetic structures.[21–23]

In this chapter, we show that AF-coupled magnetic MLs offer the possibility to engineer chiral magnetic textures. A very interesting theoretical[24–27] and experimental[28,29] body of work showed that a chiral antiphase domain wall can be induced in MLs with an even number of layers by means of surface spin-flop (SSF). However, this chiral texture was found experimentally to disappear in the absence of an applied magnetic field.[30,31] Nevertheless, this is an excellent starting point for the engineering of a controllable spin texture that runs in the vertical direction.[32] Here we will show numerically that by tuning the ratio of the uniaxial anisotropy to the interlayer coupling in such MLs, it is possible to induce antiphase domain walls—or topological magnetic kink solitons—which are stable at room temperature.[33,34] Furthermore, we show that they are mobile and that it is possible to control their propagation using an external magnetic field. Finally, we show how they interact with each other and we discuss their experimental realization. These are the main requirements of an information vector for a three-dimensional spintronic system and so could lead to a dramatic shift in data storage technologies and digital logic devices.

12.2 TOPOLOGICAL KINK SOLITONS IN MAGNETIC MULTILAYERS

Figure 12.1a shows a schematic of a ML made of in-plane magnetic material (light gray) AF coupled through a nonmagnetic spacer layer (dark gray). The strength of the coupling is defined by a coupling field $J^* = J_{AF}/M_S t_M$, where M_S is the saturation magnetization, t_M is the thickness of the magnetic layer considered, and J_{AF} is the interaction energy. The layers are assumed to have a uniaxial anisotropy K_u (anisotropy field $H_u = 2K_u/M_S$) in the direction indicated by the double arrow. The minimum energy configuration of such a system has adjacent layers AP and pointing along the easy anisotropy axis. There are two equivalent ways of realizing this ground state, as shown in Figure 12.1a, the white and gray arrows show the direction of the magnetization in each layer. We characterize these configurations by calculating the quantity $\Phi_n = (-1)^n \cos(\theta_n)$, where θ_n is the angle between the magnetization in the nth layer and the x axis. Φ_n describes the AF phase of the nth layer, where all the layers in one of the two ground states take either the value $+1$ or -1. When these two ground states, or domains, meet, as shown in Figure 12.1b, the magnetization in the two layers on each side of the meeting point (indicated by *) points in the same direction, causing frustration as the AF coupling is not satisfied.

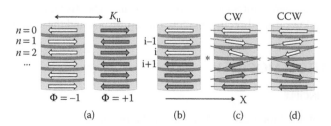

FIGURE 12.1 (a) Schematics of the two antiparallel configurations. (b, c) Meeting of two domains with opposite Φ. (b) $H_u > J^*$, the frustration is sharp and achiral. (c, d) $H_u < J^*$, the frustration is chiral and both layers forming the frustration splay in one of two directions.

If J^* is weak compared to H_u, the magnetization in both frustrated layers will point in the easy axis direction despite the AF coupling and the transition area will be sharp (Figure 12.1b); if H_u is weak compared to J^*, then both layers will lower the interaction energy at the expense of the anisotropy and splay away from each other, with the transition area extending to the adjacent layers as the coupling increases. This splaying can happen either clockwise (when going through the transition area upwards [Figure 12.1c]), or counterclockwise (Figure 12.1d), defining the chirality of the transition region.

We have performed zero-temperature macrospin simulations using a steepest descent minimization of the energy of the system. In this model, the magnetization is assumed to be uniform across each individual layer i and the total reduced energy density $e = E/M_SVJ^*$ is:

$$e = \sum_{i=1}^{N-1} \cos(\theta_{i+1} - \theta_i) - \sum_{i=1}^{N} \left[\frac{h_u}{2} \cos^2(\theta_i) + h\cos(\theta_i - \theta_H) \right] \quad (12.1)$$

where θ_H is the angle between the applied magnetic field and $+ x$, V is the volume of a magnetic layer, $h = H/J^*$, and $h_u = H_u/J^*$. Note that e depends only on reduced field parameters.

Figure 12.2a shows the calculated spin configuration across a transition in a stack with $h_u = 0.5$. The magnetization directions are represented as if viewed from the top of the stack, that is, the angles represent the actual deviation of the magnetization in the (x, y) plane. Figure 12.2b shows the evolution of Φ_n across a transition as a function of layer number n, for $h_u = 2$ and 0.5. For $h_u = 2$ the transition is sharp, that is, Φ goes directly from -1 to $+1$ between two layers; for $h_u = 0.5$ the transition is gradual. Figure 12.2c shows the width of the transition, defined as the number of layers it takes to go from 10% to 90% of the Φ transition, as a function of h_u. For high $h_u \geq 1.3 \pm 0.1$, the transition is sharp and the soliton is achiral. Then, following a sharp rise around $h_u \sim 1$ where the transition acquires a chirality, the width monotonously increases as h_u decreases. Figure 12.2d shows the same data as a function of $j^* = 1/h_u$.

There is an analogy between these systems and domain walls in soft ferromagnetic nanowires. Both are topologically locked, that is, their existence is a consequence of the conditions at the domains boundaries (topological kink solitons). Furthermore, the size of domain walls also depends on the balance between the anisotropy and the ferromagnetic exchange interaction of the material.[35] The discrete synthetic system studied here allows us to directly engineer the anisotropy and the coupling.

12.3 SOLITON MOBILITY

12.3.1 PRINCIPLE

Propagation of a soliton occurs when one of the two layers at its center reverses: if layer i on Figure 12.1b switches, then the soliton moves one layer up the stack to in-between layer $i - 1$ and i; if layer $i + 1$ switches, the soliton moves one layer down. The difference between these two layers at the center of a soliton and layers inside

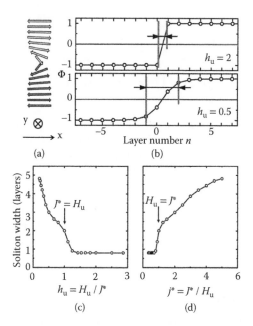

FIGURE 12.2 (a) Macrospin configuration in a stack with $h_u = 0.5$. (b) Parameter Φ for $h_u = 2$ and 0.5. (c) 10%–90% width of the transition region versus h_u. (d) Same data as (c) versus j^*.

the bulk of a domain (AP state) is that the latter are stabilized on both sides by AP neighbors whereas soliton layers have only one AP neighbor, the other neighbor being parallel (zero interaction field if we neglect the splay between layers at the center of the soliton). They will therefore be the easiest to reverse. Unidirectionality of soliton propagation is imposed by selecting which of these two layers will switch first. For a chiral soliton, this can be achieved using a rotating magnetic field.

We simulated the application of a global in-plane field of amplitude $h = H/J^*$ to stacks containing a single chiral soliton. Snapshots of the spin configuration for $h_u = 0.5$ and $h = 1$ are shown in Figure 12.3a through j; the field sequence is shown at the bottom of each snapshot (black arrows). The white and gray arrows at the bottom show the magnetization of the two layers at the center of the soliton. The field first increases from zero to h in the $+x$ direction, causing the magnetization in the two layers at the center of the soliton to point closer to the field direction (Figure 12.3b), but still symmetrically about the field. As the field starts rotating, clockwise in this case, this symmetry is broken and the white layer follows the field (Figure 12.3c and d) and switches (Figure 12.3e), becoming part of the bottom gray domain. In Figure 12.3f, the soliton has moved one layer up and the initial symmetry is restored. As the field keeps rotating with the same sense (Figure 12.3g and h), the white layer follows the field and switches (Figure 12.3i), becoming part of the bottom gray domain as well. After a full 360° rotation of the field (Figure 12.3j), the soliton has moved two layers above its original position and the magnetization in the central layers of the soliton points in the initial direction. If the sense of rotation of the field is reversed, the soliton moves in the opposite direction, and for the same rotating sense a soliton with opposite chirality (counterclockwise) moves in the opposite direction (down). It is worth

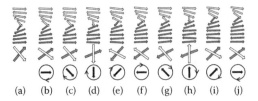

FIGURE 12.3 Macrospin simulations showing the effect of applying a clockwise in-plane rotating magnetic field to a stack containing a clockwise soliton ($h_u = 0.5$, $h = 1$). The field direction is displayed at the bottom; the middle panel shows the directions of the two layers forming the soliton. Both are shown top–down.

emphasizing that correct soliton propagation happens in discrete steps (two layers per field cycle), which is not the case for domain walls in ferromagnetic nanowires where the shape of the nanowire has to be engineered to impose the synchronization between the applied field and the domain-wall movement.[3]

12.3.2 OPERATING MARGIN

In the same way that ferromagnetic nanowires are characterized by their field-driven domain-wall conduit properties,[36] that is, how easy it is to move domain walls for a substantially lower field amplitude than necessary to nucleate new domains, these ML stacks also have soliton conduit properties. Figure 12.4a and b show a reduced rotating field amplitude diagram illustrating the different possible behavior for a soliton stack. The same set of data is plotted in both graphs—only the normalization factor differs. Figure 12.4a shows a reduced rotating field amplitude diagram $h = H/J^*$ for synchronous soliton propagation as a function of $h_u = H_u/J^*$. If h is too low (*No prop.* area below the squares), the soliton does not propagate; if h is too high (*Nucleation* area above the circles), the barrier constituted by the $2J^*$ stabilizing field in the bulk plus the anisotropy field H_u is overcome and the bulk reverses, erasing the soliton. Synchronous soliton propagation arises in the middle of these two boundaries. However, directional rotating-field-driven propagation of solitons is achieved only for the low field—low anisotropy light gray *Correct prop.* area below the triangles. We saw in the previous section that to propagate a soliton using a rotating field, the former has to be chiral. Sharp achiral solitons cannot couple to a rotating field. The consequence of this is that solitons in MLs with $h_u \geq 1.3 \pm 0.1$ (i.e., with j* ≤ 0.77) (see Figure 12.2c and d) are not expected to be controllable using rotating fields. A nonchiral soliton layer is still easier to switch than a bulk layer, but the propagation is not unidirectional anymore. Which of the top or bottom layer of the soliton follows the rotating field and switches first is now a consequence of the presence of random defects rather than a controlled process.

The dark gray *Non-unidirectional prop.* area, however, extends below $h_u = 1.3$ if the applied field is high enough. This is due to the fact that, as illustrated in Figure 12.3b, a magnetic field will pull the soliton layers toward its direction. The field necessary to fully saturate both soliton layers, and thereby erase the chirality of the soliton, increases as h_u decreases. This is reflected in the triangle line in Figure 12.4a.

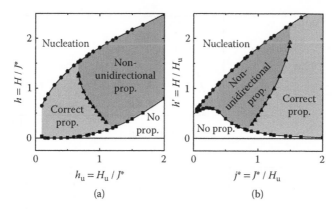

FIGURE 12.4 (a) Rotating field amplitude $h = H/J^*$ versus $h_u = H_u/J^*$ diagram. (b) Same set of data plotted as $h' = H/H_u$ versus $j^* = J^*/H_u$. The light gray area corresponds to synchronous unidirectional propagation, the dark gray area to non-unidirectional synchronous propagation.

Above that line, an otherwise chiral soliton becomes achiral and loses its ability to couple to a rotating field and propagate unidirectionally.

Stack edge effects were not included in the calculation of the diagram. Even in the absence of dipolar interactions, the propagation properties of broad solitons (low h_u) are affected by the proximity of the edges of the ML. This interaction between a soliton and stack boundaries is of the same nature as the soliton—soliton interaction, which will be discussed later. The calculations were performed on a stack containing enough layers that the properties were independent of the number of layers (50–100). Furthermore, the top and bottom layers were fixed to avoid edge-induced nucleation of domains.

Figure 12.4b shows the same data with a different normalization (H_u) and as a function of j^*. This figure will be used in Section 12.6.1.

12.4 SOLITON STABILITY

It is possible to calculate the energy barrier to propagation of a soliton within the macrospin model. For a clockwise soliton (Figure 12.1c), the energy barrier corresponding to forward propagation is obtained by a variational method where the energy of the stack is minimized as layer i is artificially rotated clockwise. This energy is shown in Figure 12.5a (dotted line) for $h_u = 1$ as a function of the rotation angle with respect to the $+x$ direction (the zero-field equilibrium angle for these parameters is indicated by a full dot). The energy increases with the layer angle and reaches a maximum when layer i is pointing along the hard axis. For a stack with $H_u = 1300\,\mathrm{Oe}, h_u = 1, M_S = 1400\,\mathrm{emu/cm^3}, t_M = 1.4\,\mathrm{nm}$ (corresponding to Fe on Cr (211) with a Cr thickness of about 1.3 nm)[37] and lateral dimensions of $200 \times 200\,\mathrm{nm^2}$, this barrier corresponds to $72\,k_BT$ at room temperature.

To calculate the energy barrier to backward propagation, the energy of the stack is computed as layer i is rotated counterclockwise, the energy profile is also shown in Figure 12.5a (solid line). In that case, the energy first increases as the magnetization

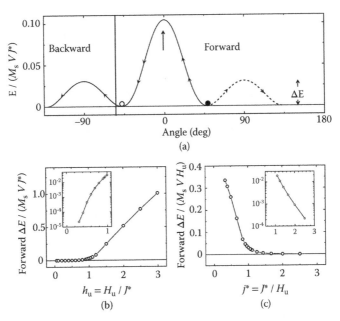

FIGURE 12.5 (a) Energy barrier to forward and backward propagation of a soliton in a stack with $h_u = 1$. The arrow indicates the barrier related to soliton chirality flipping. (b) Height of the forward energy barrier to propagation $\Delta E/M_S VJ^*$ as a function of h_u. Inset: same low h_u data in a semi-Log scale. (c) Height of the forward energy barrier to propagation $\Delta E/M_S VH_u$ as a function of j^*. Inset: same high j^* data in a semi-Log scale.

of layer i aligns more closely with that of layer $i + 1$, due to the increase in coupling energy. It reaches a maximum as both layers point in the same direction along the easy axis (corresponding to 1500 k_BT at room temperature for the parameters cited previously), then decreases as both layers splay away from each other again. At the minimum (open dot), the chirality of the soliton has flipped from clockwise to counterclockwise. The second leftmost barrier corresponds to backward propagation of a counterclockwise soliton and is identical to the one corresponding to forward propagation of a clockwise soliton.

The height of the energy barrier to forward propagation $\Delta E/M_S VJ^*$ is plotted in Figure 12.5b as a function of h_u. The inset shows the same data for $0 < h_u < 1$ with a logarithmic vertical scale to highlight the appearance of a nonzero ΔE. For $h_u < 0.25 \pm 0.05$, ΔE is zero; solitons are unstable and any perturbation such as thermal fluctuations or edge effects will expel them from a finite stack. For $h_u > 0.25 \pm 0.05$, ΔE increases monotonically with h_u. Figure 12.5c shows the same ΔE data, this time normalized by $M_S VH_u$.

Unidirectional propagation is ensured by the high energy barrier to chirality flipping. However, when a field is applied, the Zeeman energy tends to decrease the splay between the magnetizations of the two layers at the center of the soliton, that is, this energy barrier to chirality flipping is modified by the presence of the field. The line formed by the triangles in Figure 12.4a shows the boundary above which the energy barrier to chirality flipping goes to zero. Above this line, the field erases the chiral identity of the soliton, so that both layers are equally likely to switch during the subsequent half-field cycle and unidirectional propagation is lost.

It was found in Velthuis et al.[30] and Meersschaut et al.[31] that the antiphase domain wall obtained after SSF—which is essentially a chiral soliton of the kind that we are describing here—was not stable on reduction of the applied field. This is confirmed by our results. Their system has an h_u of less than 0.25; as can be seen in Figure 12.5b, it corresponds to a soliton with a zero propagation energy barrier.

12.5   SOLITON–SOLITON INTERACTION

Because solitons of the same chirality propagate in the same direction under the same globally applied field, it is possible to unidirectionally propagate several of solitons in one ML stack. However, solitons in the same ML interact; solitons of the same chirality repel; solitons of opposite chirality attract (and annihilate if close enough). This is illustrated in Figure 12.6 where two solitons of opposite chirality (Figure 12.6a) and of the same chirality (Figure 12.6b) are placed in close proximity in the same stack. When the chiralities are opposite, the system can lower its energy by rotating counterclockwise the layer in-between the two solitons (marked with *) to be more AP to the layers below and above; that is, it is energetically favorable for both solitons to merge their *tails*. If the coupling is strong enough, or if the solitons are close enough, the layers in-between the solitons (layer * plus layers above and below) will act as a block and switch collectively, resulting in the annihilation of both solitons. In the case where the solitons have the same chirality (Figure 12.6b), the equilibrium position for layer * is to point exactly along the easy axis. This acts on the layer above and below, forcing them closer to the easy axis as well, which in turn acts on the layer forming the top of soliton 1 and the layer forming the bottom of soliton 2, pushing them closer to the hard axis. If the coupling is strong enough, these layers can reverse, causing the two solitons to move away from each other. These interactions are directly related to the width of the solitons: the wider the soliton is (the smaller h_u) the longer the range of the interaction is. To be more quantitative, for $h_u = 1$, two solitons of the same chirality can have only three layers in-between (see Figure 12.6c), but if h_u is reduced to 0.7, this minimum number of layers increases to seven (see Figure 12.6d). The diagrams shown in Figure 12.4 will be modified if several solitons are to propagate in the same ML, however, this is beyond the scope of this paper.

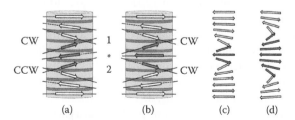

FIGURE 12.6   (a, b) Schematics of a multilayer stack with two solitons of (a) opposite chiralities and (b) same chiralities. (c, d) Simulated equilibrium configuration for two solitons of the same chirality in a stack with (c) $h_u = 1$ and (d) $h_u = 0.7$.

12.6   EXPERIMENTAL REALIZATION

The experimental realization of soliton-carrying heterostructures requires control of the anisotropy within individual magnetic layers as well as of the interlayer coupling. Control of the anisotropy is the most restrictive. Some methods allow a well-defined anisotropy direction to be induced in polycrystalline samples.[38–42] However, control of the strength of the anisotropy is limited. Controlling the interlayer coupling to reach the desired h_u ratio is more effective. RKKY coupling can be tuned by varying the spacer material or the thickness of the spacer layer,[8,43] or by changing the thickness of the magnetic layer.[44] A challenging point is to ensure the homogeneity of the coupling and the anisotropy over a large number of repeat layers, however, the large range of parameters for which solitons are stable (see Figure 12.4) should ensure their tolerance against defects along the ML.

More specifically, there are two main assumptions to our model: the dipolar interactions have been neglected, and all our calculations have been performed in the macrospin approximation. In the following, we look at the validity of these two simplifications.

12.6.1   Effect of Dipolar Interaction

The previous calculations neglect the effect of dipolar interactions between layers. In other words, they are valid in the case of infinite films only. However, as an extended ML is patterned into a device of reduced lateral dimensions, the stray field produced by each layer on its neighbors becomes nonnegligible. The stray field B produced by a cuboid structure of volume V magnetized in the (x, y)-plane, averaged inside the same volume V placed a center-to-center distance z away is directly proportional to M_S and creates an AF interaction the strength of which decreases with increasing z. In the far field, $B \propto V/z^3 = t_M w^2/z^3$, where t_M is the thickness of the cuboid and w its lateral size. Although it is more complicated, the increase of B with t_M is still valid in the near field regime. The dependency of B on w, however, is different in the far field and in the near field regime: for $w > z$ (near field), B decreases with increasing w.

In the absence of dipolar interactions, the results presented so far only depend on the ratio between anisotropy and coupling (h_u or j^*) and do not depend on the saturation magnetization M_S or on the ML parameters (magnetic and spacer thicknesses t_M and t_S). Only the total energy E depends on M_S and $t_{M,S}$, but even then the dependency is straightforward (see Equation 12.1). This ceases to be the case if dipolar interactions are included. Because of the nontrivial dependence of dipolar fields on the exact geometry of the element considered (t_M, w, and t_S), the main consequence of including dipolar interactions into our calculations is to complicate the dependency of the previously calculated quantities on M_S, t_M, and t_S. For each ($h_u = H_u/J^*$), which previously was enough to describe the whole behavior of the system, we now have six more degrees of freedom: t_M, t_S, H_u, M_S, w, and lateral shape. Therefore, in the following, we show the effect of dipolar interactions on a particular set of parameters corresponding to experimentally realistic materials.

Figure 12.7a shows the effect of the dipolar coupling at different lateral widths onto the $h_u = j^* = 1$ boundary fields of Figure 12.4. The parameters of Fe on Cr (211) were used again (H_u = 1300 Oe, M_S = 1400 emu/cm^3, t_M = 1.4 nm, t_S = 1.3 nm).[37] The calculations were performed in the macrospin approximation on square structures.[45] The dashed lines and the arrows indicate where the boundaries lie in the absence of dipolar coupling.

We now try to qualitatively understand the trends of the graphs, remembering that in a ML magnetized in-plane, dipolar interactions between layers are AF. The boundary to nucleation (circles) is the most straightforward to understand, as all layers in the bulk of the AP domains where this is relevant point along the easy axis. A layer in the bulk of the device (see Figure 12.7b) is more stable in the presence of dipolar coupling than in the absence of dipolar coupling, as it sees an overall stabilizing field from its neighbors (stabilizing field from first neighbors, lower destabilizing field from second neighbors, yet lower stabilizing field from third neighbors, etc.). A bulk layer will therefore reverse at a higher field in the presence of stronger dipolar interactions. This is in agreement with the observed increase of the field at which nucleation occurs as the width of the device decreases (see Figure 12.7a, circles).

The fact that the other two boundaries involve chiral solitons rather than bulk layers makes the contribution of dipolar interactions more complicated to estimate. Indeed, in that case, layers at a soliton and around point away from the easy anisotropy axis each at a specific angle that is the result of the balance between anisotropy, coupling, and dipolar interactions. Although dipolar interactions create an AF coupling between layers, the fact that they are long range prevents us from assuming *a priori* that they will be equivalent to a higher effective nearest neighbor AF coupling. Figure 12.8a and b show the variation of the angle α between one of the soliton layers and the easy axis (see Figure 12.8a as a function of h_u and Figure 12.8b as a function of j^*, in the absence of dipolar interactions [full circles] and in the presence of dipolar interactions at 100 nm width [open squares] and at 25 nm width [open triangles]). For low h_u (high j^*), α is close to 90° and virtually independent of the device width. As h_u increases, α decreases until the soliton is achiral (α = 0). The presence of stronger and stronger dipolar interactions as the width goes from 100 to 25 nm slows

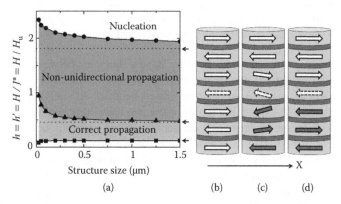

(a)   (b)   (c)   (d)

FIGURE 12.7 (a) Effect of dipolar interactions as a function of structure size on the phase diagram of a soliton stack with h_u = 1. The dotted lines and arrows indicate where the boundary lies in the absence of dipolar interactions. (b–d): schematic of (b) a bulk antiparallel domain, (c) a chiral soliton, and (d) a sharp soliton.

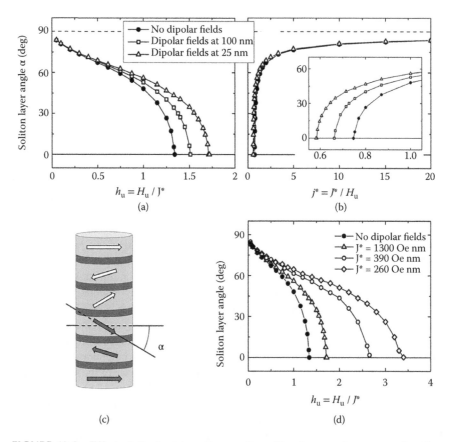

FIGURE 12.8 Effect of dipolar interactions on the soliton layer angle α. α as a function of (a) h_u and (b) j^*, without dipolar interactions (full circles), and in the presence of dipolar interactions at a lateral width of 100 nm (open squares) and 25 nm (open triangles). $J^* = 1300$ Oe·nm. (c) Schematic illustrating how α is defined. (d) α as a function of h_u without dipolar interactions (full circles) and with dipolar interactions at 25 nm lateral width, for $J^* = 1300$ Oe·nm (open triangles), $J^* = 390$ Oe·nm (open circles), and $J^* = 260$ Oe·nm (open diamonds).

down the decrease of α with h_u. In other words, the splay between soliton layers increases with increasing dipolar interactions; dipolar interactions do indeed have the same effect as an increased effective nearest neighbor AF coupling.

This is enough to explain the trends of the two remaining boundaries (triangles and squares) in Figure 12.7a: the way the corresponding switching fields change with structure size in Figure 12.7a is exactly reflected in the way the same boundaries vary with j^* in the vicinity of $j^* = 1$ in Figure 12.4b. In other words, increasing dipolar interactions has the same effect as increasing the effective AF coupling. More intuitively, the boundary between *Correct propagation* and *Non-unidirectional propagation* in Figure 12.7a (triangles) highlights the field at which the soliton goes from being chiral (Figure 12.7c) to being achiral (Figure 12.7d). As the width of the device decreases, and therefore as the splay between soliton layers increases, the chiral to achiral transition occurs at higher fields, in agreement with the observed increase of the field at which propagation becomes non-unidirectional as the width of the device decreases (triangles). The boundary between *no propagation* and *correct propagation* in Figure 12.7a (squares) indicates the field that is sufficient to overcome the energy barrier to forward propagation. This energy barrier decreases as j^* increases (see Figure 12.5c).

We started this section by pointing to the fact that the simple h_u or j^* dependency of the ML's behavior was lost in the presence of dipolar interactions. We just illustrated how the ML's geometrical parameters (width and thicknesses) become relevant in the calculation of the phase diagram and the soliton angle for instance. Figure 12.8d illustrates some of the complexity inherent to dipolar interactions, particularly the fact that in their presence the absolute value of J^* now matters. Figure 12.8d shows the soliton layer angle α as a function of h_u in the absence of dipolar interactions (full circles) and with dipolar interactions at 25 nm width (open symbols), for $J^* = 1300$ Oe·nm (triangles), 390 Oe·nm (circles), and 260 Oe·nm (diamonds). The latter three sets of data were calculated using the corresponding H_u (so that the explored range of h_u remains the same). This might not correspond to any realistic material parameter, however, it illustrates how the simple scaling in the absence of dipolar fields, where α only depends on h_u, is now lost. As can be seen on this figure, the effect of dipolar interactions at a fixed structure size is weaker for higher J^*. This can be understood by the fact that dipolar interactions for a given width introduce an absolute field scale into the system, against which the nearest neighbor exchange coupling J^* dominates or not. This will not be developed here, but by the same token, the width dependency of the $h_u = 1$ phase boundary shown Figure 12.7a also depends on the absolute value of J^* and H_u.

In conclusion, dipolar interactions present in any usefully small device do modify the functioning of the soliton shift register. Dipolar interactions create a long range AF coupling between layers that has the same qualitative effect as a higher effective nearest neighbor AF coupling. However, they introduce an absolute field scale, so that the simple scaling of the shift register's properties with the ratio between nearest neighbor AF coupling field J^* and anisotropy field H_u, j^*, or h_u, does not hold anymore. For the same structure size and for the same h_u, the effect of dipolar interactions is stronger for lower J^*. Nevertheless, for the parameters explicitly studied here (Fe on Cr [211]), the margin for correct soliton propagation is actually larger in the presence of dipolar interactions.

12.6.2 Full Micromagnetic Calculations

After dipolar interaction, the next obvious simplification of our calculations is the fact that we assume a macrospin model for the magnetization. However, how much of the behavior that we have just described would change if the full micromagnetic configuration was considered is unknown. Even the possibility of individual magnetic layers breaking into a multidomain state cannot be ruled out. There is no energetic gain for the bulk of the AP domain to break into vertically coupled AP domains, nonetheless, the layers at the center of a soliton could potentially break into domains during reversal. Even if the layers do not break into domains, the possible distortions of the magnetization allowed in full micromagnetics are likely to change their reversal behavior. There is no straightforward answer as to whether or not the soliton scheme would still work in full micromagnetics. As we have seen in Section 12.6.1, assuming specific geometrical parameters for our ML leads to setting a specific energy scale via the dipolar interactions. Furthermore, how close the dimensions of the device are from the single domain state will obviously make a difference: the smaller the device is, the better the macrospin approximation is. Because of the increased computation time, we have performed full micromagnetic simulations using the OOMMF software[46] on a ML comprising a restricted number of layers, and for a small subset of parameters for which we have run the corresponding macrospin calculations. The micromagnetic simulation parameters are the following: anisotropy energy density $K_u = 9.1 \times 10^4$ J/m^3, $M_S = 1400$ emu/cm^3, AF coupling energy density $K_{AF} = 1.36 \times 10^{-4}$ J/m^2, corresponding to the previously used parameters for Fe on Cr (211), and damping constant $\alpha = 0.5$. The ML stack was made of eight layers, with $t_M = t_S = 1.5$ nm and a square lateral footprint of width = 200 nm. The cell size was $5 \times 5 \times 1.5$ nm^3.

It has to be noted that, because of edge effects, the behavior of a ML with such a reduced number of layers is likely to differ from the virtually infinite stack we have used for the macrospin simulations. However, the aim of this section is not to describe these differences. Rather, the aim of this section is to assess the limits of the macrospin model by directly comparing its predictions on a system with a reduced number of layers, which is tractable using full micromagnetic calculations.

The simulations were initialized with a left-handed soliton pointing toward $+x$ straddling layers L3 and L4 by imposing an initial angle of $\pm45°$ on the soliton layers. An in-plane magnetic field of various amplitude and initially pointing in the $+x$ direction was applied and subsequently rotated clockwise in steps of 5° for two complete cycles, leading, if the field amplitude is correct, to the propagation of the soliton in the upward direction. Different behaviors are observed, depending on the amplitude of the rotating field, which will not be detailed here. For a restricted range of applied fields, between 245 ± 5 Oe and 1175 ± 5 Oe, the soliton propagates one layer up the stack during the first half of the field cycle, from straddling L3–L4 to straddling L4–L5, then one more layer up during the second half of the first field cycle, from straddling L4–L5 to straddling L5–L6, finally another layer up to straddle L6– L7 during the first half of the second field cycle, immediately followed by the flipping of layers L7 and L8 and hence the annihilation of the soliton at the top of the stack. Figure 12.9b through 12.9f shows a few snapshots of the calculated magnetization configuration as the left-handed soliton propagates during the second half of the first field cycle. The top panel shows an xz-view of the ML, the bottom panels show an xy-view of layers L4, L5, and L6. Note the different scales in the xy-plane and along the z axis. The middle panel shows the direction of the applied rotating magnetic field. When the field points toward $-x$ (Figure 12.9b), the soliton has already propagated one layer up during the first half of the first cycle (it straddles layers L4 and L5) and the soliton layers point toward $-x$. As the field rotates further (Figure 12.9c), the magnetization in the top layer of the soliton (L5) rotates, until somewhere between 280° (Figure 12.9d) and 285° (Figure 12.9e), the x-component of the magnetization in layer L5 changes sign and the soliton propagates one layer up to straddle layers L5 and L6.

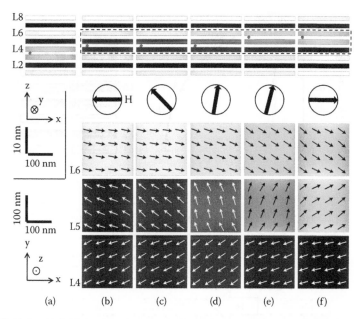

FIGURE 12.9 (a) Initial magnetic state for the micromagnetic simulations. The contrast ranges from black ($-x$) to white ($+x$). A soliton pointing in the $+x$ direction straddles layers L3 and L4. (b–f): Snapshots of the micromagnetic configuration for an eight layers multilayer where a soliton propagates from between L4 and L5 to between L5 and L6 under an applied rotating field of 500 Oe. The soliton is marked by an asterisk. Top panels: (z, x)-view; bottom panels: (x, y)-view. Note the different scale bars in the z direction and in the (x, y)-plane shown in the bottom panels of (a). The black arrow inside the circle indicates the field direction. As measured from the $+x$ axis: (b) 180°; (c) 225°; (d) 280°; (e) 285°; (f) 0°.

Figure 12.10a shows a polar plot of the field amplitude/angle at which the previously described transition occurs (controlled propagation of the left-handed soliton from straddling L4 and L5 to straddling L5 and L6, see Figure 12.10b). For both macrospin and micromagnetic simulations, the transition point is determined as the angle for which, under a given field, the average magnetization in the layer changes sign. For $H = 500$ Oe, we have checked that this transition indeed corresponds to soliton propagation, that is, that if the applied field is turned off from the configuration shown in Figure 12.9d, the soliton stays in between layers L4 and L5, whereas if the applied field is turned off from the configuration shown in Figure 12.9e, the soliton stays in between layers L5 and L6.

Micromagnetic simulations are shown in gray; the corresponding macrospin simulations are shown in black. Two different behaviors are observed in the macrospin calculations. For low fields (open black squares, below 675 ± 5 Oe), the same controlled propagation of the soliton up the stack is seen. For field amplitudes above 675 Oe (full black circles), the macrospin model shows that the soliton ceases to propagate unidirectionally, in close quantitative agreement with the results of Figure 12.7 at 200 nm width, despite the fact that the number of layers was a lot larger. In contrast, the micromagnetic simulations show correct unidirectional propagation for

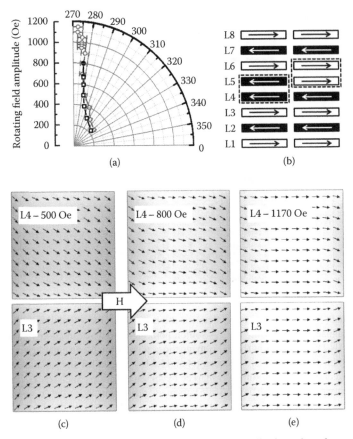

FIGURE 12.10 (a) Polar plot showing the applied field amplitude and angle at which L5 switches as a soliton propagates from between L4 and L5 to between L5 and L6. The angle is measured clockwise from the $+x$ axis. (b) Schematic of the transition considered. (c, d, e) Micromagnetic configurations of layers L3 and L4 when carrying a soliton under various field amplitudes. The field is applied in the $+x$ direction, as indicated by the large white arrow.

fields up to 1175 Oe. Loss of unidirectionality with increasing field was attributed to a loss of chirality as the field becomes high enough so that both soliton layers point exactly in the direction of the field. However, when looking into details at the micromagnetic configuration of the soliton layers, it is clear that, although the splay between soliton layers is greatly reduced as the applied field increases, the memory of the chirality is not totally erased as it is in the macrospin model (see Figure 12.10c through e). At high fields, the soliton layers are in an S-state, where most of the magnetization points along the field direction, apart from the two edges perpendicular to the field, which point closer to the direction set by the initial soliton chirality.

For field amplitudes higher than 1175 Oe, however, on initializing the soliton between L3 and L4, the soliton moves backwards down to in between L2 and L3 before annihilating at the bottom of the ML. In this case also, the field is actually not strong enough to completely erase the chiral identity of the soliton, which is still imprinted in the direction of the magnetization in the soliton layers along the edge of the structure, as in Figure 12.10e. This is beyond the scope of this paper, but we have observed that in a homogeneous ML such as the one studied here, the edge of a ML acts as a well for solitons. In our case here, this attraction is strong enough and the chiral identity of the soliton is weak enough that we observe a flipping of the chirality of the soliton, followed by its downward motion.

In brief, the qualitative agreement between macrospin and micromagnetic simulations for these particular material parameters and at 200 nm lateral width is excellent. Both show that controlled propagation of the soliton occurs at a field angle closer and closer to the hard axis for increasing field amplitudes. Quantitatively, the agreement is very good at low fields. At high field, the macrospin model underestimates the field at which the soliton will lose its chirality and unidirectional propagation will be lost, so that the calculated operating margin is larger in micromagnetics that in macrospin. Again, this constitutes in no way a general rule, and different material parameters might behave differently.

12.7 CONCLUSION

In conclusion, we have presented the results of macrospin simulations that show the existence of mobile chiral topological kink solitons in AF-coupled MLs with in-plane uniaxial anisotropy. We have demonstrated that in the absence of dipolar interactions, the behavior of these systems is solely described by the anisotropy to coupling ratio h_u. We showed that they can couple to an external in-plane rotating magnetic field and propagate synchronously in the vertical direction. As opposed to our previously demonstrated soliton-carrying ML scheme with perpendicular anisotropy,[34] the propagation of in-plane solitons such as the ones studied here is bidirectional and depends on the relative chirality of the soliton and the rotating field. We showed that several solitons can propagate controllably in the same ML stack and we described the soliton–soliton interaction mechanism. We have studied how including dipolar interactions into the macrospin calculations modifies the behavior of a ML. We showed that the main consequence of including dipolar interactions is to lose the simple dependency of the behavior of soliton-carrying stacks on h_u. Indeed, dipolar interactions depend on the geometrical parameters of the ML (lateral size and thicknesses) and therefore their inclusion adds several independent degrees of freedom. The effect of dipolar interactions is qualitatively similar to an increase in the nearest neighbor AF coupling. However, the fact that dipolar interactions are long-range means this is only a coarse approximation. We have also shown that the strength of the modifications caused by dipolar interactions depends on the absolute strength of the AF coupling J^*. Dipolar interactions calculated for a given set of geometrical parameters have a stronger effect for a small J^* than for a large J^*. We have also studied the limits of the macrospin approximation by performing full micromagnetic simulations for a particular set of parameters and a small number of layers. We showed that for this particular set of material parameters and dimensions, the qualitative agreement between the macrospin model and micromagnetics was excellent. Discrepancies were found quantitatively, in that case showing that the range of field for which the soliton propagates correctly is larger in the case of micromagnetic simulations than in the macrospin case.

Because of their properties as stable, mobile objects formed in synthetic magnetic systems, solitons in magnetic MLs are ideal candidates to code and manipulate digital information in a three-dimensional spintronic device.[47]

ACKNOWLEDGMENTS

This work was supported by the European Community under the Seventh Framework Program ERC contract no. 247368 (3SPIN) and grant agreement no. 309589 (M3D). A. Fernández-Pacheco acknowledges support by a Marie Curie IEF within the Seventh European Community Framework Program no. 251698 (3DMAGNANOW). R. Lavrijsen is gratefully acknowledged for his critical reading of parts of this manuscript.

REFERENCES

1. S. A. Wolf, D. D. Awschalom, R. A. Buhrman, J. M. Daughton, S. von Molnár, M. L. Roukes, A. Y. Chtchelkanova, and D. M. Treger, Spintronics: A Spin-Based Electronics Vision for the Future, *Science* 294, 1488 (2001).
2. R. P. Cowburn and M. E. Welland, Room Temperature Magnetic Quantum Cellular Automata, *Science* 287, 1466 (2000).
3. D. A. Allwood, G. Xiong, C. C. Faulkner, D. Atkinson, D. Petit, and R. P. Cowburn, Magnetic Domain-Wall Logic, *Science* 309, 1688 (2005).
4. A. Imre, G. Csaba, L. Ji, A. Orlov, G. H. Bernstein, and W. Porod, Majority Logic Gate for Magnetic Quantum-Dot Cellular Automata, *Science* 311, 205 (2006).
5. S. I. Kiselev, J. C. Sankey, I. N. Krivorotov, N. C. Emley, R. J. Schoelkopf, R. A. Buhrman, and D. C. Ralph, Microwave oscillations of a nanomagnet driven by a spin-polarized current, *Nature* 425, 380 (2003).
6. V. Baltz, B. Rodmacq, A. Bollero, J. Ferré, S. Landis, and B. Dieny, Balancing interlayer dipolar interactions in multilevel patterned media with out-of-plane magnetic anisotropy, *Appl. Phys. Lett.* 94, 052503 (2009).
7. I. Tudosa, J. A. Katine, S. Mangin, and E. E. Fullerton, Perpendicular spin-torque switching with a synthetic antiferromagnetic reference layer, *Appl. Phys. Lett.* 96, 212504 (2010).
8. S. S. P. Parkin, N. More, and K. P. Roche, Oscillations in exchange coupling and magnetoresistance in metallic superlattice structures: Co/Ru, Co/Cr, and Fe/Cr, *Phys. Rev. Lett.* 64, 2304 (1990).
9. S. W. Cheong and M. Mostovoy, Multiferroics: a magnetic twist for ferroelectricity, *Nature Mater.* 6, 13 (2007).
10. Y. Togawa, T. Koyama, K. Takayanagi, S. Mori, Y. Kousaka, J. Akimitsu, S. Nishihara, K. Inoue, A. S. Ovchinnikov, and J. Kishine, Chiral Magnetic Soliton Lattice on a Chiral Helimagnet, *Phys. Rev. Lett.* 108, 107202 (2012).
11. A. N. Bogdanov and D. A. Yablonskiĭ, Thermodynamically stable "vortices" in magnetically ordered crystals. The mixed state of magnets, *Sov. Phys. JETP* 68, 101 (1989).
12. U. K. Rosler, A. N. Bogdanov, and C. Pfleiderer, Spontaneous skyrmion ground states in magnetic metals, *Nature* 442, 797 (2006).
13. S. Mühlbauer, B. Binz, F. Jonietz, C. Pfleiderer, A. Rosch, A. Neubauer, R. Georgii, and P. Böni, Skyrmion Lattice in a Chiral Magnet, *Science* 323, 915 (2009).

14. X. Z. Yu, Y. Onose, N. Kanazawa, J. H. Park, J. H. Han, Y. Matsui, N. Nagaosa, and Y. Tokura, Real-space observation of a two-dimensional skyrmion crystal, *Nature* 465, 901 (2010).
15. S. Heinze, K. von Bergmann, M. Menze, J. Brede, A. Kubetzka, R. Wiesendanger, G. Bihlmayer, and S. Blügel, Spontaneous atomic-scale magnetic skyrmion lattice in two dimensions, *Nature Physics* 7, 713 (2011).
16. I. Raicević, D. Popović, C. Panagopoulos, L. Benfatto, M. B. Silva Neto, E. S. Choi, and T. Sasagawa, Skyrmions in a Doped Antiferromagnet, *Phys. Rev. Lett.* 106, 227206 (2011).
17. I. Dzyaloshinskii, A thermodynamic theory of "weak" ferromagnetism of antiferromagnetics, *J. Phys. Chem. Solids* 4, 241 (1958).
18. T. Moriya, Anisotropic Superexchange Interaction and Weak Ferromagnetism, *Phys. Rev.* 120, 91 (1960).
19. A. P. Malozemoff and J. C. Slonczewski, *Magnetic Domain Walls in Bubble Materials*, Academic Press, New York, NY (1979).
20. C. Moutafis, S. Komineas, C. A. F. Vaz, J. A. C. Bland, T. Shima, T. Seki, and K. Takanashi, Magnetic bubbles in FePt nanodots with perpendicular anisotropy, *Phys. Rev. B* 76, 104426 (2007).
21. A. Yoshimori, A New Type of Antiferromagnetic Structure in the Rutile Type Crystal, *J. Phys. Soc. Jpn.* 14, 807 (1959).
22. K. Marty, V. Simonet, E. Ressouche, R. Ballou, P. Lejay, and P. Bordet, Single Domain Magnetic Helicity and Triangular Chirality in Structurally Enantiopure $Ba_3NbFe_3Si_2O_{14}$, *Phys. Rev. Lett.* 101, 247201 (2008).
23. Y. Hiraoka, Y. Tanaka, T. Kojima, Y. Takata, M. Oura, Y. Senba, H. Ohashi, Y. Wakabayashi, S. Shin, and T. Kimura, Spin-chiral domains in $Ba_{0.5}Sr_{1.5}Zn_2Fe_{12}O_{22}$ observed by scanning resonant x-ray microdiffraction, *Phys. Rev B* 84, 064418 (2011).
24. D. L. Mills, Surface Spin-Flop State in a Simple Antiferromagnet, *Phys. Rev. Lett.* 20, 18 (1968).
25. D. L. Mills and W. M. Saslow, Surface Effects in the Heisenberg Antiferromagnet, *Phys. Rev.* 171, 488 (1968).
26. A. S. Carrico, R. E. Camley, and R. L. Stamps, Phase diagram of thin antiferromagnetic films in strong magnetic fields, *Phys. Rev. B* 50, 13453 (1994).
27. N. Papanicolaou, The ferromagnetic moment of an antiferromagnetic domain wall, *J. Phys.: Condens. Matter* 10, L131 (1998).
28. R. W. Wang, D. L. Mills, E. E. Fullerton, J. E. Mattson, and S. D. Bader, Surface spin-flop transition in Fe/Cr(211) superlattices: Experiment and theory, *Phys. Rev. Lett* 72, 920 (1994).
29. S. Rakhmanova, D. L. Mills, and E. E. Fullerton, Low-frequency dynamic response and hysteresis in magnetic superlattices, *Phys. Rev. B* 57, 476 (1998).
30. S. G. E. te Velthuis, J. S. Jiang, S. D. Bader, and G. P. Felcher, Spin Flop Transition in a Finite Antiferromagnetic Superlattice: Evolution of the Magnetic Structure, *Phys. Rev. Lett.* 89, 127203 (2002).
31. J. Meersschaut, C. Labbé, F. M. Almeida, J. S. Jiang, J. Pearson, U. Welp, M. Gierlings, H. Maletta, and S. D. Bader, Hard-axis magnetization behavior and the surface spin-flop transition in antiferromagnetic Fe/Cr(100) superlattices, *Phys. Rev. B* 73, 144428 (2006).
32. A. Fernández-Pacheco, D. Petit, R. Mansell, R. Lavrijsen, J. H. Lee, and R. P. Cowburn, Controllable nucleation and propagation of topological magnetic solitons in CoFeB/Ru ferrimagnetic superlattices, *Phys. Rev. B* 86, 104422 (2012).
33. V. G. Baryakhtar, M. V. Chetkin, B. A. Ivanov, and S. N. Gadetskii, *Dynamics of Topological Magnetic Solitons: Experiment and Theory*, Springer-Verlag, Berlin, Germany (1994).

34. R. Lavrijsen, J. H. Lee, A. Fernndez-Pacheco, D. Petit, R. Mansell, and R. P. Cowburn, Magnetic ratchet for three-dimensional spintronic memory and logic, *Nature* 493, 647 (2013).

35. S. Chikazumi, *Physics of Magnetism*, John Wiley & Sons, New York, NY (1964).

36. D. A. Allwood, N. Vernier, Gang Xiong, M. D. Cooke, D. Atkinson, C. C. Faulkner, and R. P. Cowburn, Shifted hysteresis loops from magnetic nanowires, *Appl. Phys. Lett.* 81, 4005 (2002).

37. E. E. Fullerton, M. J. Conover, J. E. Mattson, C. H. Sowers, and S. D. Bader, Oscillatory interlayer coupling and giant magnetoresistance in epitaxial Fe/Cr(211) and (100) superlattices, *Phys. Rev. B* 48, 15755 (1993).

38. A. T. Hindmarch, A. W. Rushforth, R. P. Campion, C. H. Marrows, and B. L. Gallagher, Origin of in-plane uniaxial magnetic anisotropy in CoFeB amorphous ferromagnetic thin films, *Phys. Rev. B* 83, 212404 (2011).

39. H. Raanaei, H. Nguyen, G. Andersson, H. Lidbaum, P. Korelis, K. Leifer, and B. Hjörvarsson, Imprinting layer specific magnetic anisotropies in amorphous multilayers, *J. Appl. Phys.* 106, 023918 (2009).

40. T. J. Klemmer, K. A. Ellis, L. H. Chen, B. van Dover, and S. Jin, Ultrahigh frequency permeability of sputtered Fe–Co–B thin films, *J. Appl. Phys.* 87, 830 (2000).

41. W. Yu, J. A. Bain, W. C. Uhlig, and J. Unguris, The effect of stress-induced anisotropy in patterned FeCo thin-film structures, *J. Appl. Phys.* 99, 08B706 (2006).

42. R. P. Cowburn, Property variation with shape in magnetic nanoelements, *J. Phys. D: Appl. Phys.* 33 R1 (2000).

43. S. S. P. Parkin, Oscillations in giant magnetoresistance and antiferromagnetic coupling in $[Ni_{81}Fe_{19}/Cu]_N$ multilayers, *Appl. Phys. Lett.* 60, 512 (1992).

44. I. G. Trindade, Antiferromagnetically coupled multilayers of $(Co_{90}Fe_{10}\ t_F/Ru\ t_{Ru}) \times N$ and $(Ni_{81}Fe_{19}\ t_F/Ru\ t_{Ru}) \times N$ prepared by ion beam deposition, *J. Magn. Magn. Mat.* 240, 232 (2002).

45. G. Akoun and J. P. Yonnet, 3D analytical calculation of the forces exerted between two cuboidal magnets, *IEEE Trans. Mag.* 20, 1962 (1984).

46. OOMMF, M.J. Donahue, and D.G. Porter, Interagency Report NISTIR 6376, National Institute of Standards and Technology, Gaithersburg, MD (Sept. 1999), http://math.nist .gov/oommf.

47. Patent numbers GB0820844.9, Cowburn, Russell Paul. Magnetic data storage, U.S. Patent 20100128510, filed November 13, 2009, and issued May 27, 2010.

13 Soft-Error-Aware Power Optimization Using Dynamic Threshold

Selahattin Sayil

CONTENTS

13.1 INTRODUCTION

The increasing speed and complexity of today's designs implies a significant increase in the power consumption of very-large-scale integration (VLSI) chips. The use of portable devices and popular demand on battery life has made power consumption even more important in modern complementary metal–oxide–semiconductor (CMOS) designs. To meet this challenge, various techniques extending from the circuit level to the architectural level have been proposed aimed at reducing the consumed energy. Some circuit-level mechanisms include adaptive substrate biasing, dynamic supply scaling, dynamic frequency scaling, and supply gating [1,2].

Another important issue a circuit designer faces today is the circuit reliability. With advances in CMOS technology, circuits become increasingly more sensitive to transient pulses caused by single-event (SE) radiation particles. Under favorable conditions, these transients can cause incorrect data storage resulting in soft errors. The reduced circuit node capacitances, supply voltages, and transistor drive currents coupled with increasingly denser chips are raising soft-error rates (SERs) and making them an important design issue. As circuit designers try to address the excessive power consumption problem, they need to be aware of the impact of the power optimizations on circuit soft-error robustness [3–5]. This is especially important for mission critical applications where the reliability is most important objective over the cost and the performance.

One common technique for reducing power is to reduce the supply voltage. This should be accompanied by threshold voltage reduction to keep circuit speed and high current drive to avoid performance penalties. However, the threshold voltage reduction results in great amount of increase on the subthreshold leakage current [1].

One solution to this problem is a dynamic threshold MOS (DTMOS) operation that applies an active body-bias to Metal Oxide Semiconductor Field Effect Transistors (MOSFETs) [6]. Because of low threshold voltage during the logic transition and high threshold voltage during the off-state, the dynamic threshold circuit operates at high speed with low power. The DTMOS technique allows the use of ultralow voltages (0.6 V and below) and is considered a promising candidate for low-power and high-speed circuit devices as it can improve the circuit speed without increasing the standby power consumption.

This chapter presents an analysis on various DTMOS schemes for soft-error tolerance using many benchmark circuits including AM2901 microprocessor bit-slice. The experimental results indicate that all DTMOS configurations increase circuit robustness to SE-induced soft errors and delay effects because of increased transistor current drive compared to body tie configuration.

Our initial analysis on threshold voltage has shown that decreasing threshold voltages increase the critical charge of logic circuits, thus providing more robustness to radiation transients. In a normal DTMOS scheme, the body–source junction is "forward biased" (at less than 0.6 V), forcing the threshold voltage to drop and hence this lower threshold effect can be exploited for SE transient mitigation.

Therefore, this chapter also introduces a soft-error mitigation technique where standard DTMOS configuration can be combined with driver sizing to mitigate SE transients with lot more area efficiency than driver sizing taken alone.

13.2   SINGLE-EVENT SOFT ERRORS

Terrestrial soft errors in memory have been a very well-known problem [7]. However, because of increasing clock frequencies and diminishing transistor sizes, soft errors are now affecting CMOS logic. It has been predicted that for technologies below 65 nm, the majority of the observed radiation-induced soft failures will be due to transients that will occur in combinational logic (CL) circuits [8].

At ground level, soft errors are mainly induced by three different radiation mechanisms: (1) α particles emitted from trace radioactive impurities in the device materials, (2) interaction of low-energy thermal neutrons with certain boron isotopes in the device, and (3) the reaction of high-energy cosmic neutrons (>1 MeV) with silicon and other device materials. Neutron-induced soft errors are generated by secondary charged particles that are created in neutron–silicon atom collisions [7].

When an incident charged particle strikes the sensitive area within a combinational circuit such as the depletion region of transistor drains or a reverse-biased p–n junction, many hole-electron pairs are created because of ionization mechanism (Figure 13.1). These free carriers can later drift under the electric field creating a transient voltage pulse. This transient is also named as a single-event transient (SET).

The amount of the charge collected at a particular node (Q_{col}) depends on combination of factors, including the gate's size, the biasing of the various circuit nodes, material type, the device doping level, and characteristics of the particle hit such as energy, trajectory, and its position within the device [9].

When the collected charge Q_{col} at a given node exceeds the critical charge Q_{crit} of that node, the generated SET can propagate and may reach to storage elements under certain conditions. If the generated pulse arrives at the storage element during its latching window, incorrect data can be stored resulting in soft error. This error is also termed as single-event upset (SEU). The following must be satisfied before generation of such errors:

The transient pulse generated should have sufficient amplitude and width such that it propagates along the succeeding gates without significant attenuation. Hence, *electrical masking* should not be present.

The logic path the pulse takes should be enabled by CL inputs. In another words, there should not be any *logical masking*.

The latching clock edge should be present during the presence of the SET pulse at the input of the storage element. This means no *temporal masking* should exist.

In Figure 13.2, all these criteria have been satisfied, that is, first a sufficient transient pulse is generated at the particle site such that it propagates through many stages without any attenuation. In addition, there is no logical masking in this circuit as the second input of NAND2 gate is tied to logic 1. Finally, the pulse arrives during latching edge of the clock pulse causing generation of a soft error.

With advances in technology scaling, CMOS circuits are increasingly more sensitive to transient pulses caused by SE particles. The reduced circuit node capacitances, supply voltages, and transistor drive currents coupled with increasingly denser chips are raising SERs and making them an important design issue.

In addition to SETs, radiation-induced soft delay errors (SDEs) are also on the increase in CMOS logic designs. Gill et al. [10] describe SE-induced soft delay as

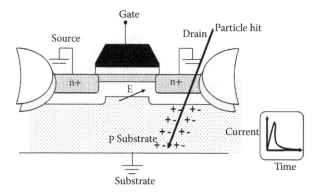

FIGURE 13.1 Transient current pulse generation due to particle hit on sensitive node.

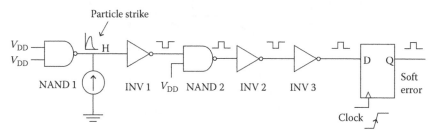

FIGURE 13.2 Propagation of an SE transient and generation of a soft error.

the amount of delay induced on a CMOS gate because of high-energy particle strike on its sensitive region, which happens only during signal switching. Incorrect data storage may occur if the delayed signal violates the timing requirements of the storage elements. Hence, an SDE is created.

As an example, consider the circuit given in Figure 13.3. A rising input pulse has been connected to the input of the first inverter. When output V_{o1} is undergoing a falling phase of transition, the P-type MOS (PMOS) transistor turns OFF and become susceptible to a particle strike. If a high-energy particle strikes node V_{o1} during this phase of transition, the generated current on PMOS transistor drain because of the SE hit (shown with the current source) can pull the signal in positive direction causing longer transition times. The delay effect is observable at the output of the succeeding gate(s), if the path is logically enabled. The maximum delay effect occurs if particle hit occurs near the 50% point of voltage transition.

It has been reported that the soft delay effect will become more pronounced in newer technologies because of reduced circuit node capacitances [10]. For high-reliability designs, SDEs must also be considered in the analysis in addition to SETs as the two error mechanisms are different in the way they are masked: Soft delay effects can only be masked by logical and latch window masking effects. As soft delay is not a voltage glitch, electrical masking does not apply. If a signal is once delayed, then it can propagate to the circuit output through functionally sensitized path without any attenuation of the delay.

Gill et al. [10] have compared SET and soft delay sensitivities of various benchmark circuits for a maximum linear energy transfer of 20 MeV. The results for some benchmark circuits indicated that a higher number of nodes were sensitive to soft delay effects than SET because of absence of electrical masking effect for SDE.

The rate at which soft errors occur is called the SER. The SER is usually defined by the failure-in-time (FIT)—a FIT being 1 failure in 1 billion device hours of operation. It has been reported that advanced processors with large multimegabit-embedded static random access memory (SRAM) can easily have soft failure rates in excess of 50,000 FIT at terrestrial level [9]. The same error rate can also be achieved for standard high-density application-specific integrated circuit designs at 90 nm and below [11].

For single-chip consumer applications, this error rate may not still be important for most designers, but for high-reliability systems composed of multichip assemblies

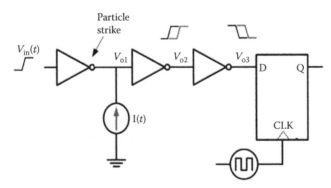

FIGURE 13.3 Soft delay effect causing delay effects on succeeding gates.

such a rate becomes intolerable [12]. Hence, for mission-critical or high-reliability applications such as military, avionics [13], medical systems [12], and so forth, where reliability is as important as energy efficiency, designers need to make clever design choices that reduce static power consumption and improve soft-error reliability of the newer designs [14].

13.3 SOFT-ERROR-AWARE POWER OPTIMIZATION VIA DYNAMIC THRESHOLD TECHNIQUES

There has been some work to examine the effect of power optimizations on soft errors. Degalahal et al. [15] have analyzed the effect of increasing threshold voltage (widely used for reducing static power consumption) on circuit SERs and have found that increasing threshold voltages can cause devices to slow down, which in turn increase CL circuit susceptibility to SETs.

For increasing CL soft-error tolerance, Dhillon et al. [3] have proposed a technique that used optimal assignment of gate sizes, supply voltages, and threshold voltages while meeting timing constraints. Later work by Choudhury et al. [4] has suggested an algorithm that uses voltage assignment (dual-V_{DD}) for SEU robustness. However, both techniques obtained potentially large power overheads.

Finally, the work by Wu and Marculescu [5] has presented a power-aware methodology for soft-error hardening using dual-supply voltages. In this methodology, a higher supply voltage is assigned to the gates that have large error impact and contribute most to the overall SER.

To address increasing power consumption challenge, operating the transistors of a digital logic in the subthreshold region has been proposed as a solution. However, the power supply voltage reduction should be followed by threshold voltage reduction to avoid performance penalties and to maintain high current drive. Subthreshold leakage current, on the other hand, increases as threshold voltages decrease. One solution proposed to solve this problem is the dynamic threshold technique that applies an active body-bias to MOSFETs [6,16]. Because of low-threshold voltage during the logic transition and high-threshold voltage during the off-state, the dynamic threshold circuit operates at high speed with low power. However, dynamic threshold technique can only be used for low voltage (0.6 V and below) VLSI circuits.

In a standard DTMOS logic gate, all transistor gates are tied to their substrates (Figure 13.4). The high-speed operation is provided by forward bias to switching transistors, whereas low leakage is obtained by applying zero bias to other transistors. Specifically, the body–source junction is "forward biased" (at less than 0.6 V), forcing the threshold voltage to drop.

Various DTMOS inverter schemes have been proposed by researchers to improve standard DTMOS logic inverter. To reduce standby leakage current, Chung et al. [17] have suggested a new scheme with minimum size small subsidiary transistors. The subsidiary transistors increase the current drive by managing the body bias as shown in Figure 13.5a. The input load of the inverter circuit is reduced as the output charges are used to raise the body potential of the main transistors. Later, Gil et al. [18] proposed another scheme with subsidiary transistors (Figure 13.5b) that achieved better performance in terms of speed.

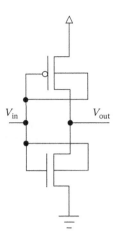

FIGURE 13.4   Standard DTMOS logic inverter.

Figure 13.5c shows another scheme of DTMOS subsidiary transistors, but the gates of subsidiary devices are tied to main transistor's drain instead of gate [19]. This configuration performs best in terms of power-delay product when compared to previous schemes, namely the standard DTMOS scheme, and circuits of Figure 13.5a and b. In the last configuration shown in Figure 13.5d, transistor drains are tied to substrates. Soleimani et al. [20] reported even better power efficiency using this style of DTMOS, although stability with respect to temperature was an issue compared to standard DTMOS configuration.

Before examining various dynamic threshold–based schemes for their vulnerability to SETs and SDEs, it is imperative that the effect of varying transistor thresholds on soft error and delay effects is studied. For this, a string of six inverters shown in Figure 13.6 were used in the analysis. For measuring SETs, the input of the first inverter is tied to logic 1, and for soft delay measurements, to a switching input waveform. In our analysis, we have considered 65 nm technology with parameters obtained using the predictive technology model [21].

An SE hit was simulated at the output of the first inverter using a double-exponential current pulse [7] that is given by

$$I(t) = \frac{Q}{\tau_\alpha - \tau_\beta}(e^{-t/\tau_\alpha} - e^{-t/\tau_\beta}) \tag{13.1}$$

where Q is the charge (positive or negative) deposited by the particle strike, τ_α the collection time constant of the p–n junction, and τ_β the ion-track establishment time constant.

The time constants τ_α and τ_β are dependent on process technology and are taken as 100 and 5 picoseconds (ps), respectively, based on the work of Hutson et al. [22] Alternatively, the single-exponential current model given by Freeman [23] may also be used. Zhou and Mohanram [24] and Ness et al. [25] mention that the single-pole model can be replaced by double-exponential current pulse model for SEUs without loss of generality. In this work, the double-pole representation has been used as an approximation to waveforms seen in mixed-mode simulations.

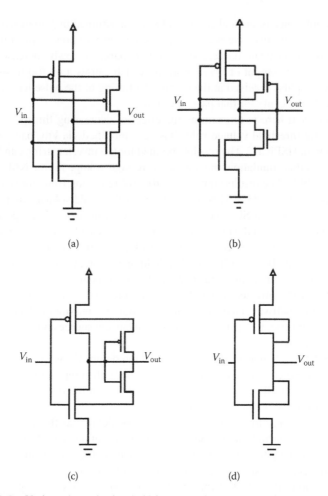

(a)   (b)

(c)   (d)

FIGURE 13.5 Various dynamic threshold inverter schemes reported. (a) DTMOS scheme with subsidiary devices that uses output charge to raise body potential. (b) Another scheme with subsidiary transistors that uses active body bias. (c) A scheme where the gates of subsidiary transistors are tied to the main transistor's drain. (d) DTMOS with drains connected to substrate.

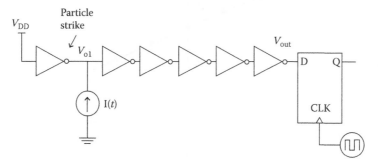

FIGURE 13.6 The inverter string used in HSPICE simulations.

In all simulations for critical charge, it has been assumed that no masking effects occur, hence the circuits have been set up in such a way to prevent masking effects. To determine the critical charge, the deposited charge is slowly increased until an SET appears at the output of inverter string, which causes a bit flip in storage element. For each pulse received at the output, the timing of the pulse has been varied to see if latching occurs for that particular pulse.

Since our goal here is to examine the effect of decreasing threshold on circuit SET, transistor threshold voltage value has been modified via Vth_0 parameter using *delvto* option in HSPICE®. The results obtained for the inverter string can be seen in Figure 13.7. In this simulation, the nominal threshold voltage for the NMOS transistor is $V_{th} = 0.29$ V. The results show that a 100 mV reduction from normal threshold value increases Q_{crit} by 33%. This indicates that if threshold voltage can be reduced, the circuit robustness to SETs could be increased as the critical charge value of a node becomes larger. On the other hand, increasing threshold values decrease circuit tolerance to SETs. This confirms the results of Degalahal et al. [15].

We have also studied the effect of threshold on soft delay effect for the first time. For soft delay measurement, a pulse signal has been connected to the input of the first inverter in the inverter string shown in Figure 13.6. A radiation hit has been injected roughly halfway during a falling transition (for maximum delay) at the output of the first inverter using double-exponential current source and the delay change has been observed.

For soft delay calculation, the 50% delay at the output V_{out} is first recorded in the presence of an SE charge. The delay measurement is then repeated with SE current source removed (no SE charge). The difference between the two delays is recorded as a *soft delay* at the output of the inverter string.

The critical charge for an SDE, $Q_{crit-delay}$ can be defined as the minimum charge collected due to a particle strike that produces sufficient delay such that the delayed signal arrives during the setup time of the storage element. To determine $Q_{crit-delay}$, the deposited charge is slowly increased until latching occurs at the storage element. Figure 13.8 shows the effect of threshold voltage on the critical charge $Q_{crit-delay}$ for

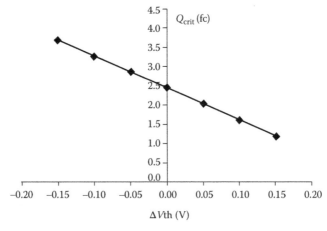

FIGURE 13.7   Change in Q_{crit} with change in device threshold.

the inverter circuit. The result shows that circuit robustness to soft delay effect can be increased with decreasing threshold voltage.

Based on these results, various dynamic threshold–based schemes have been examined for their vulnerability to SET and soft delay effects. These methods are widely used for high-speed and low-power operations. Results are then compared to that of conventional configuration where transistor body terminals are connected to source terminal.

In the comparison, five different circuits including benchmark circuits have been utilized. These circuits are a six-stage inverter chain, a c17 ISCAS-85 and an ISCAS-85 c432 benchmark circuit, a full-adder module in ISCAS-85 c6288, and finally the arithmetic and logical unit (ALU) module from AM2901 4-bit microprocessor bit-slice. Although it is dated, the ALU module alone contains 83 gates, 12 inputs and 10 outputs, and 276 SE vulnerable nodes [26].

In these configurations, the power supply voltage was taken as 0.6 V and the gate sizes have been assumed as minimum size. The hit locations for these circuits were selected at nodes close to primary inputs. For example, in the c17 benchmark circuit given in Figure 13.9, node "n7" was selected as a hit location and an erroneous signal

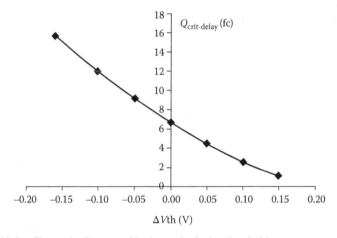

FIGURE 13.8 Change in $Q_{\text{crit-delay}}$ with change in device threshold.

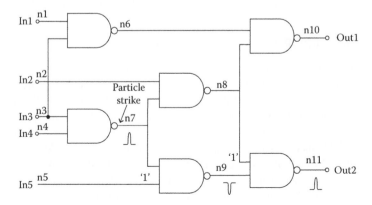

FIGURE 13.9   SET propagation in ISCAS-85 c17 circuit.

because of an SET was propagated to the output for observation. The critical charge value, Q_{crit} was determined using the procedure explained before.

The minimum (critical) soft delay required at the output to produce a SDE normally varies based on signal arrival time at storage element input. In simulations for $Q_{crit\text{-}delay}$, a critical delay of 200 ps was assumed for simulation convenience as we compared vulnerability of different DTMOS configurations. The minimum charge that created the critical delay has been recorded as critical charge $Q_{crit\text{-}delay}$.

The results for the critical charge simulation can be seen in Table 13.1. The first row shows the critical charge values for soft errors SEUs and SDEs when using a normal body tie design. For all four circuits considered, the body tie design achieves the smallest critical charge amongst all techniques and hence becomes the most vulnerable in terms of SET and soft delay effects. Compared to normal body tie design, all techniques improve critical charge, yet the standard DTMOS technique shows superior characteristics in terms of SEU robustness because of highest critical charge in all cases.

According to Table 13.1, the critical charge value needed for soft errors in standard DTMOS configuration is 50% more than what is required for normal body tie scheme for most cases, and hence is more robust in terms of SEU tolerance.

The reason can be explained as follows: Referring to Figure 13.4, when the input to the DTMOS inverter is low, the body for PMOS transistor is low and hence, the body of the PMOS transistor gets a forward bias with respect to source terminal. As this increases the drivability of the PMOS transistor, a negative SET at the output can be easily dissipated because of increased PMOS transistor current drive. A similar explanation can be made for the case that the input is high and a positive SET is present at the inverter output.

For soft delay error, the critical charge values for the standard DTMOS scheme is more than 60% higher almost in all cases compared to the normal body tie configuration. This can also be explained by referring to the inverter shown in Figure 13.4. If we assume the input waveform is a rising from logic low to high, a positive charge deposited by an SE particle on the output may result in a delay increase or soft delay on the output falling waveform. When input rises from low to high, the body of the NMOS transistor follows the input and as a result it gets a forward bias. This causes the output waveform to switch more rapidly and result in a reduced soft delay.

13.4   SOFT-ERROR HARDENING USING DTMOS

Many circuit-level techniques have been proposed by researchers to mitigate SETs in CL: Spatial redundancy techniques such as triple modular redundancy circuits triplicate the CL to be protected and then use a voting circuit to filter out the transient [27–29]. Temporal methods sample the data with different delays and produce the output to a voting circuit [30–32] to eliminate an SET. The driver sizing technique proposed in [24] increases device capacitance and drive current to decrease device vulnerability to SEUs. Larger drive strengths of NMOS and PMOS transistors quickly wOther techniques such as the one given in the work of Baze et al. [33] use different circuit layouts at the gate and transistor levels to create SEU hardened circuits.

TABLE 13.1

Critical Charge Values for Various Dynamic Threshold Inverter Schemes

Configuration	Six-stage Inverter		c17 ISCAS-85		ISCAS-85 c432		ISCAS-85 c6288		AM2901 ALU Module	
	Q_{crit} (fC)	$Q_{crit\text{-}delay}$ (fC)	Q_{crit} (fC)	$Q_{crit\text{-}delay}$ (fC)	Q_{crit} (fC)	$Q_{crit\text{-}delay}$ (fC)	Q_{crit} (fC)	$Q_{crit\text{-}delay}$ (fC)	Q_{crit} (fC)	$Q_{crit\text{-}delay}$ (fC)
Normal body tie	2.46	6.51	1.70	2.07	2.63	6.51	2.58	6.34	2.03	1.99
Standard DTMOS	3.72	11.05	2.64	4.56	3.76	10.30	3.79	10.86	3.01	3.74
Fig. 5a [17]	3.27	7.96	2.31	2.65	3.42	8.14	3.48	7.82	2.87	2.53
Fig. 5b [18]	3.53	8.33	2.50	2.55	3.59	7.94	3.64	7.98	3.44	2.56
Fig. 5c [19]	2.66	10.00	1.80	3.55	2.84	9.42	2.84	9.46	2.16	2.84
Fig. 5d [20]	3.22	9.39	2.12	3.24	3.34	9.26	3.36	8.81	2.56	3.02

Hardening against SDEs is also necessary in addition to SET hardening as soft delay effects will also be more pronounced in newer technologies because of reduced circuit node capacitances. Gill et al. [10] suggested the use of driver sizing technique in mitigating the soft delay effects, but this happens with the burden of increasing area and power penalties.

Our proposed hardening technique is based on the combined use of the standard DTMOS scheme along with driver sizing [34]. This combined approach results in considerable area saving compared to driver sizing alone. This is possible because a standard DTMOS gate is more SE robust compared to a conventional one.

In sizing simulation, we first consider the six-stage inverter chain consisting of conventional inverters (Figure 13.6) and apply various deposited charges in between 10 and 150 fC that are terrestrial level [24]. For each charge level, the necessary hit inverter size was determined to eliminate the soft error and soft delay effect. For SET-induced errors, the gate transistors are sized up until the soft error at the output is completely eliminated. For soft delay reduction, the hit driver has been sized up such that the delay is reduced to less than 200 ps. The same process has then been carried out for the standard DTMOS inverters.

Finally, the above-mentioned procedure was repeated for all four example circuits, and necessary gate sizes were determined to eliminate soft error and soft delay error effects. Table 13.2 shows the results obtained for all four circuits examined. For simplicity, the sizing has been done using whole driver sizes.

For better visualization, Figure 13.10 shows a histogram of data shown in Table 13.2, which indicates the average driver sizes (of all five circuits) needed to mitigate soft errors and SDEs in conventional and DTMOS configurations. For each error effect, the dark bar represents the sizing needed for a conventional driver gate and the light bar (shown next to it) represents the needed driver size in DTMOS configuration.

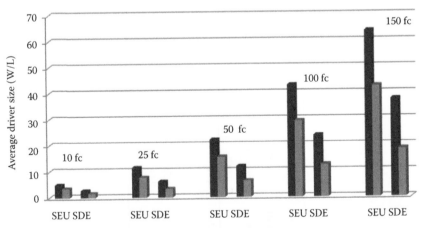

FIGURE 13.10 Average driver sizes needed for single-event upset and soft delay error mitigation in conventional and DTMOS driver configurations: Dark bar to the left represents the gate size needed for conventional driver and the light bar shown next to it shows the gate size needed for DTMOS driver configuration.

TABLE 13.2
Gate Sizes Required for Soft Error and Soft Delay Mitigation at Various Deposited Charge

Deposited Charge	Six-stage Inverter		c17 ISCAS-85		ISCAS-85 c432		ISCAS-85 c6288		AM2901 ALU Module	
	SEU Normal/DTMOS	SDE Normal/DTMOS	SEU Normal/DTMOS	SDE Normal/DTMOS	SEU Normal/DTMOS	SDE Normal/DTMOS	SEU Normal/DTMOS	SDE Normal/DTMOS	SEU Normal/DTMOS	SDE Normal/DTMOS
10 fC	5X/**3X**	2X/**1X**	7X/**5X**	4X/**2X**	4X/**3X**	2X/**1X**	4X/**3X**	2X/**1X**	5X/**4X**	4X/**3X**
25 fC	11X/**7X**	3X/**2X**	17X/**11X**	9X/**4X**	10X/**7X**	4X/**2X**	9X/**6X**	4X/**3X**	11X/**8X**	11X/**6X**
50 fC	21X/**15X**	6X/**4X**	34X/**22X**	17X/**8X**	19X/**14X**	7X/**4X**	17X/**11X**	7X/**5X**	20X/**16X**	23X/**11X**
100 fC	42X/**29X**	12X/**7X**	67X/**44X**	34X/**15X**	38X/**28X**	13X/**9X**	33X/**17X**	13X/**9X**	36X/**29X**	47X/**23X**
150 fC	62X/**43X**	18X/**11X**	100X/**66X**	50X/**22X**	57X/**42X**	19X/**13X**	50X/**21X**	19X/**14X**	51X/**42X**	83X/**33X**

Note: The bold entries represent DTMOS gate sizes.

Results show that the standard DTMOS technique can be used along with gate sizing in mitigating the SETs and soft delays using considerably less area overhead than conventional driver sizing. Compared to the conventional driver sizing technique given by Zhou and Mohanram [24], this combined approach saves about 30% in circuit area in SET mitigation and results in approximately 50% area savings in soft delay error mitigation.

13.5   CONCLUSION

As designers address reduction of static power consumption via optimizations, they need to be aware of the impact on SEU robustness. This chapter presented an analysis on various DTMOS schemes for their soft-error tolerance using various benchmark circuits including AM2901 microprocessor bit-slice. The analysis results indicate that all DTMOS configurations increase circuit robustness to SE-induced soft errors and delay effects because of increased transistor current drive. The standard DTMOS configuration, however, showed superior characteristics in terms of SEU robustness because of highest critical charge in all cases. To exploit this effect, this technique can be combined with driver sizing technique to mitigate SETs and soft delay effects with lot more area efficiency than driver sizing technique used alone. Compared to the conventional driver sizing technique, this combined approach saves about 30% in circuit area in SET mitigation and results in approximately 50% area savings in soft delay error mitigation.

REFERENCES

1. K. Roy, S. Mukhopadhyay, and H. Mahmoodi-Meimand, Leakage current mechanisms and leakage reduction techniques in deep-submicrometer CMOS circuits, *Proceedings of the IEEE*, 91(2): 305–327, 2003.
2. L. Wei, K. Roy, and V. K. De. Low voltage low power CMOS design techniques for deep submicron ICs, In *Proceedings of the 13th International Conference on VLSI Design*, pp. 24–29, Calcutta, India, 2000.
3. Y. S. Dhillon, A. U. Diril, A. Chatterjee, and A. D. Singh, Analysis and optimization of nanometer CMOS circuits for soft-error tolerance, *IEEE Transactions on Very Large Scale Integration* (VLSI) *Systems*, 14(5): 514–524, 2006.
4. M. R. Choudhury, Q. Zhou, and K. Mohanram, Design optimization for single-event upset robustness using simultaneous dual-VDD and sizing techniques, In *Proceedings of International Conference on Computer Aided Design (ICCAD)*, pp. 204–209, San Jose, CA, 2006.
5. K-C. Wu and D. Marculescu, Power-aware soft error hardening via selective voltage scaling, In *Proceedings of International Conference on Computer Design (ICCD)*, pp. 301–306, Lake Tahoe, CA, 2008.
6. F. Assaderaghi, D. Sinitsky, S. Parke, J. Bokor, P. K. Ko, and C. Hu, A dynamic threshold voltage MOSFET (DTMOS) for ultra-low voltage operation, In *IEDM Technical Digest*, pp. 809–812, San Francisco, CA, 1994.
7. P. E. Dodd and L. W. Massengill, Basic mechanisms and modeling of single-event upset in digital microelectronics, *IEEE Transactions on Nuclear Science*, 50(3): 583–602, 2003.
8. S. Mitra, T. Karnik, N. Seifert, and M. Zhang, Logic soft errors in sub-65 nm technologies design and CAD challenges, In *Proceedings of Design Automation Conference*, pp. 2–3, Anaheim, CA, 2005.

9. R. C. Baumann, Radiation-induced soft errors in advanced semiconductor technologies, *IEEE Transactions on Device and Material Reliability*, 5(3): 305–316, 2005.

10. B. S. Gill, C. Papachristou, and F. G. Wolff, Soft delay error effects in CMOS combinational circuits, In *Proceedings of 22nd VLSI Test Symposium*, pp. 325–330, Napa Valley, CA, 2004.

11. A. Lesea and P. Alfke, Xilinx FPGAs Overcome the Side Effects of Sub-90 nm Technology–a White Paper, Technical Report WP256 (v1.2), Xilinx corporation, San Jose, CA, Jan. 2007.

12. E. Normand, Single-event effects in avionics, *IEEE Transactions on Nuclear Science*, 43(2): 461–474, 1996.

13. P. D. Bradley and E. Normand, Single event upset in implantable cardioverter defibrillators, *IEEE Transactions on Nuclear Science*, 45(6): 2929–2940, 2004

14. D. Zhu and H. Aydin, Reliability-aware energy management for periodic real-time tasks, *IEEE Transactions on Computers*, 58(10): 1382–1397, 2009.

15. V. Degalahal, R. Ramanarayanan, N. Vijaykrishnan, Y. Xie, and M. J. Irwin, The effect of threshold voltages on the soft error rate, In *Proceedings of 5th International Symposium on Quality Electronic Design (ISQED 04)*, pp. 503–508, San Jose, CA, 2004.

16. F. Assaderaghi, D. Sinitsky, S. Parke, J. Bokor, P. K. Ko, and C. Hu, A dynamic threshold voltage MOSFET (DTMOS) for ultra-low voltage VLSI, *IEEE Transactions on Electron Devices*, 44(3): 414–422, 1997.

17. I. Chung, Y. Park, and H. Min, A new SOI inverter for low power applications, In *Proceedings of 1996 IEEE International SOI Conference*, pp. 20–21, Sanibel Island, FL, 1996.

18. J. Gil, M. Je, J. Lee, and H. Shin, A high speed and low power SOI inverter using active body-bias, In *International Symposium Low Power Electronics and Design*, pp. 59–63, Monterey, CA, 1998.

19. A. Drake, K. Nowka, and R. Brown, Evaluation of dynamic-threshold logic for low-power VLSI design in 0.13um PD-SOI, In Proceedings of 2003 VLSI-SOC, pp. 263–266, Darmstadt, Germany, 2003.

20. S. Soleimani, A. Sammak, and B. Forouzandeh, A novel ultra-low-energy bulk dynamic threshold inverter scheme, In *Proceedings of IMECS*, pp. 505–508, Hong Kong, China, 2009.

21. Predictive Technology Model (PTM), 2012. http://www.eas.asu.edu/~ptm

22. J. M. Hutson, V. Ramachandran, B. L. Bhuva, X. Zhu, R. D. Schrimpf, O. A. Amusan, and L. W Massengill, Single event induced error propagation through nominally-off transmission gates, *IEEE Transactions on Nuclear Science*, 53(6): 3558–3562, 2006.

23. L. B. Freeman, Critical charge calculations for a bipolar SRAM array, *IBM Journal of Research and Development*, 40: 119–129, 1996.

24. Q. Zhou and K. Mohanram, Gate sizing to radiation harden combinational logic, *IEEE Transactions on Computer-Aided Design of Integrated Circuits and Systems*, 25(1): 155–166, 2006.

25. D. C. Ness, C. J. Hescott, and D. J. Lilja, Improving nanoelectronic designs using a statistical approach to identify key parameters in circuit level SEU simulations, In *Proceedings of the 2007 IEEE International Symposium on Nanoscale Architecture*, pp. 46–53, San Jose, CA, 2007.

26. L. W. Massengill, A. E. Baranski, D. O. Van Nort, J. Meng, and B. L. Bhuva, Analysis of single-event effects in combinational logic-simulation of the AM2901 bitslice processor, *IEEE Transactions on Nuclear Science*, 47(6): 2609–2615, 2000.

27. S. Buchner and M. Baze, Single-Event Transients in Fast Electronic Systems, In *2001 NSREC Short Course*, Vancouver, BC, Canada, 2001.

28. R. D. Schrimpf and D. M. Fleetwood, *Radiation Effects and Soft Errors in Integrated Circuits and Electronic Devices*, World Scientific, Singapore, 2004.

29. R. Oliveira, A. Jagirdar, and T. Chakraborty, A TMR Scheme for SEU mitigation in scan flip-flops, In *8th International Symposium on Quality Electronic Design (ISQED'07)*, pp. 905–910, San Jose, CA, 2007.

30. D. G. Mavis and P. H. Eaton, Soft error rate mitigation techniques for modern micro-circuits, In *Proceedings of International Reliability Physics Symposium*, pp. 216–225, Dallas, TX, 2002.

31. M. Nicolaidis, Time redundancy based soft-error tolerance to rescue nanometer tech-nologies, In *Proceedings of the 17th IEEE VLSI Test Symposium*, pp. 86–89, Dana Point, CA, 1999.

32. S. Krishnamohan and N. R. Mahapatra, A highly-efficient technique for reducing soft errors in static CMOS circuits, In *Proceedings of the IEEE International Conference on Computer Design*, pp. 126–131, San Jose, CA, 2004.

33. M. P. Baze, S. P. Buchner, and D. McMorrow, A digital CMOS design technique for SEU hardening, *IEEE Transactions on Nuclear Science*, 47(6): 2603–2608, 2000.

34. Sayil, S. and N. B. Patel, Soft error and soft delay mitigation using dynamic threshold technique, *IEEE Transactions on Nuclear Science*, 57(6): 3553–3559, 2010.

14 Future of Asynchronous Logic

Scott C. Smith and Jia Di

CONTENTS

14.1 INTRODUCTION

The 2012 International Technology Roadmap for Semiconductors (ITRS) [1] states that asynchronous circuits currently account for 22% of logic within the multibillion dollar semiconductor industry, and predicts that this percentage will more than double over the next 10 years. Asynchronous logic has been around for the past 50+ years, but, until recently, synchronous circuits have been good enough to meet industry needs, so asynchronous circuits were primarily utilized for niche markets and in the research domain. However, as transistor size continues to decrease, asynchronous circuits are being looked to by industry to solve power dissipation and process variability issues associated with these emerging sub-90 nm circuits. This chapter details the state of the art of asynchronous logic, how asynchronous circuits are currently being utilized in the industry, and the future of asynchronous logic.

14.2 MODERN ASYNCHRONOUS LOGIC

Asynchronous circuits were originally designed using a bounded-delay model, which assumes that delays in both gates and wires are known, or at least bounded [2]. This requires delays to be added based on worst-case scenarios to avoid hazard conditions, which leads to extensive timing analysis to ensure correct circuit operation. In addition, bounded-delay asynchronous circuits are not suited to datapath design. Hence, micropipelines [3] were developed to avoid these issues by utilizing a bounded-delay datapath along with delay-insensitive control. Micropipelines utilize the bundled data convention, where the datapath consists of standard Boolean logic with an extra ready signal that matches the worse-case delay in the datapath to ensure that data is valid before being latched at the subsequent register. However, micropipelines still require timing analysis and delay matching, resulting in worse-case delay. Therefore, asynchronous logic research turned to completely delay-insensitive circuits, which assume that delays in both gates and wires are unbounded; however, this assumption severely limits circuit practicality [4]. So, isochronic wire forks [5,6] are assumed within basic components, such as a full-adder, meaning that wire delays within a component are assumed to be much less than the logic element delays within the component. Wires connecting components do not have to adhere to the isochronic fork assumption, making this a valid supposition even in current and future nanometer technologies. Delay-insensitive circuits with this isochronic fork assumption are referred to as quasi-delay-insensitive (QDI), and are the basis for most modern day asynchronous circuits. QDI circuits operate in the presence of indefinite arrival times for the reception of inputs, and utilize completion detection of output signals to provide handshaking control of input wave fronts. QDI circuits therefore require very little, if any, timing analysis to ensure correct operation (i.e., they are correct-by-construction), and also yield average-case performance rather than the worse-case performance of bounded-delay and synchronous paradigms.

Figure 14.1 shows a generalized block diagram of a QDI system, where QDI combinational logic is sandwiched between QDI registers, which is very similar to synchronous circuits, with the difference being that synchronous circuits utilize a clock signal to simultaneously latch data at the registers, whereas QDI circuits utilize completion detection of the combinational logic signals to generate handshaking signals that are used to latch data at a register. Note that some QDI paradigms combine combinational logic and registration together into single components, and the method of generating handshaking signals also varies.

Synchronous, bounded-delay asynchronous, and micropipelines all use a single wire to encode one bit of data: GND is encoded as Logic 0 and V_{DD} is encoded as Logic 1; the outputs of a computation are then sampled after waiting a predetermined amount of time that is long enough to ensure that the computation has been completed. On the other hand, QDI circuits do not rely on timing to determine when a computation has finished, so they require a different way of encoding data to determine computation completion. Typically, a 1-hot encoding scheme is utilized, such as dual-rail logic, depicted in Table 14.1, where two wires, D^0 and D^1, are used to encode one bit of data. D^0 being asserted is referred to as DATA0 and depicts a Logic 0; D^1 being asserted is referred to as DATA1 and depicts a Logic 1; and neither D^0 nor D^1 being asserted is referred to as the spacer, or NULL, state, which depicts that the computation has not yet finished. Note that D^0 and D^1 are mutually exclusive, such

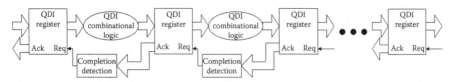

FIGURE 14.1 Quasi-delay-insensitive system framework: input wave fronts are controlled by local handshaking signals and completion detection instead of by a global clock signal. Feedback requires at least three quasi-delay-insensitive registers in the feedback loop to prevent deadlock.

TABLE 14.1
Dual-Rail Signal

	DATA0	DATA1	NULL	Illegal
D^0	1	0	0	1
D^1	0	1	0	1

that both rails can never be asserted simultaneously; this state is illegal, and will never occur in a properly operating system.

Using dual-rail logic, timing is no longer needed to determine computation completion. Starting with the system in an all NULL state, DATA is input to the system, which flows through the combinational logic, eventually transitioning all of the outputs to DATA, which depicts that the computation has finished. The system must then transition back into the all NULL state before the next DATA wave front is input to the system. NORing the two rails of a dual-rail signal generates a Logic 0 when DATA and a Logic 1 when NULL, which can be used as an acknowledge signal. To signal when all outputs are DATA or NULL, to determine when the computation or return to NULL has finished, respectively, the acknowledge signals must all be combined together using what is referred to as a C-element [7]. A C-element only changes state when all inputs are the same; when all inputs are asserted, the output is asserted, which then remains asserted until all inputs are deasserted, which requires internal feedback within the C-element gate. The C-element output can then be used to request the next DATA wave front input when asserted, and the return to NULL when deasserted.

Of the various implementations of QDI circuits, there are two main paradigms that have been primarily utilized in industry: NULL convention logic (NCL) [8] and precharge half buffers (PCHB) [9].

14.2.1 NULL CONVENTION LOGIC

NCL circuits are composed of 27 fundamental gates, as shown in Table 14.2, which constitute the set of all functions consisting of four or fewer variables. Because each rail of an NCL signal is considered a separate variable, a four-variable function is not the same as a function of four literals, which would consist of eight variables for dual-rail logic (e.g., a literal includes both a variable and its complement, F and F', whereas NCL rails are never complemented, such that a dual-rail NCL signal, F, consists of two variables, F^1 and F^0, where F^0 is equivalent to F'). The primary type of threshold gate, shown in Figure 14.2a, is the THmn gate, where $1 \le m \le n$. THmn gates have n inputs, and at least m of the n inputs must be asserted before the output will become asserted. In a THmn gate, each of the n inputs is connected to the rounded portion of the gate; the output emanates from the pointed end of the gate; and the gate's threshold value, m, is written inside of the gate.

Another type of threshold gate is referred to as a weighted threshold gate, denoted as THmnW$w_1w_2...w_R$. Weighted threshold gates have an integer value, $m \ge w_R > 1$, applied to *inputR*. Here $1 \le R < n$; where n is the number of inputs; m is the gate's threshold; and $w_1, w_2, ... w_R$, each > 1, are the integer weights of *input1*, *input2*, ... *inputR*,

TABLE 14.2

Twenty-Seven Fundamental NULL Convention Logic Gates

NCL Gate	Boolean Function	Transistor Count (Static)	Transistor Count (Semistatic)
TH12	A + B	6	6
TH22	AB	12	8
TH13	A + B + C	8	8
TH23	AB + AC + BC	18	12
TH33	ABC	16	10
TH23w2	A + BC	14	10
TH33w2	AB + AC	14	10
TH14	A + B + C + D	10	10
TH24	AB + AC + AD + BC + BD + CD	26	16
TH34	ABC + ABD + ACD + BCD	24	16
TH44	ABCD	20	12
TH24w2	A + BC + BD + CD	20	14
TH34w2	AB + AC + AD + BCD	22	15
TH44w2	ABC + ABD + ACD	23	15
TH34w3	A + BCD	18	12
TH44w3	AB + AC + AD	16	12
TH24w22	A + B + CD	16	12
TH34w22	AB + AC + AD + BC + BD	22	14
TH44w22	AB + ACD + BCD	22	14
TH54w22	ABC + ABD	18	12
TH34w32	A + BC + BD	17	12
TH54w32	AB + ACD	20	12
TH44w322	AB + AC + AD + BC	20	14
TH54w322	AB + AC + BCD	21	14
THxor0	AB + CD	20	12
THand0	AB + BC + AD	19	13
TH24comp	AC + BC + AD + BD	18	12

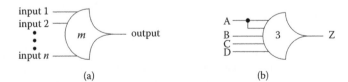

(a)   (b)

FIGURE 14.2 NULL convention logic threshold gates: (a) THmn Gate; (b) TH34w2 Gate: Z = AB + AC + AD + BCD.

respectively. For example, consider the TH34w2 gate, whose $n = 4$ inputs are labeled A, B, C, and D, shown in Figure 14.2b. The weight of input A is therefore 2. Because the gate's threshold, m, is 3, this implies that in order for the output to be asserted, either inputs B, C, and D must all be asserted, or input A must be asserted along with any other input, B, C, or D.

NCL threshold gates are designed with hysteresis state-holding capability, such that after the output is asserted, all inputs must be deasserted before the output will be deasserted. Hysteresis ensures a complete transition of inputs back to NULL before asserting the output associated with the next wave front of input data. Hence, a THnn gate is equivalent to an n-input C-element, and a TH1n gate is equivalent to an n-input OR gate. NCL threshold gates may also include a *reset* input to initialize the output. Circuit diagrams designate resettable gates by either a d or an n appearing inside the gate, along with the gate's threshold. d denotes the gate as being reset to Logic 1; n, to Logic 0. These resettable gates are used in the design of NCL registers.

NCL systems contain at least two NCL registers, one at both the input and at the output. Two adjacent register stages interact through their request and acknowledge signals, K_i and K_o, respectively, to prevent the current DATA wave front from overwriting the previous DATA wave front, by ensuring that the two DATA wave fronts are always separated by a NULL wave front. The acknowledge signals are combined in the completion detection circuitry to produce the request signal(s) to the previous register stage, as depicted in Figure 14.1. NCL registration is realized through cascaded arrangements of single-bit dual-rail registers, depicted in Figure 14.3. A dual-rail register consists of two TH22 gates (2-input C-elements) that pass a DATA value at the input only when K_i is *request for data* (rfd) (i.e., Logic 1), and likewise pass NULL only when K_i is *request for null* (rfn) (i.e., Logic 0). They also contain a NOR gate to generate K_o, which is *rfn* when the register output is DATA and *rfd* when the register output is NULL. The register depicted in Figure 14.3 is reset to NULL, because both TH22 gates are reset to Logic 0. However, the register could be instead reset to DATA0 or DATA1 by replacing exactly one of the TH22n gates with a TH22d gate.

An N-bit register stage, composed of N single-bit dual-rail NCL registers, requires N completion signals, one for each bit. The NCL completion component, shown in Figure 14.4, uses these N K_o lines to detect complete DATA and NULL sets at the output of every register stage and request the next NULL and DATA set, respectively. In full-word completion, the single-bit output of the completion component is connected to all K_i lines of the previous register stage. Because the maximum input threshold gate is the TH44 gate, the number of logic levels in the completion component for an N-bit register is given by $\lceil \log_4 N \rceil$. Figures 14.5 and 14.6 show the

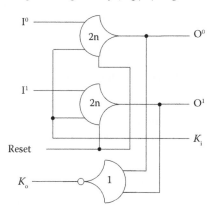

FIGURE 14.3   Single-bit dual-rail register.

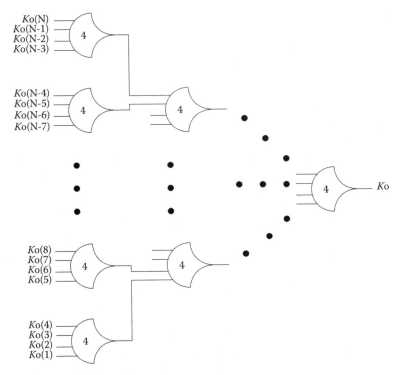

FIGURE 14.4 *N*-bit completion component.

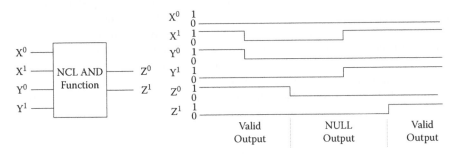

FIGURE 14.5 NULL convention logic AND function: $Z = X \cdot Y$: initially $X =$ DATA1 and $Y =$ DATA0, so $Z =$ DATA0; next X and Y both transition to NULL, so Z transitions to NULL; then X and Y both transition to DATA1, so Z transitions to DATA1.

flow of DATA and NULL wave fronts through an NCL combinational circuit (i.e., an AND function) and an arbitrary pipeline stage, respectively. The average DATA/NULL cycle time, referred to as T_{DD}, is comparable to the clock frequency of a synchronous circuit.

Because NCL threshold gates are designed with hysteresis state-holding capability, which requires internal feedback, NCL gates require both *set* and *hold* equations, where the *set* equation determines when the gate will become asserted and the *hold* equation determines when the gate will remain asserted once it has been asserted. The *set* equation determines the gate's functionality as one of the 27 NCL gates, as

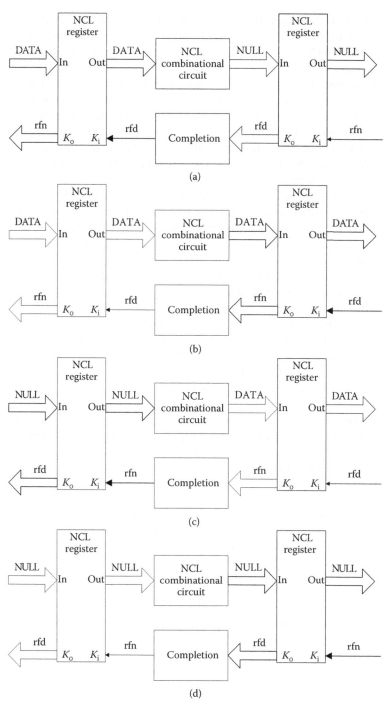

FIGURE 14.6 NULL convention logic DATA/NULL cycle. (a) DATA flowing through input register then combinational logic. (b) DATA flowing through output register then rfn flowing through completion logic. (c) NULL flowing through input register then combinational logic. (d) NULL flowing through output register then rfd flowing through completion logic.

listed in Table 14.2, whereas the *hold1* equation is simply all inputs ORed together, to ensure that the output remains asserted until all inputs are deasserted. The general equation for an NCL gate with output Z is $Z = set + (Z^- \cdot hold1)$, where Z^- is the previous output value and Z is the new value.

To implement an NCL gate using complementary metal–oxide–semiconductor (CMOS) technology, an equation for the complement of Z is also required, which in general form is: $Z' = reset + (Z^{-'} \cdot hold0)$, where *reset* is the complement of *hold1* (i.e., the complement of each input, ANDed together) and hold0 is the complement of *set*, such that the gate output is deasserted when all inputs are deasserted, and then remains deasserted while the gate's *set* condition is false. As shown in Figure 14.7, for the static realization, the equations for Z and Z', given in this paragraph and the preceding paragraph, respectively, are directly implemented in the NMOS and PMOS logic, respectively, after simplifying, whereas, the semistatic realization only requires the *set* and *reset* equations to be implemented in the NMOS and PMOS logic, respectively, and hold0 and hold1 are implemented using a weak feedback inverter.

For example, the *set* equation for the TH23 gate is $AB + AC + BC$, as given in Table 14.2, and the *hold* equation is $A + B + C$; therefore, the gate is asserted when at least two inputs are asserted, and then remains asserted until all inputs are deasserted. The *reset* equation is $A'B'C'$ and the simplified *set'* equation is $A'B' + B'C' + A'C'$. Directly implementing these equations for Z and Z', after simplification, yields the static transistor-level realization shown in Figure 14.8a. The semistatic TH23 gate is shown in Figure 14.8b. In general, the semistatic implementation requires fewer transistors, but is slightly slower because of the weak inverter. Note that TH1n gates are simply OR gates and do not require any feedback, such that their static and semistatic implementations are exactly the same.

NCL combinational logic is designed utilizing the 27 NCL gates in Table 14.2, to generate a circuit that must be input-complete and observable [10], meaning that all outputs cannot transition from NULL to DATA until all inputs have transitioned from NULL to DATA, and that all asserted gates contribute to asserting at least one circuit output, respectively. Note that, for circuits with multiple outputs, it is acceptable according to Seitz's "weak conditions" of delay-insensitive signaling [11], for some of the outputs to transition without having a complete input set present, as long as all outputs cannot transition before all inputs arrive. Figure 14.9 shows an example of an NCL full adder, and complete details of the methodology for designing NCL circuits can be found in Smith et al. [10].

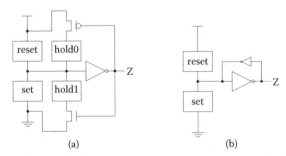

FIGURE 14.7 NULL convention logic gate realizations: (a) static implementation; (b) semistatic implementation.

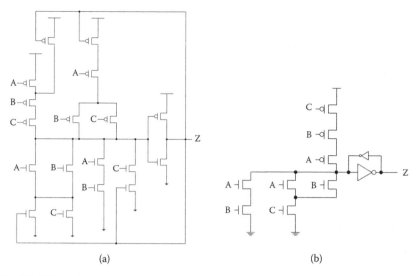

FIGURE 14.8   TH23 gate realizations: (a) static implementation; (b) semistatic implementation.

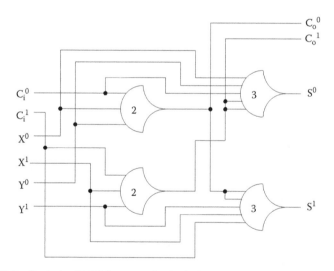

FIGURE 14.9   Optimized NULL convention logic full adder.

14.2.2   PRECHARGE HALF BUFFERS

PCHB circuits are designed at the transistor level, utilizing a style similar to domino logic, instead of targeting a predefined set of gates like NCL. PCHB circuits have dual-rail data inputs and outputs, and combine combinational logic and registration together into a single block, yielding a very fine-grain pipelined architecture.

Figure 14.10 shows an example of a PCHB NAND2 circuit. The dual-rail output is initially precharged to NULL. When request (R_{ack}) and acknowledge (L_{ack}) are both *rfd*, the specific function will evaluate when the inputs, X and/or Y, become DATA, causing the output, F, to become DATA. L_{ack} will then transition to *rfn* only

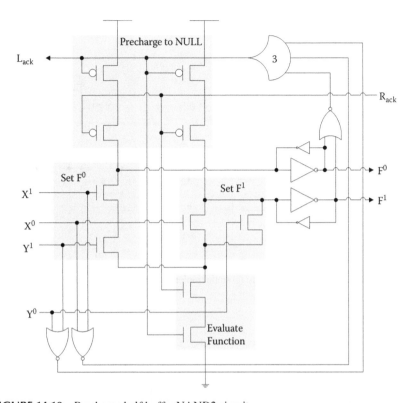

FIGURE 14.10 Precharge half buffer NAND2 circuit.

after all inputs and the output are DATA. When R_{ack} is *rfn* and L_{ack} is *rfd*, or vice versa, the output will be floating, so weak inverters must be used to hold the current output value to maintain delay-insensitivity. After both R_{ack} and L_{ack} are *rfn*, the output will be precharged back to NULL. After all inputs become NULL and the output changes to NULL, L_{ack} will change back to *rfd*, and the next DATA wave front can evaluate after R_{ack} becomes *rfd*. Note that R_{ack} and L_{ack} are equivalent to NCL's K_i and K_o, respectively, and both PCHB and NCL utilize the same protocol, referred to as four-phase return-to-zero handshaking.

Different from NCL, in PCHB, computation and registration are integrated into a single transistor-level component, which ensures observability; and both the inputs and outputs of a component are used to generate acknowledge signals, which ensures input completeness.

14.3 ASYNCHRONOUS CIRCUITS IN INDUSTRY

There have been a number of start-up companies developing asynchronous logic technology and products over the past 15 years, including Handshake Solutions, Tiempo, Theseus Logic, NanoWatt Design, Fulcrum Microsystems, and Achronix Semiconductor Corporation. Handshake Solutions was a line of business of Philips

Electronics' Technology Incubator that most notably developed low-power micro-pipeline-based bundled data asynchronous 8051 microcontrollers and ARM processors for the smart card and automotive markets; however, Handshake Solutions no longer exists. Tiempo is a fabless IC company that licenses asynchronous IP cores and accompanying automated design tools, based on their proprietary technology, which like NCL and PCHB, utilizes dual-rail logic and four-phase return-to-zero handshaking [12]. Theseus Logic, founded by the inventor of NCL, partnered with a number of companies, such as Motorola and Atmel, to develop an asynchronous microprocessor core, field-programmable gate array (FPGA), and automated design tools. Theseus Logic no longer exists; however, their NCL technology is currently being used by Camgian Microsystems to design low-power highly integrated mixed-signal Systems-on-Chip (SoCs) for wireless sensor nodes [13], and variations of NCL are being utilized by NanoWatt Design, a fabless IC company focusing on ultra-low power asynchronous design for use in mobile electronics and other devices requiring low power, such as wireless sensor nodes, utilizing their proprietary sleep convention logic (SCL) technology [14], and by Wave Semiconductor to develop a proprietary low-power, high-speed, low-cost reprogrammable device, called Azure, which is structured as a 2D mesh of byte-level processing elements that can be quickly and partially reconfigured to support expedient hardware context switching [15].

SCL, also referred to as MTNCL, combines NCL with multithreshold CMOS [16], to yield a patented and patent-pending ultra-low power asynchronous circuit design methodology [17–19]. SCL utilizes the same dual-rail logic and four-phase return-to-zero handshaking as other QDI paradigms; however, instead of propagating the NULL wave front to reset all gates to zero as in NCL, all gates in a stage are simultaneously slept to zero using the handshaking signals. This requires a modification in the overall architecture, shown in Figure 14.11, and in the gate design. The completion logic inputs now come from the register inputs, instead of register outputs, and require the completion signal from the previous stage as an additional input, which is referred to as Early Completion [20]. The new gate design, used to implement the combinational logic and completion logic, now includes a sleep input, and no longer requires internal hysteresis feedback, as shown in Figure 14.12. This allows for SCL combinational logic to no longer require input-completeness and observability, because all logic in a stage is now simultaneously reset to NULL instead of propagating the NULL wave front, which results in substantially smaller and faster circuits. Also note that select high-threshold (high-Vt) transistors (shown in dotted circles in Figure 14.12) are utilized in SCL gates to ensure that all paths

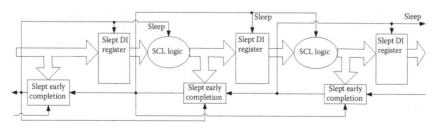

FIGURE 14.11 Sleep convention logic circuit architecture.

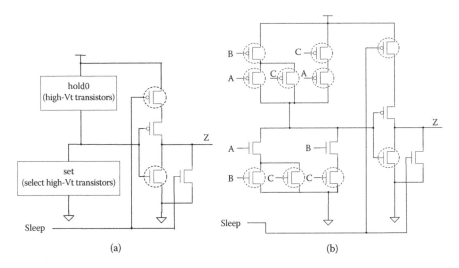

FIGURE 14.12 Sleep convention logic gate realization: (a) general implementation; (b) TH23 example.

from V_{DD} to GND include at least one high-Vt transistor that is off when the gate is deasserted, while using faster low-Vt transistors for the switching logic, to yield extremely low leakage power while the circuit is idle in the slept NULL state without substantially degrading speed.

Wave Semiconductor's patent-pending technology is similar to SCL in that it can simultaneously sleep a stage's gates to zero instead of propagating the NULL wave front, and can utilize select high-Vt transistors to reduce leakage power [21–23]. It utilizes a precharge transistor to pull up the internal node, causing the output to become zero, similar to PCHB. Like SCL, it utilizes a modified threshold gate design to build dual-rail combinational logic, combined with four-phase return-to-zero handshaking.

Fulcrum Microsystems utilized PCHB technology on a low-leakage 130-nm process to develop 4-port and then 24-port network switch chips, both of whose channels operated at 10 Gb speed with 200 ns latency through the chip, which were the fastest networking chips on the market. In 2011, Intel purchased Fulcrum Microsystems, who are now Intel Switch & Router Division, producing Ethernet switch chips with up to 64 10 Gb ports [24].

Achronix develops high-speed and low-power FPGAs that have a synchronous interface with an asynchronous core that utilizes PCHB technology [25]. Achronix FPGAs utilize Intel's 22-nm 3D tri-gate FinFET process, and include high-density versions that offer up to 1.7 million effective LUTs that utilize approximately half the power of standard commercial synchronous FPGAs, and high-speed versions that offer up to 590 thousand effective LUTs and operate at up to 1.5 GHz, three to four times faster than commercial synchronous FPGAs. Achronix FPGAs are programmed using the Achronix CAD Environment (ACE), which works in conjunction with industry-standard tools to synthesize standard Very High Speed Integrated Circuit Hardware Description Language or Verilog RTL code.

14.4 THE NEXT 10 YEARS

Ten years ago, in 2003, ITRS predicted a gradual industry shift from synchronous to asynchronous design styles to help mitigate power, robustness to process variation, and timing issues, as process technology continued to shrink [26]. The 2005 ITRS edition predicted asynchronous circuits to constitute 20% of the industry by 2012 [27]; this prediction was indeed confirmed in the latest 2012 ITRS update, which shows that asynchronous technology did account for 20% of the industry in 2012 [1]. Looking forward, ITRS predicts asynchronous circuits to account for 49% of the multibillion dollar semiconductor industry by 2024, and continue to grow to become the predominant design style [1]; hence, the future of asynchronous circuits is looking very bright.

QDI paradigms are the path forward for asynchronous circuits in the industry, because they do not rely on timing assumptions, which, due to process variation, are becoming more and more unpredictable as transistor size continues to shrink. QDI circuits are the underlying technology of all the major asynchronous companies mentioned in Section 14.3 whom are still in business, and are being utilized by the industry's leader, Intel, who purchased Fulcrum Microsystems and is currently partnering with Achronix.

Looking back, the major impediments to asynchronous design becoming more prevalent in the industry were the lack of industry-standard CAD tools for asynchronous logic and comparatively little asynchronous expertise. Because synchronous circuits were good enough to produce next generation products, asynchronous solutions received little attention. However, as synchronous circuits are becoming more and more problematic for cutting-edge process technologies, QDI paradigms are being used much more frequently, and not just by start-up companies for niche markets, but also by industry leaders, like Intel. This transition has been aided by development of mature CAD tools for asynchronous logic, such as Achronix's ACE, which hide the details of asynchronous circuits from designers, allowing them to utilize the current synchronous design style in which they are accustomed. Synchronous circuits will never go away entirely, because they are very good at performing certain tasks, such as measuring time (e.g., timers in a microcontroller). However, as asynchronous technology continues to increase its market share, comprising approximately one half of the multibillion dollar semiconductor industry within the next 10 years, circuit designers will be expected to know how to design asynchronous circuits as well as synchronous circuits; so, if you are new to asynchronous circuit design, the following references will help get you up to speed [9,10].

REFERENCES

1. ITRS, ITRS 2012 update, available at http://www.itrs.net/Links/2012ITRS/Home2012 .htm; accessed on October 2013.
2. S. H. Unger, *Asynchronous Sequential Switching Circuits*, Wiley, New York, NY, 1969.
3. E. Sutherland, "Micropipelines," *Communications of the ACM*, 32(6): 720–738, 1989.
4. A. J. Martin, "The limitations to delay-insensitivity in asynchronous circuits," *MIT Conference on Advanced Research in VLSI*, Cambridge, MA, pp. 263–278, 1990.
5. A. J. Martin, "Programming in VLSI," in *Development in Concurrency and Communication*, Addison-Wesley, Reading, MA, pp. 1–64, 1990.

6. K. Van Berkel, "Beware the Isochronic Fork," *Integration, the VLSI Journal*, 13(2): 103–128, 1992.

7. D. E. Muller, "Asynchronous logics and application to information processing," in *Switching Theory in Space Technology*, pp. 289–297, Stanford University Press, Palo Alto, CA, 1963.

8. K. M. Fant and S. A. Brandt, "NULL Convention Logic: A complete and consistent logic for asynchronous digital circuit synthesis," *International Conference on Application Specific Systems, Architectures, and Processors*, pp. 261–273, Chicago, IL, 1996.

9. A. J. Martin and M. Nystrom, "Asynchronous techniques for system-on-chip design," *Proceedings of the IEEE*, 94(6), 1089–1120, 2006.

10. S. C. Smith and J. Di, "Designing asynchronous circuits using NULL convention logic (NCL)," *Synthesis Lectures on Digital Circuits and Systems*, Morgan & Claypool Publishers, San Rafael, CA, 4(1), July 2009.

11. C. L. Seitz, "System timing," in *Introduction to VLSI Systems*, Addison-Wesley, pp. 218–262, Reading, MA, 1980.

12. Tiempo company website, available at http://www.tiempo-ic.com/, accessed on October 2013.

13. Camgian Microelectronics company website, available at http://www.camgian.com/, accessed on October 2013.

14. NanoWatt Design company website, available at http://www.nanowattdesign.com/, accessed on October 2013.

15. Wave Semiconductor company website, available at http://wavesemi.com/, accessed on October 2013.

16. S. Mutoh, T. Douseki, Y. Matsuya, T. Aoki, S. Shigematsu, and J. Yamada, "1-V power supply high-speed digital circuit technology with multithreshold-voltage CMOS," *IEEE Journal of Solid-State Circuits*, 30(8): 847–854, August 1995.

17. Jia Di and Scott Christopher Smith, "Ultra-low power multi-threshold asynchronous circuit design," U.S. Patent 7,977,972 B2, filed April 30, 2010, and issued July 12, 2011.

18. Jia Di and Scott Christopher Smith, "Ultra-low power multi-threshold asynchronous circuit design," U.S. Patent 8,207,758 B2, filed July 1, 2011, and issued June 26, 2012.

19. Scott Christopher Smith and Jia Di, "Multi-threshold sleep convention logic without NSleep," U.S. Patent Application 20130181740, filed July 18, 2013.

20. S. C. Smith, "Speedup of self-timed digital systems using early completion," *IEEE Computer Society Annual Symposium on VLSI*, pp. 107–113, Pittsburgh, PA, 2002.

21. Wave Semiconductor, Inc., "Multi-Threshold Flash NCL Circuitry," U.S. Patent Application 20130214813, filed August 22, 2013.

22. Wave Semiconductor, Inc., "Self-Ready Flash NULL Convention Logic," U.S. Patent Application 20130214814, filed August 22, 2013.

23. Wave Semiconductor, Inc., "Implementation Method for Fast NCL Data Path," U.S. Patent Application 20130249594, filed September 26, 2013.

24. Fulcrum Microsystem company website, available at http://www.fulcrummicro.com/, accessed on October 2013.

25. Achronix Semiconductor Corporation company website, available at http://www.achronix.com/, accessed on October 2013.

26. ITRS, International technology roadmap for semiconductors 2003 edition, Design, available at http://www.itrs.net/Links/2003ITRS/Design2003.pdf, accessed on October 2013.

27. ITRS, International technology roadmap for semiconductors 2005 edition, Design, available at http://www.itrs.net/Links/2005ITRS/Design2005.pdf, accessed on October 2013.

15 Memristor-CMOS-Hybrid Synaptic Devices Exhibiting Spike-Timing-Dependent Plasticity

Tetsuya Asai

CONTENTS

15.1 INTRODUCTION

One of a challenging issue in developing brain-inspired hardware is to implement nonvolatile analog synaptic devices in a compact structure. Various types of volatile (capacitor-based) analog synaptic devices have been developed [1–3], whereas types of nonvolatile analog synaptic devices are limited so far. Present nonvolatile synaptic devices use flash-based or its alternative technologies [4–6], however, they had difficulties in designing compact circuits for electron injection and ejection as well as in establishing the high reliability.

Recently, the so-called "memristor" originally introduced by Leon Chua in 1971 [7], which was claimed to be the fourth circuit element exhibiting the relationship between flux φ and charge q, has been respotlighted since Strukov et al. [8] presented the equivalent physical examples. The presented device was a bipolar ReRAM, and did not *directly* exhibit the relationship [$\varphi = f(q)$], however, the device could demonstrate its equivalent dynamics given by the temporal deviation of $\varphi = f(q)$ as

$$\frac{\mathrm{d}\phi}{\mathrm{d}t} = \frac{\partial f(q)}{\partial q}\frac{\mathrm{d}q}{\mathrm{d}t} \;\rightarrow\; v = M(q)i$$

$$M(q) \equiv \frac{\partial f(q)}{\partial q}$$

(15.1)

where $M(q)$ represents the memristance, v the voltage across the memristor, and i the current of the memristor.

Memristive devices could naturally be exploited for implementing nonvolatile synapses on electronic circuits because they are equivalent to resistors whose resistances can be held or modulated by the amount of the integrated current. Using possible memristive nanojunctions, a digital-controlled neural network has been introduced by Snider [9], and the concept has been expanded to use a nanowire-crossbar add-on as well as memristive (two-terminal) crosspoint devices such as nanowire resistive RAMs [10]. This chapter introduces yet-another analog approach, that is, a memristor (ReRAM)-based analog synaptic circuit, for possible neuromorphic computers having complementary metal-oxide-semiconductor (CMOS)-ReRAM-hybrid structures.

15.2 CMOS-RERAM-HYBRID SYNAPTIC CIRCUIT

Key ideas to implement novel synaptic circuits in this chapter are (1) to regard a memristor as a synaptic junction, and (2) to assign a membrane capacitor in one terminal of a memristor, as shown in Figure 15.1. An input node of the memristor accepts presynaptic voltage spikes (V_{pre}), and the other node is connected to membrane capacitor C_{pre}. Because this capacitor is charged or discharged by the input spike via the memristor having variable resistance (R_{mem}), a postsynaptic potential (V_{PSP}) is generated on the node where the time constant is given by $C_{pre} R_{mem}$.

At the rising edge of V_{pre} (spike onset), C_{pre} is charged via the memristor, and because of the charge flow (current) the memristor's conductance is increased, in case of the polarity settings shown in Figure 15.1. Then at the falling edge of V_{pre}, C_{pre} is discharged, and the memristor's conductance is decreased. Here it should be noticed that the absolute amount of the charge passing the memristor during the charging operation is equal to that of the discharging operation, which indicates that the memristor's conductance before applying a spike should be equal to the subsequent conductance after applying the spike, which I call "conductance-preservation law" in ideal C-memristor circuits.

Figure 15.2 shows a setup for modulating the memristor's conductance, as well as a postsynaptic-current (I_{PSC}) readout circuitry. Because a MOSFET has nonlinear characteristics between the gate voltage and the drain current, by integrating the drain currents of M_0 on the other (neuronal membrane) capacitor, one obtains different membrane potentials for different conductance of the memristor.

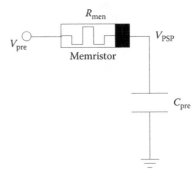

FIGURE 15.1   Fundamental unit of memristor STDP circuit.

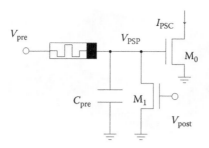

FIGURE 15.2 Conductance modulation and readout circuitry in C-memristor circuit.

If the membrane potential exceeds a given threshold voltage, a postneuron circuit (standard integrate-and-fire circuit) generates a voltage spike (V_{post} in Figure 15.2). At the same time, the PSP node is shunted by an additional MOSFET (M_1). Therefore, because of the potential difference across the memristor, the conductance is increased, which exhibits STDP in the memristor synapse (for the timing difference $\Delta t > 0$), that is, the timing difference between pre- and postsynaptic spikes results in the differential synaptic weights (differential conductance of the memristor).

15.3 EXPERIMENTAL AND SIMULATION RESULTS

Figure 15.3 shows an array of memristors (bipolar ReRAMs) with NiO thin films (Pt–NiO–Pt), which was fabricated by the author's collaborators in Osaka University, Japan. It consists of 9×9 memristor equivalents and has a common indium electrode that is connected to all the memristors. The following experimental results were obtained by using this ReRAM device.

15.3.1 EVALUATION OF MEMRISTORS

As described in Section 15.1, an original model of memristors is represented in terms of memristance, $M(q)$; however, here let us compare memristor models and physical ReRAMs by using a comprehensive model shown by D.B. Strukov et al. [8]:

$$i = g(w)v$$
$$\frac{dw}{dt} = i \tag{15.2}$$

where i represents the current of a memristor, v the voltage across a memristor, w the state variable of a memristor that corresponds to the amount of transported charges across the memristor, and $g(w)$ the monotonic increasing function representing variable conductance for w. Obviously this model explains that $g(w)$ is increased (or decreased) by positive (or negative) i. Therefore, if a constant current is given to a memristor, as shown in an inset of Figure 15.4, the conductance of memristor will increase and hence, the voltage across the memristor will decrease, as long as the model is qualitatively equivalent to physical bipolar ReRAMs. Figure 15.4 plots the transient voltage across the memristor when a constant current of 10 nA was given to

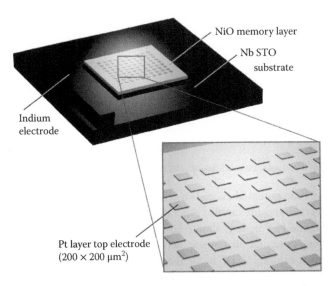

FIGURE 15.3   Array of bipolar ReRAM (memristor equivalent) (9×9 memristors on common substrate).

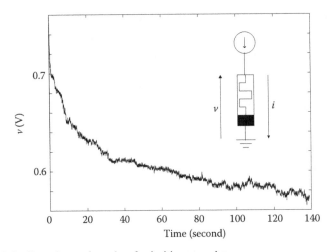

FIGURE 15.4   Experimental results of primitive memristor.

the memristor. This result showed that the voltage decreases as time increases, and hence, the model is qualitatively equivalent to the bipolar ReRAM.

A voltage bias has to be given across ReRAMs to measure their current-voltage (IV) characteristics (differential conductance of ReRAMs); however, every voltage sweep may change the conductance due to the current flow caused by the sweeping operation. Therefore, one has to decrease the voltage to avoid the unwanted change of the conductance through measurements, whereas if the voltage is too small, the current measurement becomes difficult because of thermal and environmental noises. Hence, one has to estimate moderate ranges of voltage sweeps before starting evaluations. Figure 15.5a and b plots IV characteristics of the 1st and 20th sweeps (sweep time: 640 μs) with sweeping range of 0.2 V and 1.4 V, respectively. When the range was set at 0.2 V (Figure 15.5a), no significant change of the differential conductance between the 1st and 20th measurements was observed, whereas the differential conductance of the 1st and 20th measurements was significantly changed when the range was set at 1.4 V (Figure 15.5b). Moreover, when the range was smaller than 50 mV, the IV characteristics became noisy. On the basis of these experiments, in the following experiments, the sweep range was set at 0.2 V for measuring differential conductance.

Figure 15.6 exhibits dependence of increase or decrease of differential conductance on the ReRAM's polarity. The initial IV curve is represented by (a) in the figure. Then a single current pulse (amplitude: 10 μA; pulse width: 640 μs [microseconds]) was given to the ReRAM where the current direction is indicated by an arrow of I_{ReRAM}. The resulting IV curve is represented by (b) where

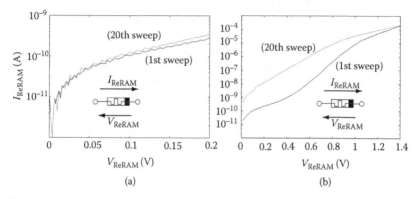

FIGURE 15.5 Experiments to determine voltage sweep ranges. (a) 0–0.2 V sweep and (b) 0–1.4 V sweep results.

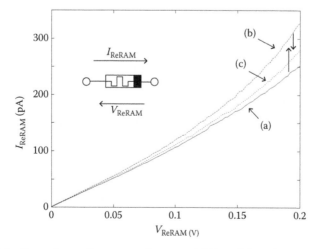

FIGURE 15.6 Dependence of increase or decrease of differential conductance on ReRAM's polarity.

the differential conductance was increased by the single current pulse. Then the same current pulse was applied to the ReRAM in opposite direction (of an arrow of I_{ReRAM}), and the resulting IV curve is represented by (c) where the differential conductance was decreased by the pulse. The results clearly show the dependence of increase or decrease of differential conductance on the ReRAM's polarity. It should be noticed that there exists small difference between curves (a) and (c) although the same current pulses were applied in different directions. This might be due to asymmetrical structure of the ReRAMs, and in case of symmetrical ReRAMs, for example, nanowire crossbars of memristive materials, the curves will be the same.

In practical ReRAMs, when large voltage or current is forwardly (or reversely) applied, the conductance will be fixed to large (or small) values. This is typical behavior of ReRAM to be operated as binary memory devices; however, this property is not suitable for their analog applications. In fact, experiments shown in Figure 15.6 were performed by setting the initial differential conductance to its minimum, to avoid the conductance saturation to its maximum. Therefore, one has to estimate appropriate amplitudes and widths of current pulses that do not make ReRAM's conductance saturated. To estimate the amplitudes and widths, differential conductance were measured by repeating the same experiments as in Figure 15.6.

Figure 15.7 shows the results. In Figure 15.7a, five current pulses were given to the ReRAM (current direction is indicated by an arrow of I_{ReRAM}), whereas three pulses in opposite directions were applied to the ReRAM in Figure 15.7b. As the conductance were not saturated within these range of pulses, as long as one uses the current pulse (amplitude: 10 µA; pulse width: 640 µs), the single pulse will not make the ReRAM's conductance saturated.

Figure 15.8 represents IV plots of sequential modifications (increase and decrease) of differential conductance by applying unidirectional current pulses. The minimum conductance is represented by (a), whereas (b) represents the intermediate

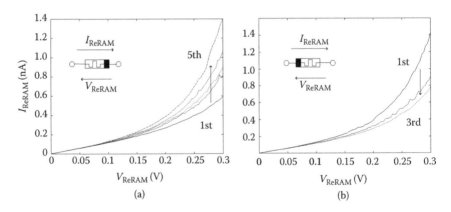

FIGURE 15.7 Dependence of differential conductance on ReRAM's polarity. (a) Forward sweep. (b) Backward sweep.

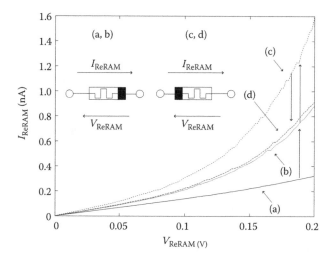

FIGURE 15.8 IV plots of sequential modifications (increase and decrease) of differential conductance.

(nonsaturated) conductance. This result clearly shows that the conductance can be modified from the intermediate (nonsaturated) state of ReRAMs.

15.3.2 Experimental Results of ReRAM-CMOS-Hybrid Synaptic Devices

Through experiments in Section 15.3.1, now we are ready to start the experiments of ReRAM-CMOS-hybrid synaptic devices. First, the ReRAM-C circuit (shown in Figure 15.1) was assembled inside a shield box of semiconductor parameter analyzer with a discrete capacitor ($C = 0.5$ nF). The result is shown in Figure 15.9. In this experiment, voltage pulse V_{pre} was applied to the circuit, and V_{PSP} and current of the capacitor (current flowing into the ground) were measured. Then the IV curve was calculated by the data because one cannot measure IV characteristics when the pulse is raised. In Figure 15.9, (a) represents the IV curve before the raise of V_{pre}, whereas (b) represents the IV curve after the fall of V_{pre}, which shows there is no significant difference between the pre- and postdifferential conductance.

Figure 15.10 shows experimental results of ReRAM-C-nMOS circuit shown in Figure 15.2 where discrete nMOSFET 2SK1398 was used and readout nMOS M_0 was omitted in this experiment. In Figure 15.10a, V_{post} was always set at 0 and a voltage pulse was applied to V_{pre} only. During the experiments, V_{PSP} and C_{pre}'s current were measured, and the IV characteristics of the memristors were reconstructed before/during/after applying the voltage spike. In the figure, plots indicated by (i) represent the initial IV characteristics of the memristor (before applying the spike), (ii) the intermediate IV characteristics (after the spike onset), and (iii) the subsequent characteristics (after the falling edge of the spike). It should be noticed that although the conductance was increased during the intermediate state, the final conductance (iii) was almost equal to the initial conductance (i), even in real experiments. Figure 15.10b shows the opposite case, that is, the resulting conductance (iii) was increased as compared with initial conductance (i) by applying a spike as

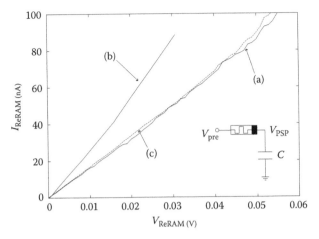

FIGURE 15.9 Experimental IV curves of memristor-C circuit. (a) Initial sweep, (b) reconstructed IV at spike onset, and (c) reconstructed IV at spike offset.

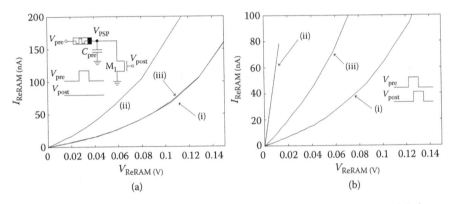

FIGURE 15.10 Experimental results of (a) conductance preservation and (b) modulation.

V_{post}. In this experiment, C_{pre} was charged by V_{pre} and then was discharged by V_{post}, which means that the absolute amount of the charge passing the memristor during the charging operation is not equal to that of the discharging operation. The result implied that the conductance could be modified by the difference of spike timings of pre- and postsynaptic neurons (Δt).

Figure 15.11 exhibits a part of STDP experiments with the synaptic device. In the experiment, two voltage spikes as illustrated in Figure 15.11a were used where the timing difference between pre- and postspike voltages (V_{pre} and V_{post}) was defined as Δt. Figure 15.11b shows the results of conductance modulation for different timing values. When Δt was large, the differential conductance (Δg) before and after applying the spike, was small as expected, whereas when Δt was decreased to 0, Δg was exponentially increased. This result clearly showed that the conductance could be controlled by the spike-timing difference (Δt).

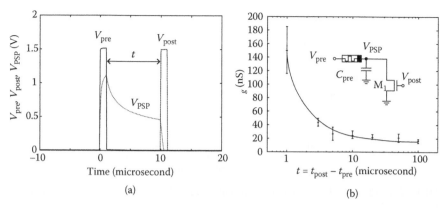

FIGURE 15.11 Experimental results of half STDP characteristics of memristor-CMOS hybrid circuit. (a) Definition of spike timing difference Δt between pre and post spikes (V_{pre} and V_{post}). (b) Change of conductance Δg as function of spike timing difference Δt.

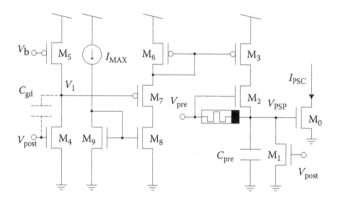

FIGURE 15.12 Proposed STDP synaptic circuit having asymmetric time window.

The circuit shown in Figure 15.2 does not modify memristor's conductance when $\Delta t < 0$ because when $V_{\text{pre}} = 0$ ($V_{\text{PSP}} = 0$), the memristor's current is always zero even if M_1 is turned on or off by V_{post}. To obtain $\Delta t < 0$ responses, the circuit has been extended where additional MOSFETs (M_2–M_5) are employed, as shown in Figure 15.12. M_4 and M_5 act as a source-common amplifier and exhibit large time delay, which can be controlled by V_b, on voltage V_1 upon the falling edge of V_{post} because C_{gd} is amplified by the mirror effect. When M_7 is turned on (just after the firing of V_{post}), I_{MAX} is mirrored to M_3 via current mirrors M_8–M_9 and M_3–M_6. At the same time, if M_2 is turned on by V_{pre}, V_{PSP} is increased. Consequently, at the falling edge of V_{pre}, conductance of the memristor is decreased.

Figure 15.13 shows the simulation results indicating asymmetric STDP characteristics. In the simulations, IV characteristics of the fabricated memristor shown in Figure 15.3 (used in previous experiments) were assumed, and models of a discrete MOS device (2SK1398 and 2SJ184) and a capacitor (0.1 µF) were used. The result

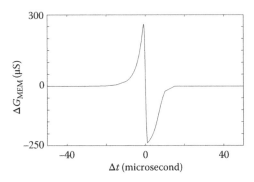

FIGURE 15.13   Simulated asymmetric STDP characteristics of proposed circuit shown in Figure 15.12.

clearly showed that the differential conductance (ΔG_{MEM}) was modified by the difference of spike timings of pre- and postsynaptic neurons (Δt).

15.4   CONCLUSION

A memristor-based STDP synaptic device was introduced. The device was built based on a combination of a capacitor and a memristor by controlling the so-called conductance-preservation law. First, a fundamental device consisting of memristor-equivalent devices called resistive RAMs as well as a discrete capacitor and a MOSFET was introduced and demonstrated by extensive experimental results. Then an STDP synaptic device was introduced where the device exhibited conductance preservation, modulation, the resulting STDP characteristics for positive timing difference between pre- and postsynaptic voltage spikes ($\Delta t > 0$). Finally, to mimic STDP characteristics in $\Delta t < 0$, the device was modified to the circuit level, by adding several MOSFETs. Through SPICE simulations, I showed that the circuit could exhibit asymmetric STDP characteristics.

Memristive devices (e.g., ReRAMs and atomic switches) offer a promising alternative for the implementation of nonvolatile analog synapses. They are applied in the CMOS molecular (CMOL) architecture [11], which combines memristive nanojunctions with CMOS neurons and their associated controllers. In ITRS 2007 [12], CMOL was introduced in terms of nanogrids of (ideally) single molecules fabricated on top of a traditional CMOS layer, but the concept has since been expanded to use a nanowire-crossbar add-on as well as memristive (two-terminal) crosspoint devices such as nanowire ReRAMs. The CMOL architecture will be further expanded to include multiple stacks of CMOS layers and crossbar layers. This may result in the implementation of large-scale multilayer neural networks, which have thus far evaded direct implementations by CMOS devices only.

ACKNOWLEDGMENTS

The author thanks Dr. Keisuke Oka, Prof. Takeshi Yanagida, and Prof. Tomoji Kawai of Osaka University for their supplying memristor-equivalent devices (array of ReRAMs). This study was supported by a Grant-in-Aid for Scientific Research on Innovative Areas [20111004] from the Ministry of Education, Culture Sports, Science and Technology (MEXT) of Japan.

REFERENCES

1. Y. Kanazawa, T. Asai, M. Ikebe, and Y. Amemiya, A novel CMOS circuit for depressing synapse and its application to contrast-invariant pattern classification and synchrony detection, *International Journal of Robotics and Automation*, 19(4): 206–212, 2004.
2. H. Tanakaa, T. Morie, and K. Aihara, A CMOS circuit for STDP with a symmetric time window, *International Congress Series*, 1301: 152–155, Elsevier, 2007.
3. H. Tanaka, T. Morie, and K. Aihara, A CMOS spiking neural network circuit with symmetric/asymmetric STDP function, *IEICE Transactions on Fundamentals*, E92-A(7): 1690–1698, 2009.
4. K. Nakajima, S. Sato, T. Kitaura, J. Murota, and Y. Sawada, Hardware implementation of new analog memory for neural networks, *IEICE Transactions on Electronics*, E78-C(1): 101–105, 1995.
5. C. Diorio, P. Hasler, B. Minch, and C. Mead, Floating-gate MOS synapse transistors, *Neuromorphic Systems Engineering, The Kluwer International Series in Engineering and Computer Science*, 447(4): 315–337, 1998.
6. S. Kinoshita, T. Morie, M. Nagata, and A. Iwata, A PWM analog memory programming circuit for floating-gate MOSFETs with 75-µs programming time and 11-bit updating resolution, *IEEE Journal of Solid-State Circuits*, 36(8): 1286–1290, 2001.
7. L.O. Chua, Memristor—the missing circuit element, *IEEE Transactions on Circuit Theory*, 18: 507–519, 1971.
8. D.B. Strukov, G.S. Snider, D.R. Stewart, and R.S. Williams, The missing memristor found, *Nature*, 453(1): 80–83, 2008.
9. G.S. Snider, Self-organized computation with unreliable, memristive nanodevices, *Nanotechnology*, 18(36): 365202, 2007.
10. S.H. Jo, T. Chang, I. Ebong, B.B. Bhadviya, P. Mazumder, W. Lu, Nanoscale memristor device as synapse in neuromorphic systems, *Nano Letters*, 10(4): 1297–1301, 2010.
11. K.K. Likharev, Hybrid CMOS/nanoelectronic circuits: Opportunities and challenges, *Journal of Nanoelectronics and Optoelectronics*, 3(3): 203–230, 2008.
12. ITRS, Emerging research architectures, In *Emerging Research Devices*, ITRS, available at http://www.itrs.net/Links/2007_ITRS_CD/Home2007.htm, 2007.

16 Very-Large-Scale Integration Implementations of Cryptographic Algorithms

Tony Thomas

CONTENTS

16.1   INTRODUCTION

Today secure storage and transfer of information as well as communication over Internet and mobile networks has become very critical. Cryptographic algorithms have been developed to provide various security services such as confidentiality, user authentication, data origin authentication, data integrity, nonrepudiation, privacy, and so on. Cryptographic algorithms are divided into two categories: symmetric-key or secret-key algorithms and asymmetric-key or public-key algorithms. Symmetric-key cryptography refers to cryptographic mechanisms in which both the sender and receiver share the same key or keys that are related in an easily computable way. In asymmetric-key cryptography, two different but mathematically related keys (not in an easily computable way) called public and private keys are used. A public-key cryptosystem is constructed in such a way that calculation of the private key is computationally infeasible from the public key.

The rapid growth of portable electronic devices with limited power and area, such as mobile phones and smart cards, has opened up the necessity to carry out the very-large-scale integration (VLSI) hardware implementations of cryptographic algorithms on low-power and compact circuits. Unlike the usual computer and network security systems that impose less stringent limitations on the area and power consumption, portable devices impose more restrictions on area and power and less on throughput. This difference in requirements necessitates a different approach in the design and implementation of the cryptographic algorithms for these devices. Thus, to be fast and feasibly practical as well as to satisfy conflicting requirements of power, area, and throughput in the security applications, cryptographic algorithms have to be implemented in hardware. Dedicated hardware devices can run encryption routines concurrently with other applications in the host computer. As hardware solutions are tamper proof, hardware implementations of cryptographic algorithms ensure confidentiality, integrity, and authentication of cryptographic keys. This makes the cryptographic keys to be safely stored inside the hardware. On the other hand, software implementations of cryptographic algorithms are prone to manipulations and a dedicated attacker may be able to reverse engineer the cryptographic keys. However, hardware is more expensive than software and memory is a constraint in hardware designs. Hence, the algorithms must be specially tailored for hardware implementations. The cryptographic algorithms should be made compact, scalable, and modular to reduce the overall cost of the hardware required.

Because, present-day multipurpose smart cards are used for a wide range of applications, they must support both symmetric-key and asymmetric-key cryptographic algorithms. Symmetric-key algorithms with high throughput are suitable for data communication, whereas asymmetric-key algorithms with lower throughput are suitable for key exchange and authentication protocols. Among various cryptographic algorithms, approved standards such as RC4, Data Encryption Standard (DES), Triple Data Encryption Standard (3DES), Advanced Encryption Standard (AES), RSA, Elliptic curve cryptosystem (ECC), SHA-3, and so on are used for these applications. RC4, for encryption of stream of data, DES/3DES, for past compatibility, and AES, for high security and throughput, are the major symmetric-key algorithms used. ECC, for high encryption efficiency, is the best candidate for asymmetric-key encryption, and SHA-3 is used in integrity applications.

This chapter focusses on the VLSI implementations of various cryptographic algorithms. The rest of this chapter is organized as follows. In Section 16.2, we introduce various cryptographic primitives and discuss the importance of VLSI implementation of such primitives. In Sections 16.3 through 16.5, we discuss the VLSI implementations of symmetric-key ciphers RC4, DES, and AES, respectively. In Sections 16.6 and 16.7, we discuss the VLSI implementations of asymmetric-key ciphers RSA and elliptic curve system. In Section 16.8, we discuss the VLSI implementations of SHA-3 hash function. Finally, the chapter concludes with Section 16.9.

16.2 VLSI IMPLEMENTATIONS OF CRYPTOGRAPHIC PRIMITIVES

In this section, we focus on the VLSI implementations of the cryptographic primitives such as stream ciphers, block cipher, asymmetric-key encryption algorithms, and hash functions.

A stream cipher is a symmetric key encryption algorithm where plaintext digits are combined with a pseudorandom keystream generated by the stream cipher. In a stream cipher, each plaintext bit or byte or block of any size is encrypted one at a time with the corresponding bit or byte or block of the keystream, to give the corresponding bit or byte or block of the ciphertext stream. Usually, the message stream is combined with the keystream with a bitwise exclusive-or (XOR) operation. The pseudorandom keystream is typically generated serially from a random seed value that serves as the key of the stream cipher. Many stream cipher algorithms are designed by combining the outputs of several independent Linear Feedback Shift Registers (LFSRs) using a nonlinear Boolean function. The VLSI area used in implementing such stream cipher systems has two components: the area used to implement the LFSRs and the area used to implement the Boolean function. The area required to implement the LFSRs is proportional to the product of the length of the LFSR and the number of inputs to the LFSR, whereas the area required to implement the Boolean function grows exponentially with the number of inputs. Thus, implementing the Boolean function is a main hurdle in this case.

Block ciphers operate on large blocks of data such as 128 bits with a fixed, unvarying transformation. A block cipher is a function that maps an n-bit plaintext blocks to an n-bit ciphertext block, where n is called the block length. The function is parameterized by a key K. The encryption operation involves several rounds of primitive operations such as substitution, permutations, key mixing, shifting, swapping, lookup table (LUT) reading, and so on. The decryption operation involves reversing the previously mentioned operations. The encryption and decryption algorithms use the same key. The fundamental difference between a block cipher and stream cipher is that in a block cipher, the same key is used for encrypting each block of data, whereas in a stream cipher, no key is repeated and thus different keys are used for encrypting different blocks of data. Five modes of operation are used for the block ciphers for encrypting data that is longer than a block length. Among these, three of the modes, namely Cipher Block Chaining mode (CBC), Cipher Feedback mode, and Output Feedback mode (OFB), are feedback modes, whereas the remaining two, namely

Electronics Code Book mode (ECB) and Counter mode (CTR), are non-feedback modes. In the feedback modes, encryption of each block of data starts only after the completion of encryption of the previous block of data. As a result, in feedback modes all the blocks must be encrypted sequentially. On the other hand, in the non-feedback modes each block of data can encrypted independently with each other. So, in non-feedback mode, many blocks of data can be encrypted in parallel. FPGA devices are highly promising for implementing block ciphers. The fine granularity of FPGAs matches well with the operations such as permutations, substitutions, LUT reading, and the Boolean functions used by these algorithms. Furthermore, FPGA can efficiently exploit the inherent parallelism of the block cipher algorithms.

Asymmetric encryption or public-key encryption refers to a cryptographic algorithm that requires two separate keys, one of which is kept as secret and the other is made public. Although different, these two key pairs are mathematically linked. The public key is used to encrypt plaintext to the ciphertext, whereas the private key is used to decrypt ciphertext back to the plaintext. The term *asymmetric* refers to the use of different keys to perform these opposite functions in contrast with the conventional *symmetric* cryptography that relies on the same key to perform both the operations. Asymmetric-key cryptographic algorithms are based on *hard mathematical problems* that currently admit no efficient solutions. Examples of such problems are integer factorization problem [1], discrete logarithm problem [1,2], shortest lattice vector problem [1], conjugacy problem [3,4], and so on. The security of asymmetric key cryptography lies in the fact that it is computationally infeasible for a properly generated private key to be determined from its corresponding public key. Thus the public key can be published without compromising the security, whereas the private key must be kept secret and should not be revealed to anyone not authorized to read the messages. Asymmetric key algorithms, unlike the symmetric key algorithms, do not require a secure initial exchange of the secret key between the sender and the receiver. The main limitation of asymmetric-key algorithms is in their complexity. Asymmetric key algorithms are much more complex and slower than the symmetric key algorithms. It is widely believed that the complexity of most of the public-key algorithms make them unsuitable for resource-constrained applications. International Standard Organization has set the limit of the maximum chip area on a smart card to 25 mm. Considering the nonvolatile and volatile memories, CPU, and other peripheral circuits required on the chip, it is desirable to fit the security component in the smallest area possible.

A cryptographic hash function maps data of arbitrary length to data of a fixed length. The output values of a hash function are called hash values or hash codes or checksums or simply hashes. Hash functions are used to protect the integrity of messages using symmetric-key-based message authentication code, asymmetric-key-based digital signature, making commitments, time stamping, key updating, protecting password files, intrusion detection, and so on. Thus, besides high security, a hash function should be suitable for implementations on a wide range of applications. The hardware efficiency of a hash function will be crucial to determine its future, because hardware resources are often limited, whereas on high-end computers, it does not matter much in general. In fact, even the slowest hash function has an acceptable performance on a PC.

16.3   VLSI IMPLEMENTATION OF RC4 STREAM CIPHER

RC4 is a stream cipher designed by Ron Rivest in 1987 for RSA Security. It is a variable key-size stream cipher with byte-oriented operations. The algorithm is based on the use of a random permutation where eight to sixteen machine operations are required per output byte. RC4 is used for encryption in the wired equivalent privacy protocol (part of the IEEE 802.11b wireless LAN security standard).

16.3.1   RC4 STREAM CIPHER ALGORITHM

The RC4 algorithm is a remarkably simple cipher. A variable-length key K of size 1–256 bytes (8–2048 bits) is used to initialize a 256-byte state vector $S = (S[0], S1, ..., S[255])$. At all times, S contains a permutation of all the 8-bit numbers from 0 through 255. For encryption and decryption, a byte is selected from S in a systematic fashion. The entries in S are then once again permuted.

Initially the entries in S are set equal to the values from 0 through 255 in the ascending order, that is, $S = (0, 1, ..., 255)$. A temporary vector, T, is also created. If the length of the key K is 256 bytes, then K is copied to T. Otherwise, for a key of length keylen bytes, K is copied to the first keylen elements of T and then K is repeated as many times as necessary to fill out T.

T is then used to produce the initial permutation of S. This involves starting with $S[0]$ and going through to $S[255]$, and, for each $S[i]$, swapping $S[i]$ with another byte $S[j]$ in S where j is computed (j is initialized 0) as $j \leftarrow (j + S[i] + T[i])$ mod 256.

After the initialization of vector S, the key K is no longer used. Stream generation involves starting with $S[0]$ and going through to $S[255]$, and, for each $S[i]$, swapping $S[i]$ with another byte in S according to the following scheme dictated by the current configuration of S. First i and j are initialized to 0.

- $i = (i + 1)$ mod 256
- $j = (j + S[i])$ mod 256
- Swap $(S[i], S[j])$
- $t = (S[i] + S[j])$ mod 256
- $k = S[t]$

The stream output is k, which is the key. After $S[255]$ is reached, the process continues, starting over again at $S[0]$. Encryption is performed by XORing the stream output with the next byte of the plaintext. Decryption is carried out by XORing the same stream output with the next byte of the ciphertext. The stream generation by the RC4 cipher is illustrated in Figure 16.1.

16.3.2   RC4 IMPLEMENTATION

In Kitsos et al. [5], a hardware implementation of the RC4 stream cipher is presented. The implementation is parameterized to support variable key lengths (key length could be 8 up to 128 bits). This implementation requires three clock cycles per byte generation in the key setup phase, and three clock cycles per byte generation in the

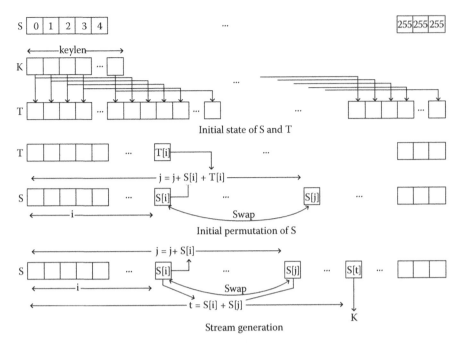

FIGURE 16.1 Stream generation by RC4 cipher.

encryption phase. The total number of operations required is $768 + 3n$ clock cycles, where n is the number of bytes of the plaintext/ciphertext.

The proposed architecture in Kitsos et al. [5] consists of a control and a storage unit. The storage unit is responsible for the key setup and generation of the keystream. There are two 256-byte arrays, S-box and K-box. The S-box is filled with the state array and the K-box consists of the key, repeating as necessary, to fill the array. The storage unit contains memory elements for the S-box and the K-box, along with 8-bit registers, adders, and one multiplexer.

The S-box RAM consists of three 256 bytes RAM blocks with same signals of clock and reset. Each RAM block has four inputs and one output. The two inputs are the read and write signals whereas the other two are the address and the data signals. When a reset signal occurs, the blocks are initialized linearly; when the write signal occurs, new data are stored in the address position; and when the read signal occurs, the data in the address position are available on the output of the block.

At the first clock cycle, the value of the counter i is used as address in the first RAM block. The value of the register $S[i]$ is used for the computation of the new value of j. Two adders are used for the computation of the new value of j. The input to the adders is the values of $T[i]$ and $S[i]$. At the second clock cycle, the new value j is used as the address for the second RAM block. The value stored in this address is temporarily stored in the $S[j]$ register. At the third cycle, the contents of the $S[i]$ register and $S[j]$ register are written in the registers at the j-th and i-th address locations, respectively.

The operations in the key setup phase are as follows. First, the value of i is used as address in the first RAM block. Then the new value of j is computed. At the second step, the new value of j is used as address of the second RAM block and the value of $S[j]$ is stored in the j-th register. Then the values of $S[i]$ and $S[j]$ are added and the result of the addition is stored in the $S[t]$ register. At the third step, the contents of the $S[i]$ register and $S[j]$ register are swapped and the value of the t-th register $S[t]$ is used as address for the third RAM block. The value of $S[t]$ is produced in the third step and this value of $S[t]$ becomes the stream-generated key byte.

The encryption (decryption) is achieved by the bitwise XORing of the keystream with the plaintext (ciphertext). The operation of the storage unit is synchronized by a control unit and the control unit is responsible for the generation of the clock and control signals.

16.4 VLSI IMPLEMENTATION OF DATA ENCRYPTION STANDARD BLOCK CIPHER

The DES was published by the National Bureau of Standards in 1977 and reaffirmed in its final form by the Federal Information Processing Standards Publication (FIPS) in 1994 [6]. DES is a block cipher with Feistel networks, which operates on data blocks of 64-bit with a key of 56 (64 with 8 parity bits) bits length. DES is a 16-round cipher. Each round is identical with different round keys.

Triple DES is built on 3 DES block cipher to support a higher security level. It operates on the Encryption–Decryption–Encryption (EDE) mode, which uses sequentially a DES encryption, then a DES decryption, and finally a DES encryption, with the support of two or three different keys.

Triple DES has been used in various cryptographic applications and wireless protocol security layers. The 3DES algorithm in the EDE scheme uses both the encryption and the decryption operations. DES encryption and decryption algorithms are identical. The only difference is that the subkeys in the 16 rounds of a DES decryption have to be generated and used in the reverse order compared to the DES in the encryption mode. To produce the subkeys in reverse order, they have to be cyclically shifted right, as opposed to left for encryption.

16.4.1 DATA ENCRYPTION STANDARD BLOCK CIPHER ALGORITHM

The DES algorithm in brief is as follows. Get a 64-bit block of data. Sixteen different round keys $K[1], \ldots, K[16]$ are first generated from the DES key K. An initial permutation is performed on the data block. The block is then divided into two halves. The first 32 bits are called the left half $L[0]$, and the last 32 bits are called the right half $R[0]$. Sixteen subkeys are then applied to the data block in 16 successive rounds. Start with $i = 1$. The 32-bit $R[i - 1]$ is expanded into 48 bits according to a bit-selection function $E (.)$. $E (R[i - 1])$ is XORed with $K[i]$. $E (R[i - 1]) \oplus K[i]$ is then broken into eight 6-bit blocks and fed into eight S-boxes (substitution boxes). Each S-box has a 4-bit long output. Outputs from all the eight S-boxes concatenated and the bits in the resulting 32 bit string is permuted. The resulting string is XORed with $L[i - 1]$ to get $R[i - 1]$. $R[i - 1]$ is assigned as the new $L[i]$. The operation in the i-th round is illustrated in Figure 16.2. This process is repeated for the 16 rounds. After the 16th round, the left and right halves are swapped. Then inverse of the initial permutation is then performed on the result of left-right swap. The ciphertext is the output of the inverse permutation operation. To decrypt, use the same process, but just use the keys $K[i]$ in the reverse order.

16.4.2 TRIPLE DATA ENCRYPTION STANDARD IMPLEMENTATION

Each DES begins with initial permutation IP and ends with the inverse of the initial permutation. These two permutations are inverse operations. When three DES are concatenated, the initial permutation of the previous DES follows the inverse initial permutation of the current DES and they cancel out. Hence, only the first IP and the last inverse IP need to be performed. As a result, the initial permutations of the second and third DES, and the final permutations of the first and the second DES, are not required to be included in the VLSI architecture. For all the following implementations of 3DES, the S-BOXes have been integrated by LUTs as well as by read-only memory (ROM) blocks.

We now describe a 48-stage VLSI architecture proposed in Kitsos et al. [7]. This architecture has the ability to process 48 independent data blocks. The subkeys for the 48 rounds are generated by three key scheduler units corresponding to the three DES components. Each key generator for a DES round produces 16 round keys. The 64-bit input key is initially permuted, and is then passed through a hardwired shifter and is finally passed through a second round permutation for each subkey. A memory of 3×64 bits is placed at the input to the key scheduling circuit. In this memory,

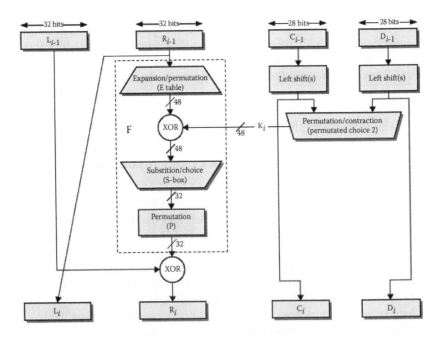

FIGURE 16.2 One round of Data Encryption Standard algorithm.

the three 64 bits keys corresponding to the three DES rounds are stored to input the appropriate key in the appropriate time. On the first clock cycle, the first key is applied on the key scheduler and the DES runs in the encryption mode. On the 17th clock cycle the second key is supplied on the key scheduler and the DES runs in the decryption mode. Finally, on the 33rd cycle the third key is supplied and the DES runs in the encryption mode.

A second architecture proposed by Kitsos et al. [7] consists of 1 DES with 16 pipeline registers between the rounds. A multiplexer is used to determine between the new data and the output of the DES from the previous stage. This architecture can process 16 independent data blocks at a time. The generation of subkeys comprises 16 rounds. This architecture is suitable for high-speed applications that support ECB or automated teller machine (ATM) counter modes of operation of the block cipher where same message blocks encrypt to same ciphertext blocks.

A third architecture proposed by Kitsos et al. [7] uses consecutive iterations. Only one basic round each is implemented for the 3DES algorithm and the key scheduler to minimize the resources allocated for the 3DES. In this case, the required hardware resources are reduced by a factor of 40, 80, and 16, respectively, compared to the previous ones.

The output of the basic round is loaded to a register and another register is used for the input plaintext. The plaintext is loaded during the initialization step into the multiplexer. Feedback is used for loading new data blocks through the multiplexer for transforming the data during the 48 rounds of the 3DES. The key scheduler has one basic round and successive round subkeys are computed by simple rotations and permutations. The key scheduler provides one subkey at every clock cycle where 48 clock cycles are needed for the execution of the whole 3DES block cipher.

Because this architecture uses the minimum hardware with feedback mode, it is suitable for area-restricted devices and for the CBC or OFB modes of the block cipher operations.

16.5   VLSI IMPLEMENTATION OF ADVANCED ENCRYPTION STANDARD BLOCK CIPHER

In January 1997, to replace the aging DES, the National Institute of Standards and Technology (NIST) invited proposals for new algorithms for the AES [8]. After two rounds of evaluation on the 15-candidate algorithms, NIST selected the Rijndael as the AES in October 2000. The AES block cipher has broad applications, in various domains such as smart cards, cellular phones, ATMs, disk security, and digital video recorders. Hardware implementations of the AES block cipher can provide better physical security as well as higher speed compared to the software implementations.

16.5.1   ADVANCED ENCRYPTION STANDARD ALGORITHM

AES, like DES, is an iterative algorithm. Each iteration is called a round. AES block cipher algorithm deals with data blocks of 128 bits using keys with three standard lengths of 128, 192, and 256 bits. The 128-bit data block is divided into 16 bytes. These bytes are mapped to a 4 × 4 array called the state, and all the internal operations of the AES algorithm are performed on the state matrix. The internal operations consist of mainly four primitive transformations: Sub-Bytes, ShiftRows, MixColumns, and AddRoundKey. AES encryption/decryption process involves these four primitive transformations executed iteratively in several rounds. The number of rounds is 10, 12, or 14, corresponding to the key sizes 128, 192, and 256, respectively.

In the encryption procedure, the input data is bitwise XORed with an initial key, and then four transformations are executed in the following order: Sub-Bytes, ShiftRows, MixColumns, and AddRoundKey for all rounds except the last one. All four transformations except the MixColumns transformation are performed in the last round (in the same order). The execution sequence is reversed in the decryption process, with the inverses of transformations such as InvSubBytes, InvShiftRows, InvMixColumns, respectively as well as the AddRoundKey operation. Because each round needs a round key, an initial key is used to generate all the round keys before the encryption/decryption operation. The block diagram of the AES encryption and the decryption algorithms is given in Figure 16.3.

16.5.2 ADVANCED ENCRYPTION STANDARD IMPLEMENTATION

Three architectural optimization approaches are usually employed to speed up the hardware implementations of AES. They are pipelining, subpipelining, and

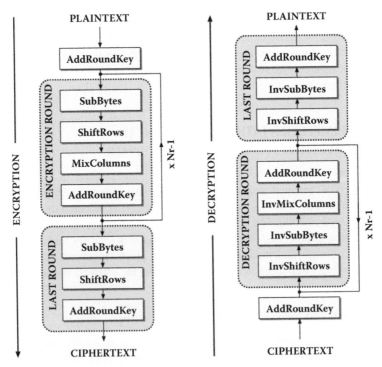

FIGURE 16.3   Advanced Encryption Standard encryption and decryption algorithms.

loop-unrolling. Among these three approaches, the subpipelined architecture can achieve maximum speedup and optimum speed–area ratio in non-feedback modes of the block cipher. The advantage of subpipelining can be enhanced further by dividing each round unit into more substages with equal delay. However, the SubBytes and the InvSubBytes operations in the AES algorithm are traditionally implemented by LUTs.

In Zhang et al. [9], the authors avoid the use of LUTs and propose the use of composite field data paths for the SubBytes and InvSubBytes transformations. They achieved high-speed subpipelined AES architectures through such data paths. Composite field arithmetic has been employed by Satoh et al. [10] to design efficient data paths. In the architecture proposed in Zhang et al. [9], the inversion in GF (2^4) is implemented by a novel approach, which leads to a more efficient architecture with shorter critical path and smaller area. The composite field arithmetic used in Rudra et al. [11] is not efficient in all the transformations used in the AES algorithm. In Zhang et al. [9], a key expansion architecture is presented, which is well suited for subpipelined designs. Further, this architecture can operate in an on-the-fly manner.

16.6   VLSI IN RSA ASYMMETRIC-KEY CRYPTOSYSTEM

RSA, which is also a standard and widely used asymmetric-key algorithm, is not considered in many hardware designs for the following three reasons [12]:

- ECC with a key size of 160 bits provides the same level of security as RSA with a key size of 1024 bits. Thus, ECC will be a better choice when implementation area is a critical factor in the design.
- RSA uses binary addition of large numbers and needs binary adders that are either slow for carry propagation or large for look-ahead carry generation.
- A larger number of bits in RSA require wider buses, which adds to the area and power consumption of the design.

However, we will quickly review a VLSI implementation of RSA algorithm.

16.6.1   RSA Encryption and Decryption Algorithms

RSA is a public-key cryptographic algorithm. It uses two keys: a public key and a private key. The public key is known to everyone and is used for encrypting messages. Messages encrypted with the public key can be decrypted only using the private key. The keys for the RSA algorithm are generated in the following way:

- Choose randomly two distinct prime numbers p and q of approximately equal bit length.
- Compute $n = pq$.
- Compute $\varphi(n) = (p - 1)(q - 1)$, where φ is the Euler's totient function.

- Choose an integer e having a short bit length and small Hamming weight such that $1 < e < \varphi(n)$ and the greatest common divisor $(e, \varphi(n)) = 1$; that is, e and $\varphi(n)$ are relatively prime. e is published as the public key exponent.
- Determine d as $ed \equiv 1 (\mathrm{mod}\ \varphi(n))$. d is securely kept as the private key exponent.

The encryption is performed as follows. Alice sends her public key (n, e) to Bob and keeps the private key d secret. To send a message M to Alice, Bob firsts converts M into an integer m, such that $0 \le m < n$. He then computes the ciphertext c corresponding to m as

$$c \equiv m^e \mod n$$

This can be done quickly using the method of exponentiation by repeated squaring. Bob then transmits c to Alice.

On receiving the ciphertext c, Alice can recover m from c by using her private key exponent d by performing the decryption operation as

$$m \equiv c^d \mod n$$

After getting m, Alice can recover the original message M by reversing the conversion operation.

16.6.2 RSA IMPLEMENTATION

LFSR can be used to generate random numbers p and q. Simultaneously, each number will be checked to see whether it is a prime number or not. The modular multiplication can be implemented using shift-add multiplication algorithm. The modular exponentiation can be implemented using a 16-bit modular exponentiation using LR binary method, where LR stands for the left-to-right scanning direction of the exponent.

Modular multiplication is a slightly complicated arithmetic operation because of the inherent multiplication and division operations. Modular multiplication can be carried out by performing the modulo operation after multiplication or during the multiplication. The modulo operation is accomplished by integer division, in which only the remainder after division is taken for further computation. The first approach requires an $n \times n$ bit multiplier with a $2n$-bit register followed by a $2n \times n$ bit divider. In the second approach, the modulo operation occurs in each step of integer multiplication. Therefore the first approach requires more hardware while the second requires more addition/subtraction computations. Different number representations such as redundant number systems and higher radix carry-save form have been used for this purpose. A carry prediction mechanism has also been used for fast calculation of modular multiplication.

Various authors [13,14,15,16] have proposed different array structures suited for VLSI implementation of modular multiplication. Vandemeulebroecke et al. [15] used a modulo after a multiplication approach using a signed digit number representation. Two arrays were used: one for multiplication and the other for integer division. Koc et al. [14] applied Blakelys algorithm [17] and use a sign estimation method by looking at the five most significant bits (MSBs) in each iteration stage. Although they obtain a bit-level systolic array structure, the latency and clock cycle are relatively long due to the control node that estimates the sign of the intermediate result in each stage [18].

Jeong et al. [18] have developed two new VLSI array architectures for modular multiplication. The idea is similar to Montgomerys algorithm [19] in which each partial product is made as a multiple of the radix to simplify the multiplication by the radix by only looking at the least significant bits. This approach requires only a post-processing step to get the final answer. In Jeong et al. [18], the MSBs are considered to remove higher bit positions while keeping the correct answer in each partial product, keeping it within a certain range. By a simple translation of a modulo operation into an addition of a precalculated complement of the modulus, the modulo during multiplication approach is used with a carry-save adder structure. Multiple XORs are used to choose the precalculated integer depending on the control that is generated in the leftmost node in each stage. Compared to other algorithms, they obtain a higher clock frequency because of the simplified modulo reduction operation.

16.7 VLSI IMPLEMENTATION OF ELLIPTIC CURVE CRYPTOSYSTEM

Elliptic curves were introduced to cryptography by Koblitz [2] and Miller [20] independently in the 1980s. Elliptic curves are described by the Weierstrass equation given below:

$$E : y^2 + a_1 xy + a_3 y = x^3 + a_2 x^2 + a_4 x + a_6$$

with $a_i, i = 1, 2, 3, 4, 6$; x; $y \in K$, where K is a field. A point $P = (x,y)$ is a valid point on the elliptic curve if it satisfies the Weierstrass equation. The basic operations performed on the elliptic curve are point addition and point doubling. Using these operations, a scalar multiplication $Q = kP$ can be defined on an elliptic curve. The set of points on an elliptic curve together with a point at infinity ∞ with point addition as an operation makes the set of points an algebraic group.

The elliptic curves based on finite fields (Galois fields $GF(p^n)$) are only used for cryptography applications. The most commonly used finite fields are prime fields $GF(p) = \mathbb{Z}_p$ and binary extension fields $GF(2^n)$. These two field types have different characteristics (p and 2) and the Weierstrass equation can be simplified and different formulas for algebraic operations can be derived for these cases. Because of the differences in the structure of those fields, the performance of both software and hardware implementations of cryptographic algorithms on them can vary significantly.

Weierstrass equation reduces to $E: y^2 = x^3 + ax + b$ for elliptic curves defined over the finite fields $GF(p)$, where $p > 3$ is prime and $a, b \in GF(p)$. Given such a curve E, the group that is employed in cryptographic protocols is a large prime-order subgroup of the group $E(GF(p))$ of $GF(p)$-rational points on E. A fixed generator of the cyclic subgroup is usually called the base point and denoted by $G \in GF(p)$.

16.7.1 Elliptic Curve Cryptography

The security of elliptic curve cryptographic systems is based on the hardness of the elliptic curve discrete logarithm problem (ECDLP). The ECDLP is as follows: given an elliptic curve E defined over $GF(p^n)$, and two points $P, Q \in E$, find an integer x such that $Q = xP$.

Given a set of domain parameters that include a choice of base prime field $GF(p)$, an elliptic curve E, and a base point G of order n on E, an elliptic curve key pair (d, Q) consists of a private key d, which is a randomly chosen nonzero integer modulo the group order n, and a public key $Q = dG$. Thus, the point Q becomes a randomly selected point in the group generated by G.

There are several different standardizations extending the basic elliptic curve Diffie–Hellman key exchange protocol. The basic Diffie–Hellman key exchange protocol is as follows. To agree on a shared key, Alice and Bob individually generate key pairs (d_A, Q_A) and (d_B, Q_B). They then exchange the public keys Q_A and Q_B over a public channel. In this case, both Alice and Bob will be able to compute the point $P = d_A Q_B = d_B Q_A$ using their respective private keys. The shared secret key is derived from P by a key derivation function, usually being applied to its x-coordinate.

The ElGamal elliptic curve cryptosystem is as follows. Suppose that E is an elliptic curve defined over a finite field $GF(p^n)$. Suppose that E, p^n, and a point $G \in E$ and the embedding $m \rightarrow P_m$ (from message to point on the elliptic curve) are publicly known. When Alice wants to communicate secretly with Bob, they proceed as follows:

- Bob chooses a random integer b, and publishes the point bG, keeping b secret.
- Alice chooses another random integer a and sends the pair of points $(aG, P_m + a(bG))$ to Bob, keeping a secret.
- To decrypt the message, Bob calculates $b(aG)$ from the first part of the pair, then subtracts it from the second part to obtain $P_m + a(bG) - b(aG) = P_m + abG - abG = P_m$, and then reverses the embedding $(P_m \rightarrow m)$ to get back the message m.

An attacker, Eve, who can only see bG, aG, and $P_m + a(bG)$, must find a from aG or b from bG to get P_m. Thus the security of the system is reduced to the ECDLP.

The Elliptic Curve Digital Signature Algorithm that was standardized in the FIPS 186-4 is as follows. A signer generates a key pair (d, Q) consisting of a private signing key d and a public verification key $Q = dG$. To sign a message m, the signer first chooses a random integer k such that $1 \le k \le n$, computes the point $kG = (x, y)$, transforms x to an integer, and computes $r = x \bmod n$. The message m is hashed to a bitstring and then transformed to an integer e. The signature of m is a pair (r, s) of integers modulo n, where $s = k^{-1}(e + dr) \bmod n$.

16.7.2 Elliptic Curve Cryptosystem Implementation

The most basic finite field algorithms are addition and subtraction. In Wenger et al. [21], algorithms for modular addition in $GF(p)$ and $GF(2^m)$ are given. Operand-scanning or product-scanning multiplication approaches are used for realizing modular multiprecision multiplications. A multiply-accumulate unit has been used to increase the efficiency of the product-scanning method. Such multiply-accumulate units can be designed for the finite fields $GF(p)$ and $GF(2^n)$. Carry-less multipliers for $GF(2^n)$ have shorter critical paths and smaller area requirements because in this case, logical XOR cells are used instead of full-adder standard cells. The power consumption for a multiplier designed out of XORs is lower compared to full adders.

16.8   VLSI IMPLEMENTATIONS OF SHA-3 HASH FUNCTION

A cryptographic hash function is a cryptographic primitive that takes an arbitrary long block of data as input and returns a fixed-length (or variable length, in the case of SHA-3) output, called the cryptographic hash value. The original data is often called the message, and the hash value is sometimes called the message digest or simply the digest. Typical lengths of hash output are 128 and 160. Hash functions are used in a wide range of communication protocols where they provide data integrity, user authentication, and in security systems such as intrusion detection systems. An ideal cryptographic hash function $H(.)$ must have the following four main properties:

- Efficiency: It is easy to compute the hash value for any given message.
- Preimage resistance: Given a hash value h, it should be infeasible to find a message m such that $h = H(m)$. In other words, $H(.)$ has to be a one-way function. That is, it is infeasible to generate a message that has a given hash value.
- Second preimage resistance: Given a message m_1, it should be infeasible to find another message m_2 such that $m_1 \neq m_2$ and $H(m_1) = H(m_2)$. That is, it is infeasible to modify a message without changing the hash value.
- Strong collision resistance: It should be infeasible to find two different messages m_1 and m_2 such that $m_1 \neq m_2$ and $H(m_1) = H(m_2)$. Such a pair is called a collision. That is, it is infeasible to find two different messages with the same hash.

In 2007, the U.S. NIST announced a public competition aiming at the selection of a new cryptographic hash function standard [22]. The motivation behind the NIST competition was to replace the two widely used hash functions MD5 and SHA-1 following a series of successful attacks [23,24,25] on them. On October 2, 2012, Keccak designed by Guido Bertoni, Joan Daemen, Michal Peeters, and Gilles Van Assche was selected as the winner of the SHA-3 hash function competition. Several applications, such as multigigabit mass storage devices and radio-frequency identification (RFID) tags, are expected to utilize SHA-3. SHA-3 hash function finds application in both high-performance and resource-constrained environments and offers good performance in terms of speed, area, and power.

16.8.1   SHA-3 Standard

Keccak Hash Function was finally selected as the SHA-3 standard. Keccak is a family of sponge functions $Keccak[r,c]$ characterized by two parameters, the bitrate r and the capacity c. The sum $r + c$ determines the width of the $Keccak - f$ permutation used in the sponge construction and can take values only in the set {25,50,100,200,400,800,1600}. For the SHA-3 standard, the designers of Keccak proposed the *Keccak*[1600] with different values for r and c for each desired length of the hash output. The *Keccak*[1600] has a 1600-bit state, which consists of 5×5 matrix of 64-bit words. For the 256-bit hash Keccak 256, the parameters are $r = 1088$ and $c = 512$. For the 512-bit hash Keccak-512, the parameters are $r = 576$ and $c = 1024$.

The compression function of Keccak consists of the following five primitive steps consisting of XOR, AND, NOT, and permutation operations:

- theta (θ)
- rho (ρ)
- pi (π)
- chi (χ)
- iota (ι)

Each compression step of Keccak consists of 24 rounds. The hash function operation consists of three phases, which are initialization, absorbing phase, and squeezing-bit phase. In the initialization phase, all the entries of the state matrix are set to zero. In the absorbing phase, each message is XORed with the current state and 24 rounds of Keccak permutation are performed. After absorbing all blocks of the input message, the squeezing phase is performed. In the squeezing phase, the state matrix is truncated to the desired output length of the hash. If a longer hash value is required, then more Keccak permutations are performed and their results are concatenated until the length of the hash values reaches the desired output length. The function is illustrated in Figure 16.4.

16.8.2 SHA-3 Implementation

Hardware implementation of SHA-3 is analyzed based on the operation speed, the circuit area, and the power consumption. The cost of an application-specific integrated circuit (ASIC) implementation of SHA-3 directly depends on the area required to realize the algorithm. The maximum available silicon area, the total number of I/O

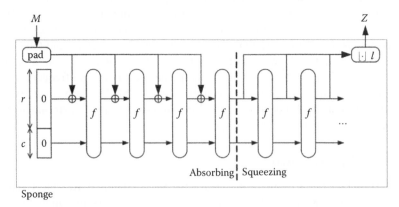

FIGURE 16.4 Sponge function.

TABLE 16.1

Performance of SHA 3

Area (kGE)	Energy (mJ/Gbit)	Throughput (Gbps)	Clock Frequency (mHz)	Clock Frequency for 20 Gbps Throughput
50	2.42	43.011	949	441

pins, and the capabilities of the test infrastructure can all limit the implementations. Further, the throughput of a hash function, which is defined as the amount of message in bits for which a hash value can be computed per second, needs to be measured. Power and energy requirements are also used in evaluations in these days. Power density can limit the circuits to comply for sub 100-nm technologies, and limit the applications in systems with scarce energy resources such as handheld devices, smartcards, RFID devices, and so on. In evaluations, the energy per bit of input information processed by the hash function may be calculated to determine the energy consumption.

In Henzen et al. [26], an ASIC hardware implementation of SHA-3 hash standard is given. The authors first developed a golden model and then used it to generate the stimuli vectors and expected responses to verify the register-transfer level (RTL) description of the algorithm written in Very high speed integrated circuit Hardware Description Language. They then used Synopsys Design Vision-2009.06 to map the RTL description to the UMC 90-nm technology using the fsd0a_a_2009Q2v2.0 RVT standard cell library from the Faraday Technology Corporation. All outputs are assumed to have a capacitive loading of 50 fF, and the input drive strength is assumed to be that of a medium strength buffer.

Postlayout performances of SHA-3 (Keccak-256) algorithms for a target throughput of 20 Gbps in the UMC 90-nm process as reported by Henzen et al. [26] are given in Table 16.1.

16.9 CONCLUSION

In this chapter, we discussed VLSI implementations of various cryptographic primitives such as stream ciphers, block ciphers, public-key ciphers, and hash functions. We included the brief descriptions of some of the major hardware implementations of these algorithms. There is a lack of standard among various implementations. With the advancement of hardware technologies, we hope that faster and efficient implementations of the cryptographic algorithms can be carried out on handheld devices.

REFERENCES

1. IEEE Standard Specifications for Public-Key Cryptography, IEEE P1363, January 2000.
2. N. Koblitz, "Elliptic Curve Cryptosystems," *Mathematics of Computation*, 48: 203–209, 1987.
3. K. H. Ko, J. W. Lee, T. Thomas, "Towards generating secure keys for braid cryptography," *Design Codes and Cryptography*, 45(3): 317–333, 2007.

4. T. Thomas, A. K. Lal, "A zero-knowledge undeniable signature scheme in non-abelian group setting," *Int. J. Net. Sec.*, 6(3): 265–269, 2008.

5. P. Kitsos, G. Kostopoulos, N. Sklavos, O. Koufopavlou,"Hardware implementation of the RC4 stream cipher," *IEEE 46th Midwest Symposium on Circuits and Systems*, 3: 1363–1366, 2003.

6. Data Encryption Standard (DES). Fed. Inf. Process. Standards Pub, FIPS 46-3, October 1999.

7. P. Kitsos, S. Goudevenos, O. Koufopavlou, "VLSI implementations of the triple-DES block cipher," *Proceedings of the 10th IEEE International Conference on Electronics, Circuits and Systems*, ICECS 2003.

8. Advanced Encryption Standard (AES). Fed. Inf. Process. Standards Pub, FIPS 197, November 2001.

9. X. Zhang, K. K. Parhi, "High-speed VLSI architectures for the AES algorithm," *IEEE Transactions on Very Large Scale Integration (VLSI) Systems*, 12(9), September 2004.

10. A. Satoh, S. Morioka, K. Takano, S. Munetoh, "A compact Rijndael hardware architecture with S-Box optimization," in *Proc. ASIACRYPT 2001*, Gold Coast, Australia, pp. 239–254, December 2000.

11. A. Rudra, P. K. Dubey, C. S. Jutla, V. Kumar, J. R. Rao, P. Rohatgi, "Efficient implementation of Rijndael encryption with composite field arithmetic," *Proc. CHES 2001*, Paris, France, pp. 171–184, May 2001.

12. Y. Eslami, A. Sheikholeslami, P. G. Gulak, S. Masui, K. Mukaida, "An area-efficient universal cryptography processor for smart cards," *IEEE Transactions on Very Large Scale Integration (VLSI) Systems*, 14(1): 43–56, January 2006.

13. S. E. Eldridge, D. Walter, "Hardware implementation of Montgomerys modular multiplication algorithm," *IEEE Trans. Comput.*, 42: 693–699, June 1993.

14. C. K. Koc, C. Y. Hung, "Bit-level systolic arrays for modular multiplication," *J. VLSI Sig. Proc.*, 3: 215–223, 1991.

15. A. Vandemeulebroecke, E. Vanzieleghem, T. Denayer, P. G. A. Jespers, "A new carry-free division algorithm and its application to a single chip 1024-b RSA processor," *IEEE J. Solid-State Circuits*, 25: 748–755, June 1990.

16. C. D. Walter, "Systolic modular multiplication," *IEEE Trans. Comput.*, 42: 376–378, March 1993.

17. G. R. Blakley, "A computer algorithm for the product AB modulo M," *IEEE Trans. Comput.*, 32: 497500, 1983.

18. Y. J. Jeong, W. P. Burleson, "VLSI array algorithms and architectures for RSA modular multiplication," *IEEE Transactions on Very Large Scale Integration (VLSI) Systems*, 5(2): 211–217, June 1997.

19. P. L. Montgomery, "Modular multiplication without trial division," *Math. Comp.*, 44: 519–521, 1985.

20. V. S. Miller, "Use of elliptic curves in cryptography," In *Advances in Cryptology - CRYPTO 85*, Springer-Verlag, Santa Barbara, CA, August 18–22, 1985.

21. E. Wenger, M. Hutter, "Exploring the design space of prime field vs. binary field ECC-hardware implementations," *Information Security Technology for Applications*, Lecture Notes in Computer Science, 7161: 256–271 2012.

22. National Institute of Standards and Technology, "Announcing request for candidate algorithm nominations for a new cryptographic hash algorithm (SHA-3) family." *Federal Register*, 72(212): 62212–62220, 2007.

23. C. D. Canniere, C. Rechberger, "Finding SHA-1 characteristics: General results and applications," In *Advances in Cryptology - ASIACRYPT 2006*, vol 4284 of LNCS, p. 120. Springer, Berlin, Germany, 2006.

24. C. D Canniere, C. Rechberger, "Preimages for reduced SHA-0 and SHA-1," In *Advances in Cryptology - CRYPTO 2008*, vol 5157 of LNCS, pp. 179–202. Springer, Berlin, Germany, 2008.
25. X. Wang, H. Yu, "How to break MD5 and other hash functions," In *Advances in Cryptology - EUROCRYPT 2005*, vol 3494 of LNCS, p. 1935. Springer, Berlin/ Heidelberg, Germany, 2005.
26. L. Henzen, P. Gendotti, P. Guillet, E. Pargaetzi, M. Zoller, F. K. Gurkaynak, "Developing a hardware evaluation method for SHA-3 candidates," *Cryptographic Hardware and Embedded Systems*, CHES 2010, Lecture Notes in Computer Science, 6225: 248–263, 2010.

17 Dynamic Intrinsic Chip ID for Hardware Security

Toshiaki Kirihata and Sami Rosenblatt

CONTENTS

17.1   INTRODUCTION

Rapid progress in science and technology has brought great convenience for business and personal life alike. In particular, evolutionary advancements in nanoscale semiconductor technology [1], while improving performance, are key contributors to the miniaturization of electronic products. These improvements, in turn, have introduced consumers to personal computing and the information era. Lately, consumer products for mobile computing and communication have been the fastest growing segments to drive the semiconductor industry. These new products further incorporate wireless, Internet, and revolutionary social network applications, allowing interpersonal communication without restrictions of time and place. However, these new products and services have created various privacy and security concerns. Although communicating with people whom one has never met is already common in a person's virtual life, it may result in a breach of privacy as well as in virus attacks to the personal system. Therefore, developing high-security products and services is an urgent task not only for business, but also for the semiconductor industry.

Security and privacy in the digital network use cryptographic protocols [2] with open and private keys, which are now commonplace in the industry. However, this high level of security cannot be realized without establishing a root of trust, which includes the use of highly reliable hardware. Thus, hardware security [3] is an equally, if not more, important concern for high-security systems. Hardware security requires identification, authentication, encryption, and decryption engines [4]. Unfortunately, a recent surge in counterfeit hardware has resulted in the occasional distribution of cheap, fake chips in the industry [5]. Distribution of these counterfeit chips not only leads to reduced product reliability, but also further increases privacy and security risks. These counterfeits are typically the result of either discarded chips reintroduced into the supply chain, or of fabrication of cheap copies that pass as authentic, reducing cost without significant security. Systems using unauthentic chips are consequently exposed to a high risk of failure in the field. More importantly, using fake chips for banking or defense systems results in a potentially unacceptable risk for national security. Thus, the reliability of the task of identifying hardware has become a necessity in contemporary security, which in turn requires a method for highly reliable chip identification and authentication using semiconductor components.

Figure 17.1 summarizes the hardware identification method. The simplest approach uses an identifier (ID) such as text that is uniquely assigned to each hardware unit. However, as the ID can be read by anyone, a counterfeit part can be created by simply writing the ID on the fake hardware. This approach is, therefore, not secure at all. Recent hardware, in particular a very-large-scale integration (VLSI) chip, uses embedded fuses for IDs that are uniquely assigned and programmed in each of the VLSI chips. This fuse approach is somewhat secure, because the ID cannot be read without breaking the module, whereas a special tool is necessary for writing the ID to the fuses. Older fuse technology uses laser [6] for blowing the fuses, which is a large and expensive process. The evolution of fuse technology has resulted in the electrically blowable fuse approach (eFUSE) [7]. This makes it possible to create a high-density one-time programmable read-only memory (OTPROM) [8,9] that can be used for chip ID, as well as for other programming elements such as redundancy replacement. Regardless of laser fuse or OTPROM using eFUSE, the fuses are visible by optical or electronic microscopy, and therefore the programmed ID information can still be obtained after delayering. Using nonvolatile memory such as embedded flash memory [10,11] makes it difficult to obtain the ID by the delayering method; however, direct readout from the chip is still possible if the read method is known.

There are two fundamental problems with the existing ID approaches. First, all of them use an ID that is extrinsic to the hardware. These extrinsic IDs are not integral to the hardware, as the information needs to be written or programmed, and they can be subsequently read by some method. Second, the ID can be cloned as long as it is

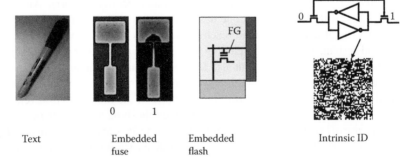

Text	Embedded fuse	Embedded flash	Intrinsic ID

FIGURE 17.1   Evolution of chip identification (ID).

known. These weaknesses make it possible to create counterfeits by emulating the ID in multiple fake chips. Therefore, the challenges to creating secure IDs are to make it difficult for unauthorized people to detect the ID and to make it impossible, if not impractical, to create a clone of the ID even if it can be made known with reasonable effort. An intrinsic chip ID uses intrinsic VLSI features, constituting an ideal solution for high-security applications and making it prohibitive to make fake copies.

In this chapter, we discuss an intrinsic chip ID using bitmaps of an embedded dynamic random-access memory (eDRAM) macro. Section 17.2 introduces the concept of intrinsic ID, followed by a review of the existing approaches. In Section 17.3, principles of eDRAM are discussed. Sections 17.4 through 17.6 are concerned with the dynamic intrinsic ID approach using retention signatures of a one-transistor and one-capacitor cell. In Sections 17.7 and 17.8, a field-tolerant method is explored, which improves success of intrinsic ID authentication in a field. Section 17.9 introduces a security and authentication enhancement method using a multi-ID approach. Section 17.10 summarizes the chapter and briefly discusses the future challenges for intrinsic ID research and development.

17.2   INTRINSIC CHIP ID

Fingerprints are widely used for secure identification of individuals. A human fingerprint is a unique and unclonable feature that each person possesses. In a similar fashion, a secure intrinsic ID exploits intrinsic features of a VLSI chip. Such features arise from random process variations, and can be used to generate an ID that cannot be reverse-engineered or easily emulated, also called a "Physically Unclonable Function (PUF)" [12–37]. This thus greatly improves chip security over the existing extrinsic ID approach. In this section, we discuss the intrinsic ID generation and authentication concept using random process variations in manufacturing, and their challenges.

An intrinsic chip ID converts process variations in manufacturing into a digital binary vector for chip identification. Figure 17.2 shows intrinsic ID generation methods using complementary metal–oxide–semiconductor (CMOS) cross-coupled inverters. In the simplest approach (a), each of the inverters drives the other's output, nodes A and B, respectively. This structure is commonly used as a temporary storage element or as a bus keeper [14] for VLSI design, and constitutes an ideal element for intrinsic ID. When used in ID generation, the cross-coupled inverter is powered up without node initialization. This power-up method, accomplished by raising a main power supply voltage (VDD), naturally determines the states of A and B as a consequence of the p-type metal-oxide-semiconductor (PMOS) and n-type metal-oxide-semiconductor (NMOS) threshold mismatch of the cross-coupled inverters. The A and B states after power-up can be used as a bit of an intrinsic ID. These states, however, may not be stable, as they may be sensitive to noise during power-up. Therefore, a preferred approach (b) includes two additional access transistors that force nodes A and B to ground (GND) level as soon as power is turned on. The states can, therefore, be determined by the PMOS threshold mismatch of the cross-coupled inverters, as nodes A and B are forced to GND level until one of the PMOS is strongly turned on, eliminating instability resulting from power-up noise. This approach (b) has two advantages. First, the threshold voltage (VT) mismatch of the cross-coupled inverters depends on the local PMOS VT mismatch, and can potentially minimize the systematic impacts of

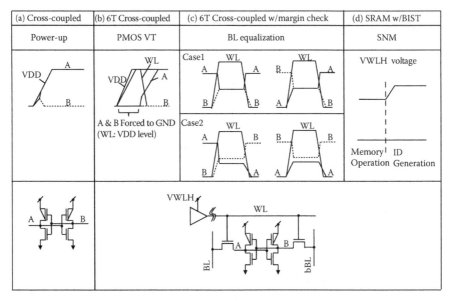

(a) Cross-coupled	(b) 6T Cross-coupled	(c) 6T Cross-coupled w/margin check		(d) SRAM w/BIST
Power-up	PMOS VT	BL equalization		SNM

FIGURE 17.2 Physically unclonable fuse (PUF) using cross-coupled inverter. (a) Cross-coupled inverter approach. The states of A and B are naturally determined after power-up. (b) 6 Transistor (6T) SRAM cell approach. The nodes A and B are forced to GND level until one of the PMOS is strongly turned on, eliminating instability due to power-up noise. (c) 6T SRAM cell approach with margin check. Nodes A and B are equalized from the predetermined 0 and 1 states, and they generate the random bit pattern by deactivating the wordline (WL). (d) Static-noise margin (SNM) approach. The fail bits are generated by reducing SNM. The bit pattern is preferably generated using a built-in-self-test (BIST) engine.

lithography, temperature, or noise. Second, this structure is identical to that of the six-transistor static random access memory (SRAM) cell, and as such enables the use of SRAM arrays and their bitmap [15–23] for intrinsic ID generation. The end result is a significant increase of density with the added benefit of allowing a product memory array to be used for intrinsic ID generation.

Approach (b) may be improved by using an ID margin detection method to eliminate unstable bits. Unlike the power-up intrinsic ID generation, approach (c) [22] generates the ID after completing a power-up sequence similar to that of conventional memory operation. Before ID generation, the A and B nodes are initially set to predetermined states ("1" and "0," or "0" and "1"). This initialization is realized by means of a SRAM write mode. Writing "1" sets nodes A and B to "1" and "0," and writing "0" sets nodes A and B at "0" and "1," respectively. Activation of the wordline (WL) opens the access transistor, whereas the bitlines (BLs) of the pair are held at GND level. This results in a short-circuit of nodes A and B, and the BL and bBL equalization levels are determined by intrinsic features such as threshold voltage mismatch or transistor strength of the cell. Cells containing balanced cross-coupled inverters allow nodes A and B to be equalized, and result in a small voltage difference between BL pairs during equalization, as shown in case 1. On the other hand, cells having largely imbalanced cross-coupled inverters remain unequalized, or flip the states in a preferred direction, as shown in case 2.

Evaluation or generation of the intrinsic ID is enabled by deactivation of the WL, which in turn disables the equalization of nodes A and B. This results in determining the node A and B states naturally. For the well-equalized BL cell (case 1), a generated bit is likely unstable. Otherwise, the generated bit is determined by the initial predetermined voltages of the nodes. On the other hand, a bit generated from the cell that is not well equalized (case 2) is always stable regardless of the initial predetermined voltages of the nodes. Generation and evaluation must be performed using both initial node states (A and B at "0" and "1" and "1" and "0") to assess the intrinsic stability of the bit.

On-chip intrinsic ID generation requires a built-in-self-test (BIST) engine [38]. Fortunately, the on-chip memory in a logic chip is typically supported by BIST, and therefore is readily available for intrinsic ID generation without additional overhead. The fourth approach (d) uses a BIST engine in the generation of a random binary vector derived from checking the static-noise margin (SNM) of a SRAM array [23]. For memory operation, the wordline high voltage (VWLH) is adjusted to have a sufficient SNM when the WL goes high. For ID generation, the VWLH voltage is increased to reduce the SNM. This results in a fail for a weak cell when the WL goes high. To weed out the weakest bits, this approach includes a feedback loop between the number of fails and the VWLH voltage supply. This feedback remains active until the fail count (FC) reasonably matches a predetermined number. The result is the generation of a stable random bit pattern comprised of the fail bit addresses, which are detected and recorded using the on-chip BIST.

An intrinsic chip ID can be further implemented using a delay-based PUF. Ring-oscillator (RO)–based PUFs [24,25] compare the delay between ROs while generating unclonable random bit strings. Similar to the RO PUF, an arbiter-based PUF [26] is composed of delay paths for signals A and B and an arbiter located at the end of the delay path. The arbiter outputs "1" when signal A arrives earlier than B, otherwise it outputs "0." Because the delay path is determined by intrinsic features in manufacturing, the output bit can be random. The challenge of this approach is to lay out the signal path symmetrically to minimize the normal delay difference between the two paths. Otherwise, the output will be skewed. Another approach [27] detects an analog voltage determined by the threshold of the MOS transistors, which is subsequently converted to a binary identification sequence using an auto-zeroing comparator. The one-time oxide breakdown PUF [28] leverages the fact that a stress condition such as high voltage can break weak cells with a higher confidence than strong cells, resulting in permanent random intrinsic ID generation. The concern with this permanent approach is that the ID bits can be detected by delayering the chip in a manner similar to the extrinsic fuse approach. Bit-string generation based on resistance variation in metals and transistors has also been reported [29].

The ultimate intrinsic ID is a fingerprint derived from product intrinsic features without need for allocating additional silicon for chip identification. VLSI chips, in particular recent multicore and multithread microprocessers [39,40], include tens or hundreds of on-chip cache memories. Using the on-chip cache as a fingerprint intrinsic ID may therefore offer an ideal solution for highly secure identification, because the memory can be used for the generation of enormous intrinsic IDs without requiring additional expensive silicon from a high-performance system. The challenges for this application reside in providing the intrinsic ID function without degrading the product chip features.

In this chapter, we discuss intrinsic chip ID generation and authentication using 32 nm HiK/metal gate eDRAM product developed for next generation multicore and multithread microprocessor. The next section describes the principle of eDRAM macro operation and the retention signature, which will be used for intrinsic ID generation and authentication to be discussed in following sections.

17.3 HIGH-PERFORMANCE EMBEDDED–DYNAMIC RANDOM ACCESS MEMORY

For nanometer technology, it is desirable and essential to integrate more functions within a single die. DRAM integration with a high-performance logic process (eDRAM) [38] not only reduces packaging cost, but it also significantly increases the memory bandwidth while eliminating input and output (I/O) electrical communication that is noisy and power hungry. Because of the smaller memory cell size, the eDRAM can be approximately 3–6 times denser than embedded-SRAM, and operates with low-power dissipation and 1000 × better soft-error rate. eDRAM macros [41] that are based on high-performance logic are extremely vital microelectronic components, making it possible to integrate 32-MB on-chip cache memory on POWER7™ processors [39].

eDRAM employs a one-transistor and one capacitor (1T1C) as a memory cell that stores a data bit for a read and write operation. To reduce the cell size, the capacitor is built using either stack [42] or trench capacitor [43] structures. A deep trench capacitor approach is the preferred structure for eDRAM because the capacitor is built before device fabrication. This facilitates the implementation of a process fully compatible with logic technology, as transistor performance does not degrade because of capacitor fabrication, and design rules for back-end-of-lines (or metal wiring) remain the same as that of the logic technology. This results in an ideal technology solution for DRAM integration on a high-performance logic chip.

Figure 17.3 shows the overview of the eDRAM macro [41] for the POWER7 microprocessor [39]. The macro consists of four 292-Kb memory arrays stacked in the Y-direction, resulting in a density of 1.168 MB. These arrays are controlled by a peripheral circuit block (IOBLOCK) arranged at the bottom of the macro. The peripheral circuit block, in turn, consists of command and address receivers, decoders, and macro I/O circuitries used to control the memory arrays with given input commands.

Each 292-Kb memory array consists of 1T1C cells arranged in a two-dimensional matrix. The memory cells in the array including row and column redundancies are accessed by 264 WLs and 1200 BLs for row and column, respectively. The architecture is optimized for a L3 cache application while taking into account the performance, power, and I/O requirements. Unlike in conventional memory, the wordline drivers are placed in the area adjacent to the global sense amplifiers (GSAs), which are in turn controlled by the global WL drivers located in the peripheral circuit block (IOBLOCK). This orthogonally segmented WL architecture [38] is the key to realize the wide I/O organization.

The 292-Kb array is organized using eight 36.5-Kb microarrays for transistor microsense amplifier architecture [41]. A total of 32 cells with an additional redundant cell (total 33 cells) are coupled to the local bitline (LBL), and read or written

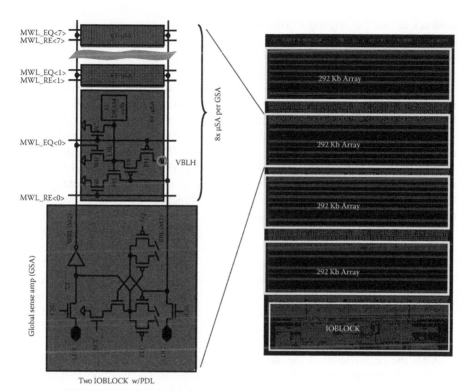

FIGURE 17.3 High-performance embedded–dynamic random access memory (eDRAM). (J. Barth, D. Plass, E. Nelson, C. Hwang, G. Fredeman, M. Sperling, A. Mathews et al., A 45 nm SOI embedded DRAM macros for POWER7™ processor 32-MByte on-chip L3 cache, *IEEE Journal of Solid-State Circuits*, 46[1], © 2011 IEEE.)

using a local microsense amplifier (µSA). Eight groups of LBLs, each one with a µSA arranged in the same column, are coupled to a GSA through the global read and write BLs (RBL and WBL). This hierarchical approach is important both to improve performance and to reduce power dissipation for a 500-MHz random cycle.

While in the stand-by state, the WL is held at a wordline low voltage (VWL), a negative voltage. This is key to turning all access transistors sufficiently off. Otherwise, the threshold of the access transistors must be increased, resulting in a higher WL boost voltage (VPP) requirement to turn the devices on. Signals MWL_ EQs and MWL_REs are, respectively, high and low, to turn the LBL precharge device (PC) on, the NMOS foot device off, and the PMOS head device (PH) off. This signal configuration precharges all LBLs to the low level (GND), while disabling the NMOS read head device (RH) and the PMOS feedback device (FB). RBL and WBL are held high and low, respectively. When the 36.5-Kb microarray is selected for WL activation, the signal MWL_EQ goes low. This disables the NMOS (PC), floating the LBL in the selected subarray. The MWL_EQ signal in other unselected arrays remains at high level, keeping the precharge device of the unselected LBLs at low level. The low-going MWL_EQ also turns on the PMOS PH, enabling the PMOS feedback device. For writing 1 to the cell, RBL goes low. This allows the LBL to go high, resulting in a high voltage written to the corresponding capacitor. For writing 0 to the cell, WBL goes high. This keeps the LBL at the low level, resulting in a low voltage written to the corresponding capacitor.

When a read command is accepted, read data are transferred from the cell to the BL when the WL goes high. For reading 1 data, LBL goes high, which turns the NMOS RH on. This results in making RBL low when the signal MWL_RE goes high and the NMOS footer device (NF) turns on. At the same time, low-going RBL turns on the PMOS FB. This results in making the LBL go high. The high voltage on the LBL is written back to the corresponding cell. For reading 0 data, LBL stays at the low level, and the NMOS RH device remains disabled. RBL therefore stays at the high level. The RBL data are subsequently multiplexed by a column select signal (CSL) and communicated to the peripheral circuit block (IOBLOCK) through a local data bus pair (LDC and LDT) and then the primary data line (PDL) for data transfer. At the same time, the GSA senses the RBL in the high state, making WBL go high. This makes the write 0 device (W0) to turn on, forcing the LBL to the low level. Therefore, the low-level voltage on LBL is written back to the corresponding cell.

Precharge operation starts when the WL falls to the VWL voltage (negative voltage). MWL_EQ and MWL_RE go high and low, respectively, precharging the LBL to the low level. This concludes one 500-MHz random access memory cycle for < 1.5 nanoseconds latency.

After the deactivation of the WL, the data bits are kept inside the capacitors. However, the charge stored in them leaks as time goes on, and therefore they should be periodically read and written back. This is a unique but important requirement for DRAMs, and is known as the refresh operation. As long as the refresh is executed before the voltage of the storage nodes drops below the detection threshold of the SA, the data bits are maintained. The time interval during which a cell can hold the data bit is called retention time. The eDRAM intrinsic chip ID uses this retention signature in order to generate unclonable random bit patterns.

17.4   RETENTION-BASED INTRINSIC CHIP ID

Figure 17.4 shows the concepts of intrinsic chip ID generation and authentication using retention signatures from eDRAM. Before ID generation, a logic 1 is written to all bit cells in an array, stored in the form of charge in the capacitors. However, the charge in each cell leaks as a function of time. Therefore, after waiting for a predetermined amount of time, the read data bits from the corresponding array may remain as 1 or change to 0. Preferably, the read operation may be managed by a BIST engine or, if not implemented, by an external tester. If the charge remains sufficiently high, the output (DQ) from the SA results in the expected value of 1, resulting in a PASS. If the charge leaks beyond the SA detection point of 1, then DQ is 0, resulting in a FAIL. Because each cell has a different retention time, the PASS and FAIL address locations in the array are random and physically unclonable. The array bit pattern (BITMAP) may then be recorded as the intrinsic chip ID in a bit-string format such that the length of the string is equal to the array size, and where a passing bit (pbit) is stored as 0, and a failing bit (fbit) is stored as 1, ordered from the first to

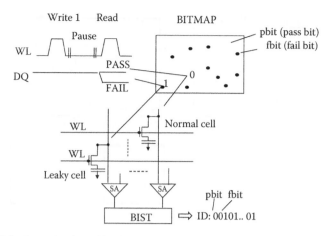

FIGURE 17.4 Retention-based intrinsic chip ID.

last logical addresses. Notice that the bits of interest for the intrinsic ID binary string are the fbits, and therefore their binary values are conveniently set to 1 as opposed to their logical 0 values after the read operation.

Because generation of fbits can be done by changing the predetermined pause time in a nondestructive manner, this retention-based intrinsic ID approach allows for use of the eDRAM array IP as is. However, the retention time depends significantly on various process or design parameters, including subthreshold leakage, junction leakage, GIDL, bitline capacitance, cell capacitance, and noise. Therefore, chip authentication using retention time becomes highly sensitive to the precise test condition used during ID generation. In addition, creating fbits at low temperatures requires a long pause time, increasing the time for ID generation.

To improve bit stability, retention-based ID generation uses a higher VWL than the product target voltage. This is equivalent to emulating a low-threshold access transistor, forcing the retention signature to be sub-threshold driven. Figure 17.5 employs a VWL controlled by the VWL generator (VWLG) [44] using a feedback loop driven by the array's FC. Using higher VWL voltage increases the subthreshold leakage of the access transistor, which in turn generates more fbits in the bitmap. This VWL voltage can therefore be used as a tunable array input parameter controlling the number of fbits in the BITMAP.

Specifically for applications, product use is optimized using a negative VWL resulting in no retention failures within a predetermined retention target. This is labeled *Memory mode* in Figure 17.5. In the *ID mode*, significantly higher VWL voltage is applied to the array such that the retention time is determined by sub-threshold leakage. To reduce the dependency on temperature and voltage conditions during ID generation, the retention-based intrinsic ID approach includes a counter for the number of fbits in the BITMAP. VWLG adjusts the VWL voltage such that the number of fbits matches the predetermined target number. Consequently, implementing feedback between the BIST-counter pair and VWLG enables the generation of an intrinsic BITMAP (or binary string) for any target FC of the retention failures (or 1 in bit string). The target FC may be given by the Original Equipment

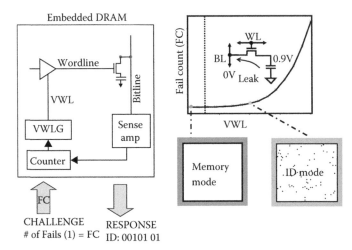

FIGURE 17.5 Method to generate the bitmap with wordline low voltage (VWL) control. Fail count (FC) is used as a CHALLENGE. VWL voltage is controlled by the VWL generator (VWLG) or an external tester such that the FC in the bitmap satisfies the target FC. The RESPONSE is the bit string created by the bitmap, where 0 and 1 are PASS and FAIL address locations in the array, respectively. (Rosenblatt, S. et al., Field tolerant dynamic intrinsic chip ID using 32 nm high-K/metal gate SOI embedded DRAM, *IEEE J. Solid-State Circuits*, 48[4], © 2013 IEEE.)

Manufacturer (OEM) as a CHALLENGE to the chip. The generated intrinsic ID vector with the given FC is recorded in the OEM database to identify the chip for subsequent authentication.

Authentication is realized by extracting the ID string from the intrinsic ID binary vector using the same target FC as used earlier during the recording phase. Once the bit string with the target FC is achieved, the chip outputs the generated ID bit string as a RESPONSE to the OEM, and the database searches for the ID from within its list until it identifies the corresponding chip.

The uniqueness of a set of IDs for a given number of parts can be calculated exactly. The probability of generating n different IDs by randomly choosing j fbits out of an array of size i is given by the following expression (see Table 17.1):

$$P_{BASE}(i,j,n) = \prod_{\alpha=0}^{n-1} \frac{\binom{i}{j}-\alpha}{\binom{i}{j}} \tag{17.1}$$

For practical implementations, the inherent information entropy contained in an ID set is already very large. For example, a set of 4 Kb strings containing 100 fbits, each will be unique up to roughly 10^{100} parts.

17.5 DYNAMIC INTRINSIC CHIP ID

The retention-based intrinsic chip ID allows for control of the number of fails in the array bit pattern (BITMAP) by means of the VWL. During generation, the number of the fails (fbits) in the ID matches a target number, and therefore that number can

TABLE 17.1

Probability of Creating n Unique Intrinsic IDs with j Fails in Array of Size i

ID Generation Turn	1	2	—	n
Failing bit addresses in ID_{GEN}	$F_1 = \{f_{11}, f_{12}, \quad f_{1j}\}$	$F_2 = \{f_{21}, f_{22}, \quad f_{2j}\}$	—	$F_n = \{f_{n1}, f_{n2}, \quad f_{nj}\}$
# Combinations available per turn resulting in a new ID_{GEN}	$\binom{i}{j}$	$\binom{i}{j} - 1$	—	$\binom{i}{j} - (n-1)$
# Total combinations per ID_{GEN} per turn	$\binom{i}{j}$	$\binom{i}{j}$	—	$\binom{i}{j}$
Probability of choosing a new ID_{GEN} per turn	$P_1 = \dfrac{\binom{i}{j}}{\binom{i}{j}} = 1$	$P_2 = \dfrac{\binom{i}{j} - 1}{\binom{i}{j}}$	—	$P_n = \dfrac{\binom{i}{j} - (n-1)}{\binom{i}{j}}$
Joint probability $P_{\text{BASE}} = P_1 \times P_2 \ldots \times P_n$	$P_{\text{BASE}}(i,j,n) = \displaystyle\prod_{\alpha=0}^{n-1} \dfrac{\binom{i}{j} - \alpha}{\binom{i}{j}}$			

be adjusted by the VWL voltage. It is not practical, however, to generate the same exact number of fails as set by the target, and even if it were, it would be too expensive or time consuming. In addition, it may not be safer from a security standpoint to use the exact same ID binary string for each authentication request. The dynamic intrinsic ID approach is introduced to eliminate the necessity to generate the exact same FC while enhancing ID security.

Figure 17.6 shows the concept of dynamic chip ID generation and authentication. It consists of a window-based authentication method using retention time in which the fbit locations corresponding to an ID with a larger FC number include the fbit locations corresponding to an ID with a smaller number. Unlike the direct intrinsic ID method of Section 17.4, the dynamic intrinsic ID method uses a pair of IDs (ID_{MIN} and ID_{MAX}). ID_{MIN} is the intrinsic ID binary string corresponding to a target minimum FC, FC_{MIN} (i.e., 10). ID_{MAX} is the intrinsic ID binary string corresponding to a target maximum FC, FC_{MAX} (i.e., 100). During ID generation, the ID_{MIN} and ID_{MAX} pair is generated by adjusting the VWL voltage while counting the fails until the respective FC_{MIN} and FC_{MAX} targets are achieved. These FC targets are used to determine the target window (i.e., 10–100) for subsequent authentication. Because of the nature of the DRAM cell, the fbits in the ID_{MAX} bit string include those in the ID_{MIN} bit string (Generation Rule).

Authentication uses the FC target FC_{AUTH} inside the target window, such that $\text{FC}_{\text{MIN}} \leq \text{FC}_{\text{AUTH}} \leq \text{FC}_{\text{MAX}}$. The FC_{AUTH} target is dynamically changed within the window at each authentication request (i.e., 30, then 50, then 70), significantly improving hardware security. By construction, ID_{AUTH}, the regenerated intrinsic ID corresponding to FC_{AUTH}, should have a FC at least larger than ID_{MIN} and at most

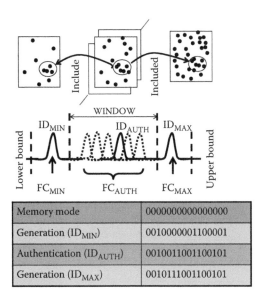

Memory mode	0000000000000000
Generation (ID$_{MIN}$)	0010000001100001
Authentication (ID$_{AUTH}$)	0010011001100101
Generation (ID$_{MAX}$)	0010111001100101

FIGURE 17.6   Concept of dynamic chip ID generation and authentication method.

smaller than ID$_{MAX}$. Again because of the nature of the DRAM cell, the fbits in ID$_{AUTH}$ (i.e., 0010011001100101) obtained from the FC$_{AUTH}$ target (i.e., 50) include those in ID$_{MIN}$ (i.e., 0010000001100001) obtained from the FC$_{MIN}$ target (i.e., 10), and are included in the fbits in ID$_{MAX}$ (i.e., 0010111001100101) obtained from the FC$_{MAX}$ target (i.e., 100). Authentication using ID$_{AUTH}$ is therefore possible as long as the generated FC is within the target window defined by the pair ID$_{MIN}$ and ID$_{MAX}$ (Authentication Rule).

As ID generation does not need to satisfy the exact FC target provided by the CHALLENGE, window-based authentication simplifies ID generation. As shown in Figure 17.6, the FC in typical hardware has a distribution centered about the target FC. It is therefore important to determine a target window suitable for authentication while considering the FC distribution, to not violate the authentication rule. One possible violation of the rule can happen if too few fails or too many fails in the ID pair reduces the uniqueness of the chip ID. Hence, it is necessary to establish a lower bound for FC$_{MIN}$ and an upper bound for FC$_{MAX}$ of the generated ID pair to avoid false positive authentication (misidentification of the ID with others), which can be regarded as a Window Rule. As an example of what can happen without boundaries, it is possible to authenticate 0 fails or 100% fails for any chip, which is not an acceptable situation. The number of fails in the ID should therefore be within a predetermined window to satisfy ID uniqueness for the target application while remaining guaranteed with an analytically chosen degree of confidence.

Figure 17.7 exemplifies the chip- and system-level architecture enabling the dynamic intrinsic ID method. For ID generation, the OEM server sends the ID generation request (CHALLENGE) with the target FC pair (FC$_{MIN}$ and FC$_{MAX}$). This pair is used one time as a CHALLENGE for generating the corresponding

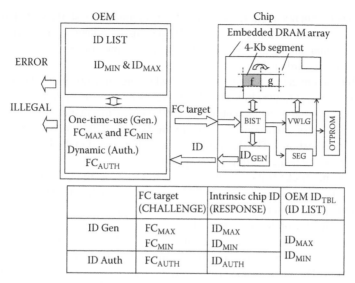

FIGURE 17.7   Chip and system architecture for dynamic chip ID.

chip ID_{MIN} and ID_{MAX} pair. On receiving the CHALLENGE, the chip starts test-ing the first 4-Kb memory segment (f) with the BIST. The test first verifies if the segment (SEG) is reliable for product use, such that the product's nominal VWL voltage produces no fails. To avoid process edge effects, the edge segments are not used. If the BIST detects a fail, the segment is not reliable. The circuit SEG then increments the segment number to choose the next segment (g) and the BIST tests it next. After confirming no fails in the selected segment, the BIST requests the VWL generator to increase the VWL voltage. This emulates reducing the thresh-old of the access transistor, resulting in a shorter retention time. The BIST retests the selected segment with the new increased VWL voltage, and checks if the FC is within the FC_{MIN} band (FC_{MIN} +/- GB), where GB is the predetermined guard band. The BIST test continues until the BIST has found the first bitmap having FC satisfying the FC_{MIN} band by incrementing the VWL voltage. The BIST test then starts searching for the second bitmap using an FC target satisfying the FC_{MAX} window while incrementing the VWL voltage. The ID generator circuit (ID_{GEN}) reformats the detected first and second bitmaps to the two binary bit strings com-posed of PASS and FAIL bits as 0 and 1 bits, respectively. This is the pair ID_{MIN} and ID_{MAX}. The chip then confirms if the locations of fbits in ID_{MAX} include all fbits in the ID_{MIN}.

If the ID_{MIN} and ID_{MAX} pair does not satisfy the Generation Rule, the circuit SEG chooses a new segment, and starts testing the new one to search for a new ID pair. The confirmed ID_{MIN} and ID_{MAX} pair is then sent to the system. The system first checks if the pair is unique (no collisions with other pairs). If the pair passes the uniqueness test, the system stores the ID pair in its list for subsequent authentication. The segment (i.e., g) used for ID generation is then stored in OTPROM within the chip for later authentication. If the pair fails the uniqueness test, the system requests new ID generation to search for a new ID pair using a new segment. In case no ID pair is found among all segments, the system outputs the error code (ERROR), resulting in a discarded chip.

The ID authentication request is enabled when the authentication CHALLENGE FC_{AUTH} is provided to the chip. The chip searches for the bitmap having FC satisfying the FC_{AUTH} band by incrementing the VWL voltage in a manner similar to the ID generation process, after confirming there are no fails with nominal VWL voltage (product specifications). The detected bitmap is reformatted to the ID_{AUTH} binary bit string. This string is then output to the system.

The system checks if the FC of ID_{AUTH} is the same as the given CHALLENGE (FC_{AUTH}). If these FCs do not match within a specified tolerance (FC_{AUTH} +/- GB), the chip may be counterfeit and spoofing a known chip ID. The dynamic intrinsic chip ID, therefore, protects against illegal use with output code ILLEGAL. This prevents against a counterfeiter using a previous CHALLENGE by preprogramming it. If the FC is valid, the system searches within its list (ID LIST) for the chip having an ID_{MIN} and ID_{MAX} pair whose fbit locations are included in and include ID_{AUTH}, respectively (in other words, satisfies the Authentication Rule).

The uniqueness of an authentication key ID_{AUTH} in recognizing a valid chip is minimized when $ID_{AUTH} = ID_{MIN}$. After all, an authentication key with $FC_{AUTH} = FC_{MIN}$ has the least number of combinations to choose from and is by construction the least unique. It follows that authentication of a pair of ID_{MIN} and ID_{MAX} using the ID_{MIN} bit string has the lowest uniqueness rating for dynamic authentication. For n parts, with a binary string of i bits, $FC_{MIN} = k$ and $FC_{MAX} = j$, an approximate analytical expression that provides a lower bound for the uniqueness of the pair of IDs is given by the following expression (see Table 17.2):

$$
P_{MIN}(i,j,k,n) = \left(\prod_{\alpha=0}^{n-2} \frac{\binom{i}{k}\binom{j}{k} - \alpha}{\binom{i}{k}} \right)^n
$$

$$
\approx \operatorname{Exp}\left(-n^2 \left(\frac{\binom{j}{k} + \frac{n}{2}}{\binom{i}{k}} \right) \right), i \gg j, \binom{i}{k} \gg n
$$

(17.2)

As seen in Figure 17.8, the analytical model shows > 99.999% ID uniqueness using a 4-Kb segment ($i = 4096$), lower bound of 12 ($k = 12$ or $FC_{MIN} = 12$), and upper bound of 128 ($j = 128$ or $FC_{MAX} = 128$) for 10^6 parts ($n = 10^6$). The actual uniqueness is characterized by the probability P_{UNIQUE}, which is typically larger than P_{MIN}. They are equal only when all the ID_{MAX} in the set have no fails in common. Figure 17.9 compares the accuracy of the analytical expressions in Equation 17.2 with Monte Carlo simulations of the dynamic intrinsic ID method for various input parameters.

TABLE 17.2
Probability P_{UNIQUE} That Retention-Based Dynamic Intrinsic ID Set Is Unique. Number of Chips = n, Array Size = i, FC_{MAX} Target = j, FC_{MIN} Target = k

Failing bit addresses in a given ID_{MAX}	$F_0 = \{f_{01}, f_{02}, \cdots f_{0J}\}$				
ID_{AUTH} generation turn		1	2	$\cdots$	$n-1$
Failing bit addresses in ID_{AUTH}		$F_1 = \{f_{11}, f_{12}, \cdots f_{1k}\}$	$F_2 = \{f_{21}, f_{22}, \cdots f_{2k}\}$	$\cdots$	$F_{(n-1)} = \{f_{(n-1)1}, f_{(n-1)2}, \cdots f_{(n-1)k}\}$
# Combinations available per turn resulting in a new ID_{AUTH} that does NOT dynamically authenticate ID_{MAX}		$\binom{i}{k} - \binom{j}{k}$	$\binom{i}{k}\binom{j}{k}-1$	$\cdots$	$\binom{i}{k}\binom{j}{k}-(n-2)$
# Total combinations per ID_{AUTH} per turn		$\binom{i}{k}$	$\binom{i}{k}$	$\cdots$	$\binom{i}{k}$
Probability of choosing new unique ID_{AUTH} per turn for a given ID_{MAX}		$P_1 = \dfrac{\binom{i}{k}\binom{j}{k}}{\binom{i}{k}}$	$P_2 = \dfrac{\binom{i}{k}\binom{j}{k}-1}{\binom{i}{k}}$	$\cdots$	$P_{(n-1)} = \dfrac{\binom{i}{k}\binom{j}{k}-(n-2)}{\binom{i}{k}}$

Probability of choosing a unique set of ID_{AUTH}
$$P_C(i,j,k,n) = \prod_{\alpha=0}^{n-2} \frac{\binom{i}{k}\binom{j}{k}-\alpha}{\binom{i}{k}}$$
$$= P_1 \times P_2 \cdots \times P_{(n-1)}$$

Total probability of unique authentication
$$P_{MIN}(i,j,k,n) = \left[\prod_{\alpha=0}^{n-2} \frac{\binom{i}{k}\binom{j}{k}-\alpha}{\binom{i}{k}}\right]^n$$
$$= P_{C1} \times P_{C2} \cdots \times P_{Cn}$$
$$= P_C^n$$
$$\approx \mathrm{Exp}\left[-n^2\left(\frac{\binom{j}{k}+\frac{n}{2}}{\binom{i}{k}}\right)\right], \quad i \gg j, \binom{i}{k} \gg n$$

$$P_{UNIQUE} \geq P_{MIN}$$

P_{MIN} excludes overlaps between chip IDs $\rightarrow$ lower count $\rightarrow$ lower probability boundary

Note: P_{MIN} is the exact expression when all ID_{MAX} have no fails in common and corresponds to a lower bound for P_{UNIQUE}.

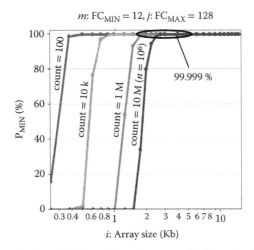

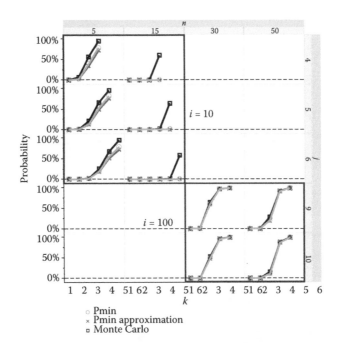

FIGURE 17.8 Calculated probability of ID uniqueness with Equation 17.2. (Rosenblatt, S. et al., Field tolerant dynamic intrinsic chip ID using 32 nm high-K/metal gate SOI embedded DRAM, *IEEE J. Solid-State Circuits*, 48[4], © 2013 IEEE.)

FIGURE 17.9 Probability of uniqueness of the retention-based dynamic intrinsic ID set using analytical expressions for P_{MIN} (exact and approximate from Table 17.2), and Monte Carlo of this probability for 10,000 simulation sets. In the Monte Carlo, the k fails used for authentication are taken from the sample of n chips. Results are shown for $i = 10$ and 100. When j is much smaller than i, all the methods of estimating the probability converge toward each other, demonstrating the accuracy of the model.

17.6 DYNAMIC INTRINSIC CHIP ID HARDWARE AUTHENTICATION

The concept of dynamic ID generation and authentication has been studied using 32 nm SOI eDRAM [45], which was a base design for an eDRAM macro IP for an IBM microprocessor [46]. Figure 17.10 shows a chip microphotograph of (a) an eDRAM array and (b) the test site of the IP used for this feasibility study along with its features. The eDRAM uses high-performance SOI eDRAM design features including 6T microsense amplifier and orthogonal WL architecture developed for POWER7 [41].

The IP used in this feasibility demonstration does include neither BIST engine nor a VWLG, and therefore the VWL voltage is controlled externally by the tester with a resolution of 5 mV. Before generating the ID, confirmation of no fails in the target 4-Kb segment is done using a nominal (within product specifications) VWL, also known as "Memory mode." In this study, CHALLENGEs of $FC_{MIN} = 25$ and $FC_{MAX} = 95$ were used to generate a pair of intrinsic IDs. The eDRAM was tested repeatedly, varying the VWL in the tester to search for a binary string with fbit count close to, but not necessarily equal to, the desired target, as long as the Generation Rule was satisfied. If a pair of 4 Kb binary strings (ID_{MIN} and ID_{MAX}) was found which:

1. Reasonably matched the corresponding CHALLENGES: $FC_{MIN} = 25$ and $FC_{MAX} = 95$
2. Satisfied the Generation Rule: the fbit locations of ID_{MAX} included the fbit locations of ID_{MIN}

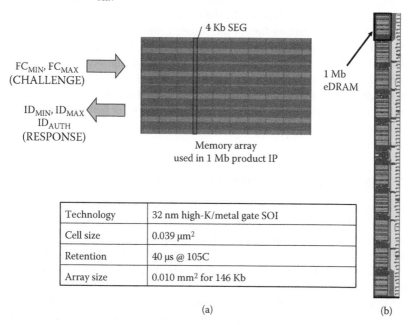

Technology	32 nm high-K/metal gate SOI
Cell size	0.039 μm²
Retention	40 μs @ 105C
Array size	0.010 mm² for 146 Kb

(a)   (b)

FIGURE 17.10 Chip microphotograph of (a) an eDRAM array and (b) the test site of the IP used for this feasibility study along with its features. (Rosenblatt, S. et al., Field tolerant dynamic intrinsic chip ID using 32 nm high-K/metal gate SOI embedded DRAM, *IEEE J. Solid-State Circuits*, 48[4], © 2013 IEEE.)

The ID pair was stored in an ID management system for authentication; else, the generation scheme was applied to successive 4 Kb SEGs of the array until a suitable area was found. Beginning with the "Memory mode" condition of no fails, the typical time to find the target number of fails required was 30 minutes per chip. The search time can be reduced significantly by means of a binary search [47].

For the purposes of authentication security and counterfeit prevention demonstration, chip authentication was realized by using different CHALLENGEs, selected as a different FC value at each authentication request, for example, a FC_{AUTH} of 65 that is between FC_{MIN} of 25 and FC_{MAX} of 95. Similar to ID_{MIN} and ID_{MAX} pair generation, VWL was adjusted across multiple tests of the 4 Kb SEG manager circuit originally used for ID_{MIN} and ID_{MAX} pair generation. The voltage was adjusted until an ID_{AUTH} binary string was generated with a number of fbits close to FC_{AUTH} of 65. The ID_{AUTH} string vector was the output of the system, and the string was compared to the list of ID_{MIN} and ID_{MAX} pairs previously stored in the system to identify the corresponding chip. A match occurred when the ID_{AUTH} string included all of the fbit locations in ID_{MIN} and was included in the fbit locations in ID_{MAX} for a pair of strings in the database (Authentication Rule from Section 17.5). If the bit string ID_{AUTH} did not follow the rule for any ID_{MIN} and ID_{MAX} pair in the list, then the chip was deemed invalid.

Figure 17.11a shows the measured ID distribution generated by CHALLENGEs (FC_{MIN} = 25, FC_{MAX} = 95, and FC_{AUTH} = 65) for 346 chips at a fixed temperature of 25°C and voltage of VDD = 0.9 V. It was confirmed that the 346 chips can be uniquely identified by the dynamic retention-based chip ID generation and authentication rules. Detailed analysis of the data shows that the mean/standard deviation of overlapping fbit locations for (b) the Monte Carlo simulation and (c) hardware are 0.55/0.73 and 0.58/0.76, respectively, showing a good correlation between simulation and hardware results. Figures 17.12a and b show the fbit overlap count by chip of 32 IDs between generation and authentication for simulation and hardware, respectively. Each ID was generated with 100 fbits and regenerated with 50 fbits in 32 different chips. The results show an excellent agreement between hardware and simulation, demonstrating a maximum of 5 overlapping fbits in any one of the 32 × 32 combinations.

17.7 FIELD-TOLERANT INTRINSIC CHIP ID

The dynamic intrinsic ID approach allows for some bit changes as long as the fbit addresses in ID_{AUTH} are included in the ID_{MAX} fbits. It may, however, result in an error if some newly generated fbits in ID_{AUTH} violate the authentication rule. This may happen if all the fbits in ID_{MIN} do not constitute any longer a subset of the fbits in ID_{AUTH}, or if all of the newly generated fbits in ID_{AUTH} cease to be a subset of the ones in ID_{MAX}. This results in false-negative authentication. Therefore, it is necessary to exert precise voltage and temperature control, so that ID_{AUTH} generation can emulate the conditions used for ID generation. This is impractical, if not very expensive, for use in the field. Even if the exact same conditions are used in the field, some unstable fbits may still change, resulting in an authentication error. A field-tolerant intrinsic ID is introduced to overcome this problem.

Figure 17.13 illustrates the concept of (a) dynamic ID and (b) field-tolerant intrinsic ID generation and authentication. As discussed in Section 17.5, the dynamic intrinsic chip ID approach (a) uses an authentication window using ID_{MIN} and ID_{MAX}. This results in successfully authenticating the corresponding chip as long as the fbits in the ID_{AUTH} vector are within the fbit location window bounded by ID_{MIN} and ID_{MAX}. Imperative to enabling the field-tolerant ID approach (b) is the guarantee of unique ID generation and successful chip recognition even if some bits change across various conditions. This can be realized by detecting common fbits between the IDs in a database and the ID for authentication. To increase the probability of common fbits, FC_{AUTH} is made larger than the FC_{GEN} used for generation of ID_{GEN} by means of a higher VWL voltage during authentication than during generation. In contrast, if FC_{AUTH} is smaller than FC_{GEN}, the number of fbits in common between ID_{AUTH} and ID_{GEN} is at most FC_{AUTH}, which by construction is a fraction of FC_{GEN}. This makes it harder to authenticate ID_{AUTH}, as there is potentially too little common fbit overlap between the two IDs as a percentage of ID_{GEN}. Therefore, as FC_{AUTH} increases

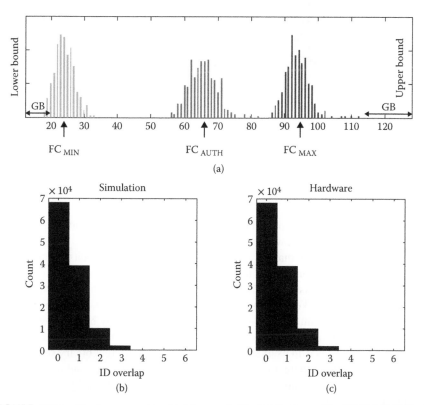

FIGURE 17.11 Hardware results. (a) Measured ID distribution with CHALLENGE of $FC_{MIN} = 25$, $FC_{MAX} = 95$, and $FC_{AUTH} = 65$. ID overlap for different IDs: (b) Monte Carlo simulation and (c) measured results. (Rosenblatt, S. et al., Field tolerant dynamic intrinsic chip ID using 32 nm high-K/metal gate SOI embedded DRAM, *IEEE J. Solid-State Circuits*, 48[4], © 2013 IEEE.)

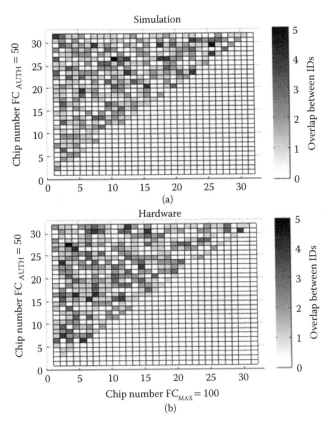

FIGURE 17.12 (a) Simulation and (b) hardware fbit overlap count by chip for $FC_{MAX} = 100$ and $FC_{AUTH} = 50$.

relative to FC_{GEN}, a predetermined target percentage of FC_{GEN} common fbits may be used to authenticate ID_{AUTH} with a required level of confidence.

Successful recognition may be achieved even if some unstable fbits change due to variations in field test conditions and/or device shifts over the product's lifetime, for example, NBTI [48]. As discussed in the dynamic ID approach (a), the field-tolerant approach (b) must also keep the number of retention fails, FC_{AUTH}, less than that of a predetermined maximum value—roughly half the total array bits—to avoid false positive authentication or ID spoofing. This limit comes from the fact that a binary string with many fbits is in itself a combination of many different binary strings with fewer fbits. Therefore, there are more chances of multiple ID_{GEN} combining to form a larger ID_{AUTH}, or of an attacker using an ID_{AUTH} with an arbitrarily large number of fbits for spoofing.

The field-tolerant approach requires choosing proper array and ID sizes to avoid false authentication. Although false negative authentication is the main goal of this approach, an increase in retention FC will also increase the probability of false-positive authentication, in which a chip is recognized as a different chip in the

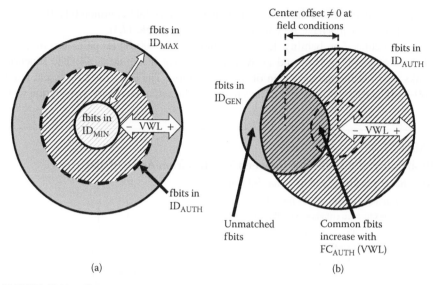

(a) (b)

FIGURE 17.13 Concept of dynamic and field-tolerant ID approaches. Dynamic ID (a) uses ID_{MIN} and ID_{MAX} to improve ID security. Field-tolerant ID (b) detects common failing bits (fbits) while accepting some unmatched fbits that may be the result of a different environment in the field. (Rosenblatt, S. et al., Field tolerant dynamic intrinsic chip ID using 32 nm high-K/metal gate SOI embedded DRAM, *IEEE J. Solid-State Circuits*, 48[4], © 2013 IEEE.)

database. For randomly generated IDs, this probability is modeled by the number of expected fbits in common within any pair of IDs belonging to the set of stored IDs. Given a string representing 100 random FAIL locations in 4096 bits, the probability that another similar random string has k fails in common with the first obeys the well-known and documented hypergeometric distribution [49]. This can be interpreted as all the ways of choosing k FAIL locations in common with the original 100 while simultaneously choosing the remaining 100-k FAIL locations from the remaining 4096-100 passing locations, and it is given by:

$$
P(k) = \frac{\begin{bmatrix} \text{\# ways for} \\ k \,''\text{successes}'' \\ \text{picked} \end{bmatrix} \times \begin{bmatrix} \text{\# ways for} \\ (100-k)\,''\text{failures}'' \\ \text{picked} \end{bmatrix}}{[\text{Total \# picks}]}
$$

$$
= \frac{\dbinom{100}{k}\dbinom{4096-100}{100-k}}{\dbinom{4096}{100}}
$$

(17.3)

Using these parameters, the mean of the distribution is 2.44 common bits; therefore, most of these strings have very little ID overlap (see Table 17.3).

The calculated probability density function $P(k)$ of 100-bit ID pairs as a function of the k overlapping fbits and array size i is shown in Figure 17.14. The function peaks near the theoretical mean of the distribution. The solid lines represent fixed 100-bit IDs, whereas the dashed lines represent the actual hardware distribution with

---

TABLE 17.3

Probability of d Collisions between Two IDs from Array of Size i

ID_{MAX} has fail count j_{max} and ID_{MIN} has fail count j_{min}

# Combinations of exactly d fbits from ID_{MIN} with j_{min} fbits that are common with fbits in ID_{MAX} with j_{max} fbits, $j_{min} \le j_{max}$	$\begin{pmatrix} j_{max} \\ d \end{pmatrix}$
# Combinations of exactly $j_{min}-d$ fbits from ID_{MIN} that are not common with ID_{MAX}	$\begin{pmatrix} i-j_{max} \\ j_{min}-d \end{pmatrix}$
# Total Combinations of j_{min} fbits in memory bank of size i	$\begin{pmatrix} i \\ j_{min} \end{pmatrix}$
Probability of d fbits from ID_{MIN} being common with ID_{MAX}	$P(d) = \dfrac{\begin{pmatrix} j_{max} \\ d \end{pmatrix}\begin{pmatrix} i-j_{max} \\ j_{min}-d \end{pmatrix}}{\begin{pmatrix} i \\ j_{min} \end{pmatrix}}$
Normalization condition	$\sum P(d) = \dfrac{\displaystyle\sum_{d=0}^{j_{min}} \begin{pmatrix} j_{max} \\ d \end{pmatrix}\begin{pmatrix} i-j_{max} \\ j_{min}-d \end{pmatrix}}{\begin{pmatrix} i \\ j_{min} \end{pmatrix}} = 1$
Hamming distance	$d_{Hamm} = j_{max} + j_{min} - d$
Expected ID overlap	$\langle d \rangle = \sum dP(d) = \dfrac{\displaystyle\sum_{d=0}^{j_{min}} d\begin{pmatrix} j_{max} \\ d \end{pmatrix}\begin{pmatrix} i-j_{max} \\ j_{min}-d \end{pmatrix}}{\begin{pmatrix} i \\ j_{min} \end{pmatrix}} = \dfrac{j_{min}j_{max}}{i}$
Expected square of ID overlap	$\langle d^2 \rangle = \sum d^2 P(d) = \dfrac{\displaystyle\sum_{d=0}^{j_{min}} d^2\begin{pmatrix} j_{max} \\ d \end{pmatrix}\begin{pmatrix} i-j_{max} \\ j_{min}-d \end{pmatrix}}{\begin{pmatrix} i \\ j_{min} \end{pmatrix}}$
Standard deviation of ID overlap	$\sigma_d = \sqrt{\langle d^2 \rangle - \langle d \rangle^2} = j_{min}\dfrac{j_{max}}{i}\dfrac{i-j_{max}}{i}\dfrac{i-j_{min}}{i-1}$
Probability of d common fbits between any pair of unique IDs within window between ID_{MIN} and ID_{MAX} (averaging)	$P(d) = \dfrac{\displaystyle\sum_{a=j_{min}}^{j_{max}}\sum_{j=j_{min}}^{j_{max}}\begin{pmatrix} a \\ d \end{pmatrix}\begin{pmatrix} i-a \\ j-d \end{pmatrix}}{\displaystyle\sum_{a=j_{min}}^{j_{max}}\sum_{j=j_{min}}^{j_{max}}\begin{pmatrix} i \\ j \end{pmatrix}}$

For $j_{max} = j_{min} = j$, the $P(d)$ expression reduces to the hypergeometric distribution of Equation 17.3. If the IDs are randomly distributed within a window between j_{min} and j_{max}, the probability of d collisions is given by multiple summations representing all the possible combinations of j values within the window.

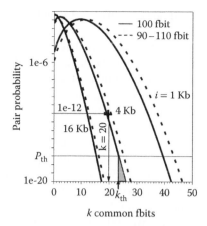

FIGURE 17.14 Calculated probability to have the corresponding ID overlap with respect to ID bit string length ($i = 1$ Kb, 4 Kb, and 16 Kb). (Rosenblatt, S. et al., Field tolerant dynamic intrinsic chip ID using 32 nm high-K/metal gate SOI embedded DRAM, *IEEE J. Solid-State Circuits*, 48[4], © 2013 IEEE.)

an average over all ID sizes in a 90- to 110-bit window (accomplished by additional averaging of the function over that range). The level of accuracy in meeting the target FC is therefore expected to affect the value of the mean (peak) and the tightness of the distribution of overlapping fails (standard deviation).

Statistical extrapolation can be used to establish a proper guard band of common bits k_{th} for false positive prevention. For a set of 10^6 parts, the total number of pairs is on the order of $10^6 \times 10^6 = 10^{12}$. Referring to the 4-Kb distribution of Figure 17.14, the odds of 1 occurrence in 10^{12} pairs correspond to a probability of 10^{-12}, which in turn occurs when the number of common fbits is $k = 20$. By construction, the probability from the model of finding a pair within the set with more than 20 common fbits decreases exponentially. Because this model is also derived from a discrete probability curve, it is reasonable to assume that the cumulative probability above this point never adds to the probability of the point itself. To illustrate this, $k_{th} = 23$ fbits in common represents a probability $P(k_{th}) = P_{th}$ that is roughly 5 orders of magnitude lower than that of 20 fbits in common. This probability also corresponds to the shaded area to the right of k_{th}. Therefore, if no ID is ever authenticated with less than 23 fbits in common, this represents a guard band providing over 10^5 lower odds ($P_{th} = 10^{-17}$) than finding 1 pair with 20 common fbits in a set of 10^6 parts. In other words, the final guard-banded threshold therefore satisfies unique chip IDs 99.999% (1.0–10^{-5}) of the time for 10^6 parts.

To ensure 100% successful authentication, hardware analysis using voltage and temperature fluctuations must be used to take into account unstable retention fails and prevent a false negative result.

17.8 FIELD-TOLERANT HARDWARE AUTHENTICATION

Field-tolerant intrinsic ID generation and authentication have been demonstrated using 32 nm SOI eDRAM product IP. The 4 Kb ID_{GEN} bit strings with target of 100 retention fails were extracted using a 4-Kb SEG from a 292-Kb array in each of 266 eDRAM chips. A nominal voltage condition of VDD = 0.9 V and 85°C was used. These ID_{GEN} bit strings identifying each one of the 266 chips were recorded in the local database.

Field-tolerant authentication was then emulated with ID_{AUTH} string generation for a target of 200 retention fails using the same 4-Kb SEGs. Two temperatures, 25°C and 85°C, and three voltage conditions of VDD = 0.9 V ± 10%, representative of voltage tolerance in the field, were used. The six-generated ID_{AUTH} from each of the 266 chips were used to search the corresponding chips with previously stored ID_{GEN} in the local database.

In searching for false positives in hardware, every ID_{AUTH} (total 6 × 266 = 1596) was compared to the stored 266 ID_{GEN} of a *different* chip as shown in Figure 17.15 for a total of approximately 420 k (= 266 × 1596) pairs. The normalized count of common fbits is overlaid with the hypergeometric model of (3) with no fitting parameters. The model assumes that the set of stored IDs have 100 random target fails (FC_{GEN} = 100) and the set of authentication keys have 200 random target fails (FC_{AUTH} = 200), with a +/–10% FC tolerance window for each set. The excellent agreement with the random fail model is a measure of the uniqueness and randomness of the set of IDs.

False negative authentication occurs when a previously stored chip ID cannot be recognized because of physical instability of the bit cells in the field. False negatives were characterized by comparing each ID_{AUTH} to the stored ID_{GEN} of the *same* chip. The results are summarized in Figure 17.16 for all 266 chips. No more than 10% of fbits in the authentication set fail to match the original ID. Most fbit loss occurs at 25°C, and as recorded ID size approaches authentication ID size. Tighter FC control can be used to limit this problem.

Whereas false negative characterization depends on the physical properties of the circuit and the test methodology, false positives rely on the randomness of the ID set and can be characterized with use of the model of (3). Conservatively, it is sufficient to use as few fbits as possible for recognition, as long as it ensures that all parts can be

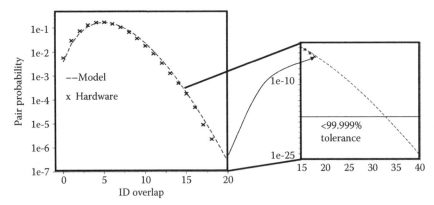

FIGURE 17.15 Probability to have the corresponding ID overlap for any of approximately 420 K pairs. A 40% overlap between a 100-fail stored ID and a 200-fail authentication key is sufficient to guarantee 99.999% ID uniqueness for 10^6 parts. Fail count windows of +/–10% are ensured with a +/–5 mV VWL window. (Rosenblatt, S. et al., Field tolerant dynamic intrinsic chip ID using 32 nm high-K/metal gate SOI embedded DRAM, *IEEE J. Solid-State Circuits*, 48[4], © 2013 IEEE.)

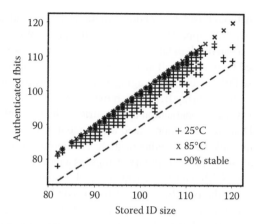

FIGURE 17.16 Measured ID overlap for each of the same 266 chips during authentication using VDD = 0.9 V and VDD = 0.9 V +/–10%, at 25°C and 85°C (total 6 conditions). The results show at least 90% of the ID string 1 bits are stable even if the test condition is changed. (Rosenblatt, S. et al., Field tolerant dynamic intrinsic chip ID using 32 nm high-K/metal gate SOI embedded DRAM, *IEEE J. Solid-State Circuits*, 48[4], © 2013 IEEE.)

successfully authenticated without misrecognition. For 10^6 parts, an approach requiring that a minimum of approximately 40% of the bits from the stored ID_{GEN} (40 fbits for 100-fail FC_{GEN} target) are common with the authentication key ID_{AUTH} of a different chip (200-fail FC_{AUTH} target) guarantees authentication with at least 99.999% success rate. On the other hand, security is enhanced when a larger number of fbits is used for recognition, as it makes counterfeiting harder. Unfortunately, false negative results are made more likely by increasing the number of fbits that must be regenerated. As no more than 90% of the stored ID_{GEN} fbits can be ensured to match the ID_{AUTH} fbits in the simulated field authentication of Figure 17.16 while avoiding a false negative, the resulting 40%–90% window of overlapping fbits contains all the redundant authentication fbits for this hardware. A compromise between false authentication boundaries can be therefore reached with a fixed 65% fbit target to be used for field-tolerant authentication.

Use of larger keys for authentication can further reduce the number of unstable fbits that are not matched by forcing those weak cells to fail eventually. As all cells will fail for sufficiently high VWL or retention time, the target value must be bounded to ensure uniqueness of the set and can be calculated using the hypergeometric model. A composite field-tolerant/dynamic key approach follows: a 200-fbit ID needs only to match roughly 65% of fbits in a 100-fbit ID, whereas having only 65% of its fbits in common with a 400-fbit maximum ID is also stored to improve security.

17.9 SECURITY AND AUTHENTICATION ENHANCEMENT FOR INTRINSIC CHIP ID

The field-tolerant approach with common fbit-detection scheme enables chip authentication even if some of the bits change in a field condition. This approach, however, may turn problematic if the ID generated for authentication includes many new fails instead of reproducing most of the old ones, because that can significantly reduce the common bit ratio. Although isolated random single-bit fails are ideally suited for the field-tolerant approach, the appearance of block fails along WLs or BLs is a major source of concern and can result in false negative authentication. Some of these fail signatures, when caused by noise or soft errors, may still be recovered by test repetition. However, when these fails are caused by aging and stress, they may not be recoverable. This results in rendering the array from which the ID is collected unrecognizable.

Figure 17.17 shows the authentication and security enhancement approach. Unlike the intrinsic chip ID approach demonstrated using one 4-Kb ID per chip, an alternative approach uses multiple IDs for authentication. The approach (a) uses two IDs (ID0 and ID1), each having 4-Kb intrinsic binary strings, as a pair. The pair (with combined binary strings totaling 8 Kb) can be generated by choosing two 4-Kb SEGs from a memory array such that the two different 4-Kb intrinsic BITMAPs can be obtained. Both ID0 and ID1 are unique relative to any other IDs from different chips. The ID0 and ID1 pair is then stored in the OEM's database. Authentication is realized by regenerating the ID0 and ID1 pair on request from the OEM. The regenerated ID0 and ID1 pair is sent to the OEM database as a response. The OEM database then checks if the IDs are the same as the respective ID0 and ID1 in the OEM database, independently. A successful authentication result is given if at least one of the IDs has a positive match. Given that odds of such block fails are very low for actual products meeting the technology reliability requirements, the odds that these failure modes may occur in

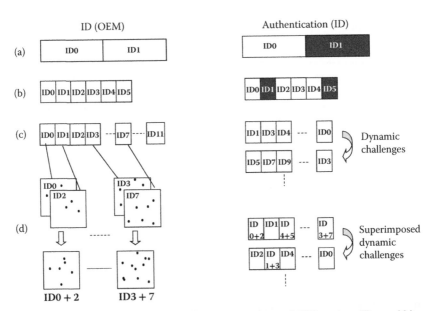

FIGURE 17.17 Security enhancement approaches using multi-IDs, micro-IDs, and bitmap superposition methods. (a) Two ID approach. Positive authentication is given if at least one of the two IDs has a positive match. (b) Micro-ID approach using six ID0-5. Positive authentication is given if at least four out of six micro-IDs (for any combination) are found in the database. (c) Dynamic chip authentication using 12 micro-IDs. Authentication is realized by providing a dynamic CHALLENGE to the chip, which identifies a selection of 8 out of 12 micro-IDs. (d) Superimposed micro-ID approach using dynamic authentication. At least two of the micro-IDs selected from a dynamic CHALLENGE are superimposed with the ID.

several segments simultaneously are vanishingly small. This fact allows one of the IDs (i.e., ID0) to serve as a backup ID for the other (i.e., ID1), successfully authenticating the chip as long as one of the IDs remains good. This ID pair authentication approach is simple enough and extendable to authentication with multiple IDs resulting in more than one backup ID per chip. However, this increases both the overhead of ID volume and the computing power required for authentication, as the number of ID searches and comparisons increases as the number of backup IDs is increased.

The micro-multiple ID approach (b) subdivides the ID into more and smaller ID segments (i.e., ID0–5). Similar to the backup approach, the OEM database stores all micro-IDs (i.e., ID0–5), each having 1-Kb binary strings. The total required ID space per chip is hence 6 Kb, which is overall smaller than that of the backup approach with two full sets of 4-Kb IDs (total 8 Kb per chip). Authentication is realized by comparing each of the 1-Kb micro-IDs with those in the OEM database. A positive authentication result is given if at least four out of six micro-IDs are found in the database. Successful authentication may happen even if two of the micro IDs do not match because of an uncoverable field failure. In addition, the ID search in the OEM database can be executed in parallel using small IDs, resulting in less computing power.

More than a backup solution, the micro-ID approach offers additional levels of security when coupled with a dynamic CHALLENGE system similar to the dynamic chip ID authentication discussed in Section 17.5. The chip is authenticated by using more than six micro-IDs (12, for example), each chip having 512b, as shown in Figure 17.17c. All 12 micro IDs (combined binary strings totaling 6 Kb) are stored in the OEM database. Authentication is realized by providing the dynamic CHALLENGE to the chip, which identifies a selection of 8 of the 512b micro-IDs. This results in approximately 20 million different 4-Kb IDs when order is taken into account, only one of which is the correct 4-Kb binary string corresponding to CHALLENGE. As the dynamic CHALLENGE changes at each authentication request, the ID used in the previous attempt is rendered invalid, consequently improving an authentication security. For example, if the CHALLENGE <1,3,4,…,0> is used in the first authentication request, a subsequent CHALLENGE <5,7,9,….3> renders the first one invalid, which prevents an attacker from attempting to spoof the ID by reusing the previous response.

The dynamic micro-ID approach can be made further secure by superimposing two or more microarray BITMAPs to create one corresponding micro-ID. For example, two of 512b BITMAPs from ID0 and ID2 are superimposed, generating a single 512b micro-ID (ID 0+2). Similarly to the micro-ID approach, a sequence of eight micro-IDs can be generated from superimposing one to at most eight of the original micro-IDs. The CHALLENGE selects the eight micro-IDs and which BITMAPs to be superimposed. The chip then sends to the OEM the string of superimposed micro-IDs as a RESPONSE, which acts as a pseudoencryption function. Simultaneously, the OEM also generates its own expected RESPONSEs from the stored micro-IDs from every chip and then compares the chip's RESPONSE to its own ones until the chip is found in the database.

Another level of security is provided as a secret ID. Instead of targeting reliability issues, this application can be used to foil an attack by a counterfeit chip. If a fake chip tries to replicate individual micro-IDs, a suspicious OEM can choose to authenticate another micro-ID that was not disclosed during a previous authentication request. The intrinsic ID approach using memory arrays can generate a large number of micro-IDs, and is overhead-free for chip design. The micro-ID and superimposed micro-ID approaches can be combined with the backup ID approach with additional micro-IDs, and may include a field-tolerant approach to overcome some unstable fbits within the micro-ID.

17.10 SUMMARY

In this chapter, intrinsic ID generation and authentication using a VLSI chip has been covered at length. As discussed, the approach may preferably use an intrinsic feature relating to retention fails using eDRAM. The DRAM cell allows for retention fails to be controlled by changing the subthreshold leakage. This results in generation of a skewed binary string, having more 0 bits than 1 bits, where 1 is associated with the memory array cell location causing a fail in the bitmap. The challenge of employing retention fails for the generation of intrinsic ID binary strings is how to generate a stable retention fail. This requires operating the device in the subthreshold domain, and can be realized by controlling the wordline low voltage with FC feedback during the retention test. One consequence of this FC feedback method is that the intrinsic ID can be changed dynamically as the FC target changes at each authentication request, resulting in enhanced hardware security. The discussion further includes a method to improve successful authentication rates in a field using a fuzzy match detection approach [15,16], as well as a multi-ID backup option, while further improving the dynamic micro-ID approach with superimposed BITMAPs.

Most of the ideas discussed in this chapter use intrinsic ID for chip identification to protect the product from counterfeiting. However, the applications of intrinsic ID are not limited to chip identification. As an intrinsic chip ID cannot be easily copied, one of its most technologically relevant applications is as a cryptographic key [34,35]. Unlike the intrinsic ID approach for chip identification, cryptographic keys require 100% stable and secure ID bit-string generation, which in turn requires innovations such as advanced circuit design for ID generation and authentication, while integrating the error correction code within the chip [36]. To further improve security, implementation of a reconfigurable PUF [37] that limits multiple use of the secret key may also be of special relevance. In addition to the cryptographic key challenge, a standardization of the intrinsic ID should be strongly encouraged to disseminate the use of intrinsic ID in various products. This requires the development of an intrinsic ID engine that can be embedded into any VLSI chip. This engine must take into consideration the system architecture to maximize the hardware security advantage [33]. Intrinsic ID is an emerging technology that can greatly benefit from tight interactions among technologists, circuit and system designers, and security engineers.

REFERENCES

1. S. Narasimha, P. Chang, C. Ortolland, D. Fried, E. Engbrecht, K. Nummy, P. Parries et al., 22 nm High-performance SOI technology featuring dual-embedded stressors, epi-plate high-K deep-trench embedded DRAM and self-aligned via 15LM BEOL, *IEEE IEDM*, San Francisco, CA, 2012.
2. W. Diffie and M. Hellman, New directions in cryptography, *IEEE Transactions on Information Theory*, 22(6), 644–654, 1976.

3. N. Potlapally, Hardware security in practice: Challenges and opportunities, *2011 IEEE International Symposium on Hardware-Oriented Security and Trust (HOST)*, pp. 93–97, San Diego, CA, 2011.

4. L. Dong, K. Chen, M. Wen, Y. Zheng, Protocol Engineering Principles for Cryptographic Protocols Design, *Eighth ACIS International Conference on Software Engineering, Artificial Intelligence, Networking, and Parallel/Distributed Computing* (SNPD), 3: 641–646, 2007.

5. R. King. (2010, March 1). *Businessweek* [Online]. Available at http://www.businessweek .com/technology/special_reports/20100302ceo_guide_to_counterfeit_tech.htm.

6. R. T. Smith, J. D. Chlifala, J. F. M. Bindels, R. G. Nelson, F. H. Fischer, and T. F, Mantz, Laser programmable redundancy and yield improvement in a 64K DRAM, *IEEE Journal of Solid-State Circuits*, 16(5): 506–514, 1981.

7. N. Robson, J. Safran, C. Kothandaraman, A. Cestero, X. Chen, R. Rajeevakumar, A. Leslie, D. Moy, D. T. Kirihata, and S. Iyer, Electrically programmable fuse (eFUSE): From memory redundancy to autonomic chips, *IEEE Custom Integrated Circuits Conference*, pp. 799–804, San Jose, CA, 2007.

8. J. Safran, A. Leslie, G. Fredeman, C. Kothandaraman, A. Cestero, X. Chen, R. Rajeevakumar et al., A Compact eFUSE Programmable Array Memory for SOI CMOS, *2007 IEEE Symposium on VLSI Circuits*, pp. 72–73, Kyoto, Japan, 2007.

9. G. Uhlmann, T. Aipperspach, T. Kirihata, K. Chandrasekharan, Y. Z. Li, C. Paone, B. Reed et al., A commercial field-programmable dense eFUSE array memory with 99.999% sense yield for 45 nm SOI CMOS, *ISSCC Digest of Technical Papers*, pp. 406–407, San Francisco, CA, 2008.

10. C. Deml, M. Jankowski, and C. Thalmaier, A 0.13 µm 2.125 MB 23.5ns Embedded flash with 2GB/s read throughput for automotive microcontrollers, *ISSCC Digest of Technical Papers*, pp. 478–479, San Francisco, CA, 2007.

11. H. Hidaka, Evolution of embedded flash memory technology for MCU, *2011 IEEE International Conference on IC Design & Technology (ICICDT)*, pp. 1–4, Kaohsiung, China, 2011.

12. R. S. Pappu. Physical one-way functions, PhD thesis, Massachusetts Institute of Technology, Cambridge, MA, 2001.

13. R. Maes, V. Rožić, I. Verbauwhede, P. Koeberl, E. van der Sluis, and V. van der Leest, Experimental evaluation of physically unclonable functions in 65 nm CMOS, *Proceedings of the ESSCIRC*, pp. 486–489, Bordeaux, France, 2012.

14. P. Simons, E. van der Sluis, and V. van der Leest, Buskeeper PUFs, a promising alternative to D Flip-Flop PUFs, *2012 IEEE International Symposium on Hardware-Oriented Security and Trust (HOST)*, pp. 7–12, San Francisco, CA, 2012.

15. G.-J. Schrijen and V. van der Leest, Comparative analysis of SRAM memories used as PUF primitives, Design, Automation & Test in Europe Conference & Exhibition (DATE), pp. 1319–1324, Dresden, Germany, 2012.

16. G. Selimis, M. Konijnenburg, M. Ashouei, J. Huisken, H. de Groot, V. van der Leest, G.-J. Schrijen, M. van Hulst, and P. Tuyls, Evaluation of 90 nm 6T-SRAM as physical unclonable function for secure key generation in wireless sensor nodes, *2011 IEEE International Symposium on Circuits and Systems (ISCAS)*, pp. 567–570, Rio de Janeiro, Brazil, 2011.

17. Y. Su, J. Holleman, and B. Otis, A 1.6pJ/bit 96% stable chip-ID generating circuit using process variations, *ISSCC Digest of Technical Papers*, pp. 406–611, San Francisco, CA, 2007.

18. D. Holcomb, W. P. Burleson, and K. Fu, Power-up SRAM state as an identifying fingerprint and source of true random numbers, *IEEE Transactions on Computers*, 58(9): pp. 1198–1210, 2009.

19. M. Cortez, A. Dargar, S. Hamdioui, and G.-J. Schrijen, Modeling SRAM start-up behavior for physical unclonable functions, *2012 IEEE International Symposium on Defect and Fault Tolerance in VLSI and Nanotechnology Systems (DFT)*, pp. 1–6, Austin, TX, 2012.

20. H. Handschuh, Hardware-anchored security based on SRAM PUFs, Part 1, IEEE Security & Privacy, 10(3): 80–83, 2012.

21. H. Handschuh, Hardware-anchored security based on SRAM PUFs, Part 2, IEEE Security & Privacy, 10(4): 80–81, 2012.

22. S. Chellappa, A. Dey, and L. T. Clark, Improved circuits for microchip identification using SRAM mismatch, *IEEE Custom Integrated Circuits Conference*, pp. 1–4, San Jose, CA, 2011.

23. H. Fujiwara, M. Yabuuchi, H. Nakano, H. Kawai, K. Nii, and K. Arimoto, A chip-ID generating circuit for dependable LSI using random address errors on embedded SRAM and on-chip memory BIST, *Symposium On VLSI Circuits*, pp. 76–77, Honolulu, HI, 2011.

24. C. Costea, F. Bernard, V. Fischer, and R. Fouquet, Analysis and enhancement of ring oscillators based physical unclonable functions in FPGAs, *2010 International Conference on Reconfigurable Computing and FPGAs*, pp. 262–267, Quintana Roo, 2010.

25. G. E. Suh and S. Devadas, Physical unclonable functions for device authentication and secret key generation, *44th ACM/IEEE Design Automation Conference (DAC'07)*, pp. 9–14, San Diego, CA, 2007.

26. D. Lim, J. W. Lee, B. Gassend, G. E. Suh, M. van Dijk, and S. Devadas, Extracting secret keys from integrated circuits, *IEEE Transactions on Very Large Scale Integration (VLSI) Systems*, 13(10): 1200–1205, 2005.

27. K. Lofstrom, W. R. Daasch, and D. Taylor, IC identification circuit using device mismatch, *ISSCC Digest Technical Papers*, pp. 372–373, San Francisco, CA, 2000

28. N. Liu, S. Hanson, D. Sylvester, and D. Blaauw, OxID: On-chip one-time random ID generation using oxide breakdown, *2010 IEEE Symposium on VLSI Circuits (VLSIC)*, pp. 231–232, Honolulu, HI, 2010.

29. J. Ju, J. Plusquellic, R. Chakraborty, and R. Rad, Bit string analysis of physical unclonable functions based on resistance variations in metals and transistors, *2012 IEEE International Symposium on Hardware-Oriented Security and Trust (HOST)*, pp. 13–20, San Francisco, CA, 2012.

30. M. Yu, R. Sowell, A. Singh, D. M'Raihi, and S. Devadas, Performance metrics and empirical results of a PUF cryptographic key generation ASIC, *IEEE Symposium on Hardware-Oriented Security and Trust* (HOST), pp. 108–115, San Francisco, CA, 2012.

31. D. Fainstein, S. Rosenblatt, A. Cestero, J. Safran, N. Robson, T. Kirihata, and S. S. Iyer Dynamic intrinsic chip ID using 32 nm high-K/metal gate SOI embedded DRAM, *IEEE Symposium on VLSI Circuits*, pp. 146–147, Honolulu, HI, 2012.

32. S. Rosenblatt, D. Fainstein, A. Cestero, J. Safran, N. Robson, T. Kirihata, and S. S. Iyer, Field tolerant dynamic intrinsic chip ID using 32 nm high-K/metal gate SOI embedded DRAM, *IEEE Journal of Solid-State Circuits*, 48(4): 940–947, 2013.

33. S. Rosenblatt, D. Fainstein, A. Cestero, N. Robson, T. Kirihata and S. S. Iyer, A self-authenticating chip architecture using an intrinsic fingerprint of embedded DRAM, *IEEE Journal of Solid-State Circuits*, 48(11): 2934–2943, 2013.

34. V. van der Leest, B. Preneel, E. van der Sluis, Soft decision error correction for compact memory-based PUFs using a single enrollment, *Cryptographic Hardware and Embedded Systems (CHES), Lecture Notes in Computer Science* 7428: 268–282, 2012.

35. V. van der Leest, E. van der Sluis, G-J. Schrijen, P. Tuyls, and H. Handschuh, Efficient implementation of true random number generator based on SRAM PUFs, *Cryptography and Security: From Theory to Applications, Lecture Notes in Computer Science*, 6805: 300–318, 2012.

36. Z. Paral and S. Devadas, Reliable PUF value generation by pattern matching, USA Patent US 2012/0183135 A, 2012.
37. S. Katzenbeisser, Ü. Koçabas, V. van der Leest, A-R. Sadeghi, G-J. Schrijen, H. Schröder, C. Wachsmann, Recyclable PUFs: Logically reconfigurable PUFs, *Cryptographic Hardware and Embedded Systems (CHES) 2011, Lecture Notes in Computer Science*, 6917: 374–389, 2011.
38. T. Kirihata, High-performance embedded dynamic random access memory in nano-scale technologies, In *CMOS Processors and Memories*, edited by K. Iniewski, pp. 295–336, Springer: Dordrecht, Heidelberg, London, New York, ISBN 978-90-481-9215-1, 2010.
39. D. Wendel, R. Kalla, J. Warnock, R. Cargnoni, S. G. Chu, J. G. Clabes, D. Dreps et al., POWER7™, a highly parallel, scalable multi-core high end server processor, *IEEE Journal of Solid-State Circuits*, 46(1): 145–161, 2011.
40. S. Rusu, S. Tam, H. Muljono, J. Stinson, D. Ayers, J. Chang, R. Barada, M. Ratta, S. Kottapalli, and S. Vora, A 45 nm 8-core enterprise XEON^R processor, *IEEE Journal of Solid-State Circuits*, 45(1): 7–14, 2010.
41. J. Barth, D. Plass, E. Nelson, C. Hwang, G. Fredeman, M. Sperling, A. Mathews et al., A 45 nm SOI embedded DRAM macros for POWER7™ processor 32 MByte on-chip L3 cache, *IEEE Journal of Solid-State Circuits*, 46(1): 64–75, 2011.
42. M. Takeuchi, K. Inoue, M. Sakao, T. Sakoh, T. Kitamura, S. Arai, T. Iizuka et al., A 0.15 μm logic based embedded DRAM technology featuring 0.425 μm² stacked cell using MIM (metal–insulator–metal) capacitor, *Symposium on VLSI Technology, Digest Technical Papers*, Kyoto, Japan, pp. 29–30, 2001.
43. G. Bronner, H. Aochi, M. Gall, J. Gambino, S. Gernhardt, E. Hammerl, H. Ho et al., A fully planarized 0.25 μm CMOS technology for 256Mb DRAM and beyond, *Symposium on VLSI Technology, Digest Technical Papers*, Kyoto, Japan, pp. 15–16, 1995.
44. J. B. Kuang, A. Mathews, J. Barth, F. Gebara, T. Nguyen, J. Schaub, K. Nowka et al., An on-chip dual supply charge pump system for 45 nm eDRAM, *ESSCIRC 2008*, pp. 66–69, Edinburgh, United Kingdom, 2008.
45. J. Golz, J. Safran, H. Bishan, D. Leu, Y. Ming, T. Weaver, A. Vehabovic et al., 3D stackable 32 nm high-K/metal gate SOI embedded DRAM prototype, *IEEE Symposium on VLSI Circuits*, pp. 228–229, Honolulu, HI, 2011.
46. J. Warnock, Y. H. Chan, H. Harrer, D. Rude, R. Puri, S. Carey, G. Salem et al., 5.5 GHz System z microprocessor and multichip module, *IEEE International Solid-State Circuits Conference (ISSCC)*, pp. 46–47, San Francisco, CA, 2013.
47. A. Al-Rawi, A. Lansari, and F. Bouslama, A new non-recursive algorithm for binary search tree traversal, *Proceedings of the 2003 10th IEEE International Conference on Electronics, Circuits and Systems (ICECS)*, 2: 770–773, 2003
48. T. Yamamoto, K. Uwasawa, and T. Mogami, Bias temperature instability in scaled p+ polysilicon gate p-MOSFETs, *IEEE Transactions on Electron Devices*, 46(5): 921–6, 1999.
49. W. Feller, The hypergeometric series, In *An Introduction to Probability Theory and Its Applications*, 3rd Ed., Vol. 1, pp. 41–45. New York: Wiley, 1968.

18 Ultra-Low-Power Audio Communication System for Full Implantable Cochlear Implant Application

Yannick Vaiarello and Jonathan Laudanski

CONTENTS

In memory of Jonathan, who passed away in an accident on May 11, 2014.

18.1   INTRODUCTION

The ear is a formidable organ capable of encoding sensory information over an incredible range of intensities of more than 12 orders of magnitude. With it, our hearing function can easily track the voice of someone speaking amidst a room full of conversations (the well-known cocktail-party effect). The ear is a fast entry point to our brain and provides a very good temporal accuracy. For instance, we are able to detect a gap of a few milliseconds or differences in timing of a few microseconds between our two ears. In comparison, our retina provides to the visual system with a relatively slow input that does not even allow us to perceive the 50-Hz flickering of a light bulb.

After describing how the complex apparel of the inner ear transforms pressure waves into nerve signals, we expose the different pathologies of the ear, their functional effect, and the social impacts. We focus then on the implanted medical devices alleviating the profound deafness, particularly the cochlear implant. In the prospect of designing a future fully implantable system, we describe in detail the only external device of this system: the microphone.

18.2   EAR AND HEARING BRAIN

18.2.1   Outer and Middle Ear: From Acoustics to Mechanics Waves

Sound enters into the ears as an acoustic wave through the air before entering through the pinna, the concha, and the external auditory meatus as displayed in Figure 18.1. This succession of elements constitutes the outer ear, which conveys acoustical waves to the tympanic membrane in the middle ear. The pinna plays an important role in amplifying and shaping the sound spectrum: it creates troughs and peaks at certain frequency [1,2]. This directional spectral amplification is critical for localizing sounds vertically and to disambiguate back from front horizontally [2].

The middle ear is composed of the tympanic membrane and three bones: the malleus, the incus, and the stapes. Its main role is the transformation of acoustical waves into hydrodynamic wave inside the cochlea (see Figure 18.1) [3]. This requires

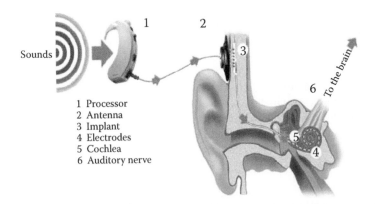

1 Processor
2 Antenna
3 Implant
4 Electrodes
5 Cochlea
6 Auditory nerve

FIGURE 18.1   Cochlear implant system.

a complex energy efficient impedance matching. The main amplification is produced by collecting pressure over the large surface of the tympanic membrane and transmitting it into the small oval window. Without it, most sound energy would be reflected by the fluid surface, and only a small percent of this energy would be transmitted.

A second amplification is produced by a lever effect. The malleus is longer than the incus by around 20% and because both are connected by elastic cartilage articulation, a small gain is produced. The elastic properties of the articulations render the gain frequency dependent [3]. Furthermore, the transmission of the mechanic energy via the ossicular chain allows a second important role for the middle ear: the protection of the inner ear against loud sounds. The stapedian muscle connects to the incus and contracts when sound level is over 70–85 dB [4,5]. This mechanism, called stapedius reflex, produces a change in the ossicular chain's impedance and can reduce its gain by up to 15–20 dB for frequencies in the range 500 Hz to 1.5 kHz [5,6].

18.2.2 Inner Ear: Transducing Hydraulic Waves into Electric Signals

The last bone of the ossicular chain, the stapes, transmits pressure changes through the oval window to the inner fluid of the cochlea. This snail-shaped organ is embedded deeply in the temporal bone and composed of three spiraling ducts of about 35 mm. These ducts are separated by membranes that are oscillating in response to the pressure waves produced by the ossicles. More specifically, a travelling wave is initiated on the basilar membrane on which lies the organ of Corti. The organ of Corti contains two types of hair cells: the inner hair cells (IHCs), which are transducers, and the outer hair cells (OHC), which act as amplifiers of the basilar membrane movement. Depolarization of the IHCs release neurotransmitter, which triggers action potentials in the auditory nerve fibers.

The motion of the stapes initiates a travelling wave in the cochlear duct. This wave can be observed on both membranes and was originally recorded stroboscopically by von Bekesy [7]. Because the basilar membrane has a varying mass and stiffness along its length, at each position the membrane possesses a specific resonant frequency, called the characteristic frequency. The cochlea thus performs a time–frequency decomposition of the hydrodynamic waveform. The time–frequency decomposition provides an accurate representation of the sound energy contained in each frequency band. This mapping of frequency to cochlear position is called the tonotopy and is a major factor used when restoring hearing using cochlear implants.

18.3 PATHOLOGIES OF HEARING AND HEARING PROTHESIS

The transformation of acoustical waves into auditory nerve discharges is a sequence of complex and intricate stages. At each of these stages, the possibility of a malfunctioning exists. Around 15% of teenagers suffer from hearing loss [8], a prevalence rate that increases to 63% for adults older than 70 [9]. For most cases, the hearing problems are linked to the aging of the auditory system. The progressive loss of OHC produces sensorineural hearing loss, a condition characterized by reduced audibility (i.e., increased threshold of hearing), reduced frequency selectivity (i.e., increased bandwidth of the basilar membrane resonance), and abnormal loudness growth. No treatment exists to restore fully the hearing function and hearing impairment is alleviated by the use of hearing aids. Hearing aids are complex amplifying devices that help perceive low-level sounds and avoid the sensation of abnormal loudness growth. This chapter deals with microphone for implantable devices prescribed in the case of more severe hearing problems.

18.3.1 MIDDLE EAR PATHOLOGIES AND MIDDLE EAR IMPLANT

There are many origins of middle ear pathologies [10]. In the case of ossicular chain malformations, or otospongiosis [11] or ossification of the articulations after repeated otitis media [12], an important decrease in audibility is observed resulting from the loss of the sound transmission. At least two devices are available depending on whether the inner ear is also affected. Replacement of one of the ossicles by an implanted device can be performed [13,14]. In the case of putative inner ear malfunction on the same side as middle ear pathology, the treatment is usually to reroute the sound to the other ear using bone-anchored hearing system [15,16].

18.3.2 INNER EAR PATHOLOGY AND COCHLEAR IMPLANTS

Irreversible loss or malfunctioning of IHCs results in the impossibility to treat deafness by amplification of vibrations irrespective of whether these are hydrodynamic or bony vibrations. Whether the pathologies are of genetic origin (connexin deficiency [17], Jervell and Lange-Nielsen's syndrome, Meniere's syndrome, etc. [18]), traumatic (loud sound exposure, temporal bone fracture), or results from an extended medical history (aging, ototoxic treatment, recurrent otitis media, etc.), IHC impairment produces profound (70 dB < loss < 90 dB) to severe hearing loss (>90 dB). However, the loss of the IHCs does not imply a pathological state for the auditory nerve and fibers may be present and fully functional. Indeed, the degeneration of nerve auditory fibers is progressive over a timescale of a few years after IHC loss.

The cochlear implant is an active device with no implanted battery (unlike pacemaking devices), which obtains both its energy and stimulating command from an outside "behind-the-hear" processor. A high-frequency magnetic transcutaneous communication link transmits a sparse representation of the sound energy present in different frequency bands. The implant then stimulates sequentially electrodes disposed along the length of the cochlea. By using the tonotopic organization of the cochlea, the implant activates the auditory nerve fibers and produces a coarse grained spectro-temporal representation of sound. The current Neurelec device has a diameter of 28 mm and a height of 4–5.5 mm. The electrode array inserted into the cochlea has 20 channels. The microphone has a dynamic range of 75 dB and a sampling frequency of 16.7 kHz. Cochlear implantation offers now speech understanding in quiet environments to implanted users, with major clinical benefits in terms of social interactions and quality of life [19].

18.3.3 FUTURE SOLUTION: FULL IMPLANTABLE DEVICE

The new device will be fully implanted. Thus, with this system, the processor and the microphones are located with the neurostimulation circuit on the patient's head.

To power this system, a rechargeable battery is inserted under the skin of the patient. The importance of the detailed module in this chapter occurs in the case of a malfunction of the implanted microphone. In order not to operate on the patient again, a microphone with radio frequency (RF) link to communicate with the implant will be provided while being as unobtrusive as possible [20]. This module, shown in Figure 18.2, will be placed in the ear canal of the patient to make it almost invisible.

18.4 SPECIFICATIONS

18.4.1 PHYSICAL DIMENSIONS AND CONSUMPTION

The integration of the system including a microphone [21], battery [22], a transmitting antenna, and a printed circuit board (PCB) containing an integrated circuit dedicated to the application shall not exceed a volume of 6 mm (diameter) by 10 mm (height).

This constraint passed on the microphone or battery requires the use of button cells battery (A10), which is the volume occupied by 5.8 mm to 3.6 mm (diameter to height) for a nominal capacity of 95 mAh. For patient comfort, it is necessary that the product has a lifetime of about 3 days with 10 hours of use per day. This constraint allows us to determine the average power consumption of the circuit should not exceed 3.2 mA at 1.2 V.

18.4.2 RADIO FREQUENCY LINK

18.4.2.1 Propagation Channel

The system will incorporate a communication from the outside to the inside of the human body. The propagation channel is composed of skin, cartilage, and fat. It connects the transmitter module placed in the ear canal and the receiver, implanted, located at the temple against the skull.

18.4.2.2 Frequency and Link Budget

Carrier frequency choice is a compromise between physical size and attenuation in the propagation channel. Indeed, the higher the frequency, the higher its integration will be easy. Nevertheless, the higher the frequency, the higher the attenuation in

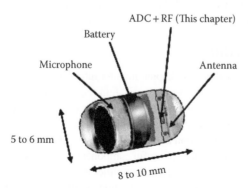

FIGURE 18.2 Description of the module.

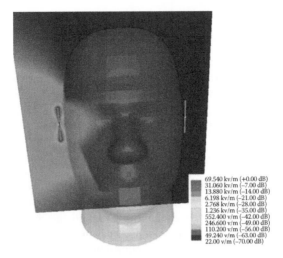

| 69.540 kv/m (+0.00 dB) |
| 31.060 kv/m (−7.00 dB) |
| 13.880 kv/m (−14.00 dB) |
| 6.198 kv/m (−21.00 dB) |
| 2.768 kv/m (−28.00 dB) |
| 1.236 kv/m (−35.00 dB) |
| 552.400 v/m (−42.00 dB) |
| 246.600 v/m (−49.00 dB) |
| 110.200 v/m (−56.00 dB) |
| 49.240 v/m (−63.00 dB) |
| 22.00 v/m (−70.00 dB) |

FIGURE 18.3 Electromagnetic simulation of a human head with the Empire software.

TABLE 18.1
Characteristics of the Three Human Tissues at 2.45 GHz

	Skin	Cartilage	Fat
Thickness (mm)	1	4	34
Relative permittivity ε_r	38.01	38.77	5.28
Conductivity σ (S/m)	44.25	52.63	8.55
Loss factor δ	0.0226	0.0190	0.1170

this environment will be important. Therefore, and considering only the band of free with permission of audio transfer, our choice fell on the 2.45 GHz Industrial, Scientific, and Medical (ISM) (radio spectrum) standard [23]. Figure 18.3 shows the simulation of the propagation channel has 2.45 GHz by considering different electromagnetic characteristics of each tissue (Table 18.1) on the Empire tools. We notice in this figure that the attenuation at the location of the implant is about 34 dB. Finally, Table 18.2 shows the link budget of our application made by these results and a literature review and showing that the radio link is possible.

18.5 ARCHITECTURE

The architecture developed in this chapter aims for ultra-low-power consumption. Thus, the signal digitization will be in two parts. First, in the transmitter, on the principle of a single-ramp analog-to digital converter (ADC), a pulse width modulation (PWM) is created by making a comparison between a preamplified audio signal and a ramp sampling. On the RF part, a direct conversion and an on-off keying (OOK) modulation have been privileged to minimize the number of useful components. Moreover, always for having a low-power system, the output of the power amplifier is controlled by the PWM signal to have the OOK modulation. Turning off the amplifier on the low state of the PWM modulating signal permits to improve the energy efficiency of 50% statistically.

TABLE 18.2
Communication Link Budget

Parameters	Values	Total	Unit	Comments
Frequency	2.45		GHz	ISM
Transmit power	−10	−10	dBm	
Tx antenna gain	−11.23	−21.23	dBi	[24]
Margin on the Tx antenna	5	−26.23	dB	
Output power		−26.23	dBm	
Propagation channel losses	−34	−60.23	dB	Empire
Margin on the losses	5	−65.23	dB	
Rx antenna gain	−18.5	−83.73	dBi	[25]
Margin on the Rx antenna	5	−88.73	dB	
Received power		−88.73	dBm	
LNA sensitivity	−90	−88.73	dBm	Trade [26]

On the receiver part, the signal carrying the dual PWM/OOK modulation at 2.45 GHz will be amplified and demodulated to baseband. To obtain a signal compatible with the 10-bit signal processing device (as digital signal processing), the PWM will be transformed into a digital signal using a counter.

18.6 WIRELESS MICROPHONE DESIGN

18.6.1 AMPLIFIER

Figure 18.4 illustrates the amplifier architecture for granting the dynamic of the audio signal with dynamic of the ramp sampling. In postlayout simulation, we obtain a gain between 14.5 and 16.5 dB (depending on the temperature and the manufacturing process) and a bandwidth of about 26 MHz for an average consumption of 130 μA. An important parameter of this block is the total harmonic distortion with ambient noise (THD + N). Equation 18.1 shows this parameter where V_F is the rms value of the fundamental harmonic and V_{HX}, the rms value of each other harmonics.

$$THD + N = 100 \left[\frac{\sum_x V_{Hx}}{V_F} \right] \tag{18.1}$$

Figure 18.5 shows the THD + N as a function of the input voltage applied to the amplifier for all manufacturing processes. In our application, the input voltage is about 30 mVp and give a THD + N of about 5.5%, which is acceptable for an understanding audio signal.

18.6.2 OSCILLATORS

Both oscillators contained in the system have similar topologies. The RF oscillator and the oscillator sampling are three stages and nine stages ring oscillators, respectively. This difference is due to the operating frequencies: 2.45 GHz and 20 kHz. Figure 18.6 shows the oscillators' topologies. Thus, we see, framed in the "3-Stages oscillator" part, the heart of the system for the oscillation. This phenomenon, being due to the propagation, through a delay τ of a voltage above or below the threshold voltage of an inverter (M13–M18). This architecture, commonly used in RF identification and in biomedical systems [27,28], has the advantage of low power consumption. The oscillation frequency, f_{osc}, the relationship between inverter delay τ and the number of stages N, can easily be estimated by Equations 18.2 through 18.4. It is possible to give an electrical model using the amplitude of the oscillations V_{osc}, the current on each node of the inverters I_{ctrl} and the different gates capacitances C_g and metal connections capacitances C_p.

$$f_{osc} = \frac{1}{2N\tau} \tag{18.2}$$

$$\text{with } \tau = \frac{V_{osc}(C_g + C_p)}{I_{ctrl}} \tag{18.3}$$

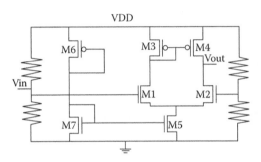

FIGURE 18.4 Amplifier topology.

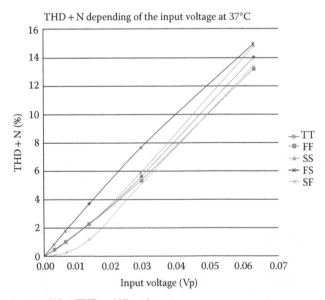

FIGURE 18.5 Amplifier THD + N Results.

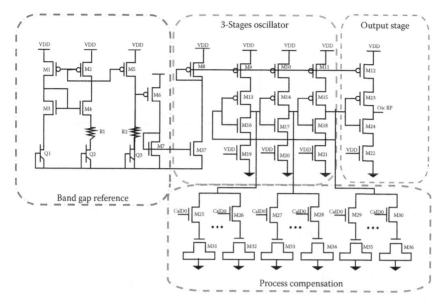

FIGURE 18.6 Oscillator topology.

$$f_{osc} = \frac{I_{ctrl}}{2NV_{osc}(C_g + C_p)} \qquad (18.4)$$

The carrier frequency, permitting the RF communication, represents 45% of the system consumption with an average of 1.28 mA. Note that this circuit gives a good fit in terms of phase noise with –74 dBc/Hz at 1 MHz from the carrier.

18.6.3 ANALOG TO TIME CONVERTER

The ramp generator is the reference element of the sampling. It is controlled by the sampling oscillator to apply a reset through the transistors M22 and M24 mounted as switches. The ramp signal has the same dynamics as the output microphone amplifier to make a comparison of these two signals.

As shown in Figure 18.7, the main part of the ramp generator is a *folded cascode OTA* operational amplifier. A study by Sansen [29] shows that this amplifier is a good compromise between dynamic output (1 V desired to obtain a comparison with the preamplifier) and power consumption. It should be noted that, to achieve maximum performance of this architecture, it is necessary that the current through each transistor of the differential pair (M12–M13) is equal to the mirror cascode current (M14–M17). These currents will then be added in the mirror M20–M21 to the DC bias of the output stage.

The postlayout simulations show that the dynamics of the ramp reaches a minimum of 91% of the full scale, that is, 1 V. In addition, the gain and bandwidth of the amplifier of this circuit are 33 dB and 900 kHz, respectively. Finally, this circuit, representing 21.4% of the system silicon surface, has a typical consumption of 340 µA (12% of the system consumption).

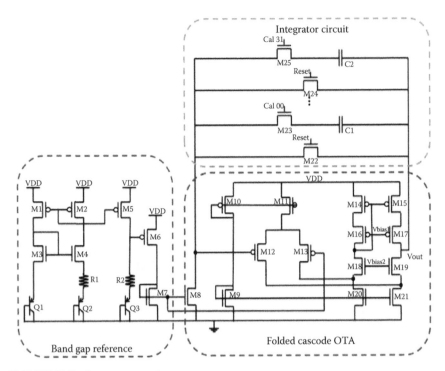

FIGURE 18.7 Integrator topology.

18.6.4 COMPARATOR

The comparator performs the PWM. Always in the context of low power circuit, we used a two-stage comparator. Thus, for a consumption of 95 µA, this circuit has the following characteristics (postlayout simulation):

- The input offset is between 250 and 320 µV.
- The rise and fall time of the output signal is less than 5.4 ns (for a 10-bit signal, LSB period is 48.8 ns).
- A minimum simulated bandwidth of 1.5 MHz (for a gain of 75 dB) is more than acceptable because our signal is sampled at 20 kHz.

18.6.5 POWER AMPLIFIER

This circuit has a double benefit from its implementation. First, it acts as an amplifier to ensure sufficient transmission power for communication. Then it is involved in the OOK modulation signal through the transistors M1 and M2 (Figure 18.8) functioning as complementary metal–oxide–semiconductor (CMOS) switches and supplying the amplifier as a function of the state of the PWM signal from the comparator. These electrical characteristics expressed below are derived from postlayout simulation. One decibel compression point is around –16 dBm while the input signal has an approximate value of –30 dBm and the noise figure (NF) varies between 2.4 and 2.9 dB depending on the manufacturing process.

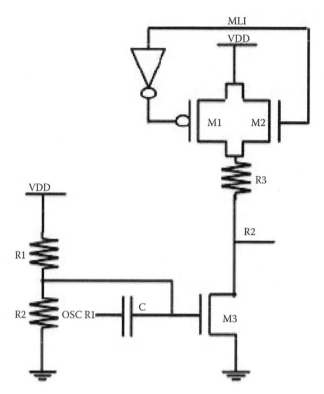

FIGURE 18.8   Power amplifier topology.

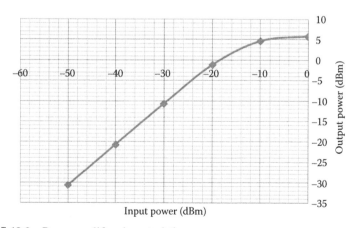

FIGURE 18.9   Power amplifier characteristics.

Figure 18.9 shows the output power as a function of input power at 37°C for a Typical (NMOS)–Typical (PMOS) (TT) process. Thus, a gain of 20 dB can be observed (at –30 dBm input) in the linear part of the power response. To improve the system consumption, the class-A power amplifier will be changed in the future by a class-C power amplifier.

18.6.6   PROCESS COMPENSATION

The main circuit of the calibration is a flash ADC. This architecture was chosen to have a fast processing time. Furthermore, a study by Allen and Holberg [30] shows that the silicon surface for a flash converter is smaller than a successive approximation register converter if the resolution does not exceed 5 bits.

However, two points are negative with this implementation. First, having to calibrate two oscillators and the simple ramp requires using three different ADC. This constraint greatly increases the silicon surface. To overcome this problem, we propose to implement a simple digital controller and an output memory system to perform the three calibrations with only one converter.

The second problem, solved by using the digital controller too, is the power consumption. The converter is used only during the calibration time. So, to reduce the consumption, it will be activated just during the first microsecond (4 clock cycles). After the conversion and the memorization of the three calibrations, an end of conversion signal (EOC) will turn off the power. The ADC consumption will be negligible compared with the consumption of the whole system.

For the calibration, we generate four control digital signals arriving one by one into a state machine. The first three signals enable the corresponding analog signal and the flip-flop storage to calibrate the three different components. The last signal is an EOC to control the digital controller.

The analog multiplexer retrieves the analog signal to compare within the ADC. And after, the conversion value is stored into the flip-flop. At the four clock cycles, the digital controller turns off the ADC. This architecture is illustrated in Figure 18.10.

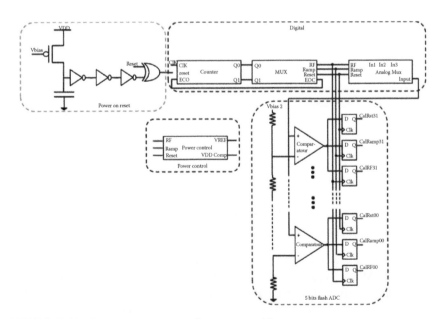

FIGURE 18.10   Low power compensation system architecture.

We will present, here, only comparison results of the dependent process block. These simulations were postlayout simulation at 37°C (typical temperature of the human body). Table 18.3 shows a comparison between the frequencies of the two oscillators with and without compensation. We can notice that, in the Slow-Fast (SF) process, the sampling frequency does not go down below 22.2 kHz due to a too important frequency shift caused by the layout (only 5 bits on the calibration converter).

18.7 IMPLEMENTATION AND EXPERIMENTAL RESULTS

This circuit has been implemented in a 130 nm CMOS technology from ST Microelectronics with an area of 1 mm². Figure 18.11 shows the layout of the chip on the left and an element representation on the right. For test, this circuit has been encapsulated in a 3 × 3 mm2 Quad Flat No-Lead 16 package and integrated into a 6 × 2 cm² PCB including a microphone, a battery, a 1.2 V LDO, and verification elements (Figure 18.12). The circuit consumption is approximately 3 mA.

First, the tests validate the transmitter. Figure 18.13 shows the validation of the audio amplifier, channel 1, a signal from the microphone, and channel 2, the same signal after amplification. Subsequently, the different internal analog signals were

TABLE 18.3
Oscillators Frequencies with and without Compensation

Process	Sampling (kHz)		Carrier (GHz)	
	With	Without	With	Without
TT	20.1	20.1	2.45	2.45
FF	20.2	20.8	2.45	2.71
SS	20.0	19.0	2.46	2.20
SF	22.2	27.4	2.45	2.45
FS	20.2	14.6	2.46	2.46

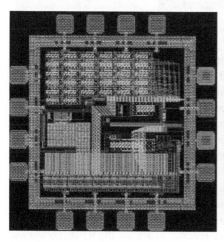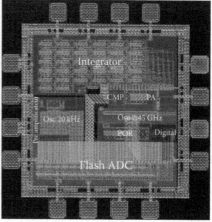

FIGURE 18.11   Transmitter layout.

FIGURE 18.12 Transmitter test printed circuit board.

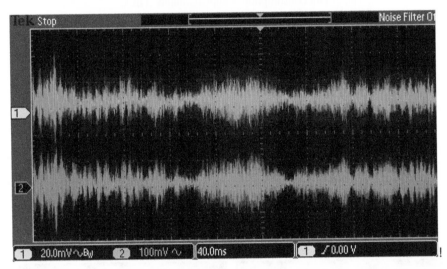

FIGURE 18.13 Microphone acquisition (channel 1) and amplifier output (channel 2).

validated. To realize the PWM signal quality, the sampling frequency was increased to 40 kHz on this test. Figure 18.14 is used to realize the result. We can observe the preamplified input signal at 2.2 kHz (channel 3), the ramp conversion at 40 kHz (channel 1), and the PWM (channel 2) resulting from the comparison of the two previous signals. We can see in this picture that the analog-time conversion is not a problem in spite of a larger sampling frequency than in the specifications.

Then, the principle of analog demodulation receiver has been validated (Figure 18.15). This measure, combining the two modules (transmitter and receiver), was made without RF communication. Thus, as shown in Figure 18.16, a 7-kHz sine was injected into the system (channel 4). Then, the comparison of this signal with the ramp sampling (channel 2), results in the PWM modulation represented in channel 1 in this figure. This signal is then injected on the input filter of the receiver to not use the RF demodulation and to highlight the final signal to be sent to the speakers (channel 3). This test highlights

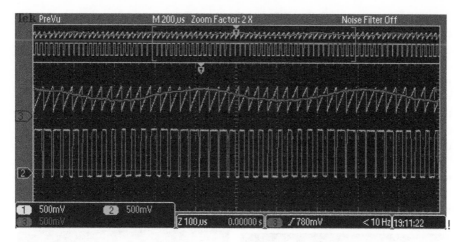

FIGURE 18.14 Internal analog signal of the transmitter.

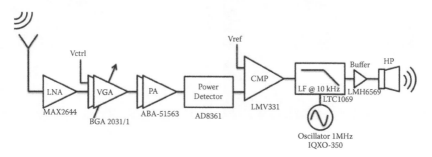

FIGURE 18.15 Receiver architecture.

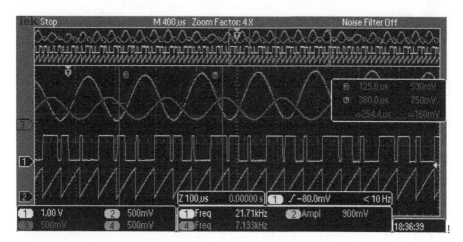

FIGURE 18.16 Analog acquisition and transmission of a 7 kHz.

the proper functioning of the analog modulation produced in the transmitter and the demodulation in the receiver. It should be noted that the output signal at 7 kHz shows no distortion introduced by the electronic transceiver.

Finally, the complete system has been tested, first, the RF emission whose spectrum is shown in Figure 18.17. Figures 18.18 and 18.19 illustrate the general function of the system. Figure 18.18 illustrates the acquisition of a 2-kHz sine (channel 1) and the signal returned to the speakers after the RF link (channel 2). The channel "M" curve shows the fast Fourier transform of the output signal whose amplitude is –28 dB at a frequency of 2 kHz and –75.1 dB at the first harmonic (4 kHz).

Then, we present in Figure 18.19 a realistic audio signal at the microphone input (channel 1), the amplified signal (channel 2) and the restored signal at the input of the speakers (channel 4). This acquisition is realized at 1-m RF communication with antennas presented at IEEE IWAT Conference [31]. We can observe on this

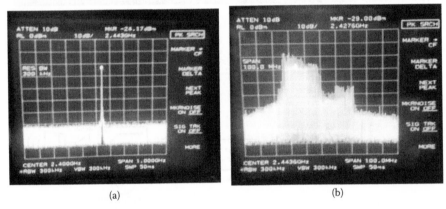

(a) (b)

FIGURE 18.17 Transmitter spectrum: (a) without pulse width modulation and (b) with pulse width modulation.

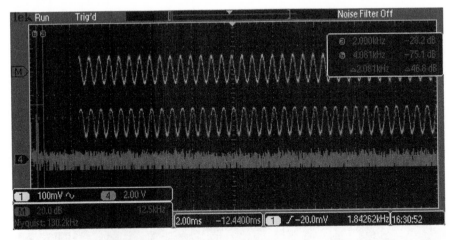

FIGURE 18.18 Top to bottom: two-kHz sine on the microphone, output signal on the speakers, and its fast Fourier transform.

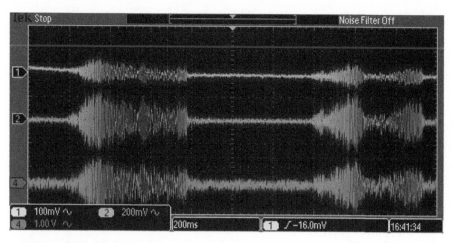

FIGURE 18.19 Top to bottom: audio signal acquisition, amplified signal, and output signal on the speakers.

oscilloscope capture the audio signal corresponding to a hissing on the microphone at different points in the system where information is purely analog. The good agreement between these curves is used to validate the chip emission and the entire demonstrator.

REFERENCES

1. J. J. Rice, B. J. May, G. A. Spirou, and E. D. Young, "Pinna-based spectral cues for sound localization in cat," *Hear Res*, 58(2): 132–152, March 1992.
2. A. D. Musicant and R. A. Butler, "The influence of pinnae-based spectral cues on sound localization," *J Acoust Soc Am*, 75(4): 1195–1200, April 1984.
3. R. Aibara, J. T. Welsh, S. Puria, and R. L. Goode, "Human middle-ear sound transfer function and cochlear input impedance," *Hear. Res.*, 152(1–2): 100–109, February 2001.
4. W. Niemeyer and G. Sesterhenn, "Calculating the hearing threshold from the stapedius reflex threshold for different sound stimuli," *Audiology*, 13(5): 421–427, 1974.
5. G. Flottorp, G. Djupesland, and F. Winther, "The acoustic stapedius reflex in relation to critical bandwidth," *J. Acoust. Soc. Am.*, 49(2B): 457–461, August 2005.
6. E. Borg and J. E. Zakrisson, "Stapedius reflex and speech features," *J. Acoust. Soc Am.*, 54(2): 525–527, August 2005.
7. G. von Bekezy, "Current status of theories of hearing," *Science*, 123(3201): 779–783, May 1956.
8. J. Shargorodsky, S. G. Curhan, G. C. Curhan, and R. Eavey, "Change in prevalence of hearing loss in U.S. adolescents," *JAMA*, 304(7): 772–778, August 2010.
9. F. R. Lin, R. Thorpe, S. Gordon-Salant, and L. Ferrucci, "Hearing loss prevalence and risk factors among older adults in the United States," *J. Gerontol. A. Biol. Sci. Med. Sci.*, 66A(5): 582–590, May 2011.
10. E. B. Teunissen and W. R. Cremers, "Classification of congenital middle ear anomalies. Report on 144 ears," *Ann. Otol. Rhinol. Laryngol.*, 102(8) Pt 1: 606–612, August 1993.
11. C. Parahy and F. H. Linthicum, "Otosclerosis and Otospongiosis: Clinical and histological comparisons," *The Laryngoscope*, 94(4): 508–512, 1984.

12. R. Dass and S. S. Makhni, "Ossification of ear ossicles: The stapes," *Arch. Otolaryngol.*, 84(3): 306–312, September 1966.

13. P. W. Slater, F. M. Rizer, A. G. Schuring, and W. H. Lippy, "Practical use of total and partial ossicular replacement prostheses in ossiculoplasty," *The Laryngoscope*, 107(9): 1193–1198, 1997.

14. T. Shinohara, K. Gyo, T. Saiki, and N. Yanagihara, "Ossiculoplasty using hydroxyapatite prostheses: Long-term results," *Clin. Otolaryngol Allied Sci.*, 25(4): 287–292, 2000.

15. C. A. J. Dun, H. T. Faber, M. J. F. de Wolf, C. Cremers, and M. K. S. Hol, "An overview of different systems: The bone-anchored hearing aid," *Adv. Otorhinolaryngol*, 71: 22–31, 2011.

16. P. Westerkull, "The Ponto bone-anchored hearing system," *Adv. Otorhinolaryngol*, 71: 32–40, 2011.

17. D. A. Scott, M. L. Kraft, E. M. Stone, V. C. Sheffield, and R. J. H. Smith, "Connexin mutations and hearing loss," *Nature*, 391(6662): 32–32, January 1998.

18. N. E. Morton, "Genetic epidemiology of hearing impairment," *Ann. NY Acad. Sci.*, 630(1): 16–31, 1991.

19. F. G. Zeng, "Trends in cochlear implants," *Trends Amplif.*, 8(1): 1–34, 2004.

20. U. Kawoos, R. V. Warty, F. A. Kralick, M. R. Tofighi, and A. Rosen, "Issues in wireless intracranial pressure monitoring at microwave frequencies," *PIERS*, 3(6): 927–931, 2007.

21. Knowles Electronics, Datasheet microphone FG-6107-C34, available at http://www.knowles.com, accessed 2012.

22. ZeniPower, Datasheet pile A10, available at http://www.zenipower.com/en/products005.html?proTypeID=100009376&proTypeName=A10%20Series, accessed 2012.

23. ISM 2.4 GHz Standard, ETSI EN 300 440-1/2, v1.1.1, 2000-07.

24. O. Diop, F. Ferrero, A. Diallo, G. Jacquemod, C. Laporte, H. Ezzeddine, and C. Luxey, "Planar antennas on integrated passive device technology for biomedical applications," *IEEE iWAT Conference*, Tucson, AZ, pp. 217–220, 2012.

25. F. Merli, L. Bolomey, E. Meurville, and A. K. Skrivervik, "Dual band antenna for subcutaneaous telemetry applications," *Antennas and Propagation Society International Symposium*, Toronto, Canada, pp.1–4, 2010.

26. Maxim Integrated, MAXIM IC, Datasheet Amplificateur faible bruit MAX2644, available at http://www.maxim-ic.com/datasheet/index.mvp/id/2357.

27. F. Cilek, K. Seemann, D. Brenk, J. Essel, J. Heidrinch, R. Weigel, and G. Holweg, "Ultra low power oscillator for UHF RFID transponder," *IEEE International Frequency Control Symposium*, Honolulu, HI, pp. 418–421, 2008.

28. R. Chebli, X. Zhao, and M. Sawan, "A wide tuning range voltage-controlled ring oscillator dedicated to ultrasound transmitter," *IEEE International Conference of Microelectronics*, Tunis, Tunisia, ICM, 2004.

29. W. M. C. Sansen, *Analog Design Essentials*, Springer, Dordrecht, the Netherlands, 2006.

30. P. E. Allen and D. R. Holberg, *CMOS Analog Circuit Design*, 2nd ed. Oxford University Press, New York, pp. 445–449, 2002.

31. O. Diop, F. Ferrero, A. Diallo, G. Jacquemod, C. Laporte, H. Ezzeddine, and C. Luxey, "Planar antennas on Integrated Passive Device technology for biomedical applications," *IEEE iWAT Conference*, Tucson, AZ, pp. 217–220, 2012.

19 Heterogeneous Memory Design

Chengen Yang, Zihan Xu,
Chaitali Chakrabarti, and Yu Cao

CONTENTS

19.1 INTRODUCTION

Memory systems are essential to all computing platforms. To achieve optimal performance, contemporary memory architectures are hierarchically constructed with different types of memories. On-chip caches are usually built with static random-access memory (SRAM), because of their high speed; main memory uses dynamic random-access memory (DRAM); large-scale external memories leverage nonvolatile devices, such as the magnetic hard disk or solid state disk. By appropriately integrating these different technologies into the hierarchy, memory systems have traditionally tried to achieve high access speed, low energy consumption, high bit density, and reliable data storage.

Current memory systems are tremendously challenged by technology scaling and are no longer able to achieve these performance metrics. With the minimum feature size approaching 10 nm, silicon-based memory devices, such as SRAM, DRAM, and Flash, suffer from increasingly higher leakage, larger process variations, and severe reliability issues. Indeed, these silicon-based technologies hardly meet the requirements of future low-power portable electronics or tera-scale high-performance computing systems. In this context, the introduction of new devices into current memory systems is a must.

Recently, several emerging memory technologies have been actively researched as alternatives in the post–silicon era. These include phase-change memory (PRAM), spin-transfer torque magnetic memory (STT-MRAM), resistive memory (RRAM), and ferroelectric memory (FeRAM). These devices are diverse in their physical operation and performance.

Figure 19.1 compares the device operation and performance of different types of memories [1]. As the figure illustrates, these nonvolatile memories (NVMs) have much lower static power consumption and higher cell density, as compared to SRAM or DRAM. But there is no winner of all: some of them have higher latency (e.g., PRAM and MRAM), and some have higher programming energy (e.g., PRAM). Moreover, akin to the scaled CMOS memory devices, one of the biggest design challenges is how to compensate for the excessive amount of variations and defects in the design of a reliable system.

This chapter introduces a top–down methodology for heterogeneous memory systems that integrate emerging NVMs with other type of memories. While PRAM and STT-MRAM are used as examples, for their technology maturity and scalability, the integration methodology is general enough for all types of memory devices. We first

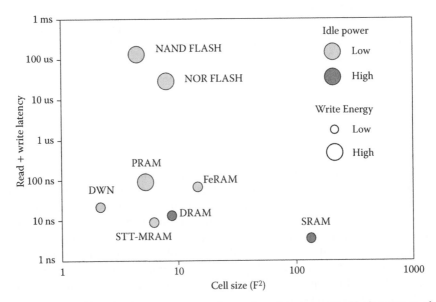

FIGURE 19.1 Diversity in memory operation and performance. (R. Venkatesan et al., "TapeCache: A high density, energy efficient cache based on domain wall memory," *ACM/ IEEE International Symposium on Low Power Electronics and Design, ISLPED'12,* © 2012 IEEE.)

develop device-level compact models of PRAM and STT-MRAM based on physical mechanism and operations. The models cover material properties, process and environmental variations, as well as soft and hard errors. They are scalable with device dimensions, providing statistical yield prediction for robust memory integration. By embedding these device-level models into circuit- and architecture-level simulators, such a CACTI and GEM5, we evaluate a wide range of heterogeneous memory configurations, under various system constraints. We also provide a survey of existing heterogeneous memory architecture that shows superior timing or energy performance. Finally, to address the reliability issues of the emerging memory technologies such as multilevel cell (MLC) PRAM and STT-MRAM, we also show how to extract error models from device-level models and review existing techniques for enhancing their reliability.

This chapter is organized as follows: Section 19.2 gives a brief introduction to PRAM and STT-MRAM devices and derives the models. The device models are embedded into CACTI, SimpleScalar, and GEM5 to analyze heterogeneous memory architectures (Section 19.3). In Section 19.4, process variability and reliability effects are studied, followed by error characterization and compensation techniques. Section 19.5 concludes the chapter.

19.2 NONVOLATILE MEMORY BACKGROUND

19.2.1 PHASE CHANGE MEMORY

The structure of a PRAM cell is shown in Figure 19.2a. It consists of a standard NMOS transistor and a phase change device. The phase change device is built with a chalcogenide-based material, usually $Ge_2Sb_2Te_5$ (GST), that is put between the top electrode and a metal heater that is connected to the bottom electrode. GST switches between a crystalline phase (low resistance) and an amorphous phase (high resistance) with the application of heat; the default phase of this material is crystalline. The region under transition is referred to as programmable region. The shape of the programmable region is usually of mushroom shape due to the current crowding effect at the heater to phase change material contact [2].

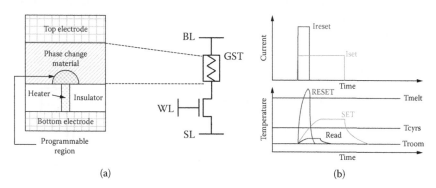

(a)   (b)

FIGURE 19.2 (a) Phase-change memory cell structure (H. S. P. Wong et al., "Phase change memory," *Proceedings of the IEEE*, pp. 2201–2227, © 2010 IEEE.) (b) Phase-change memory cells are programmed and read by applying electrical pulses with different characteristics.

During write operation of single-level cell (SLC) PRAM, a voltage is applied to the word line (WL), and the current driver transistor generates the current that passes between the top and bottom electrodes to heat the heater, causing a change in the phase of the GST material. During write-0 or RESET operation, a large current is applied between top and bottom electrodes (see Figure 19.3b). This heats the programmable region over its melting point, which when followed by a rapid quench, turns this region into an amorphous phase. During write-1 or SET operation, a lower current pulse is applied for a longer period of time (see Figure 19.3b) so that the programmable region is at a temperature that is slightly higher than the crystallization transition temperature. A crystalline volume with radius r' starts growing at the bottom of the programmable region as shown in Figure 19.3b. At the end of this process, the entire programmable region is converted back to the crystalline phase. In read operation, a low voltage is applied between the top and bottom electrodes to sense the device resistance. The read voltage is set to be sufficiently high to provide a current that can be sensed by a sense amplifier but low enough to avoid write disturbance [2].

Because the resistance between the amorphous and crystalline phases can exceed two to three orders of magnitude [3], multiple logical states corresponding to different resistance values can be accommodated. For instance, if four states can be accommodated, the PRAM cell is a 2-bit MLC PRAM. The four states of such a cell are "00" for full amorphous state, "11" for full crystalline state, and "01" and "10" for two intermediate states. MLC PRAM can be programmed by shaping the input current profile [2]. To go to "11" state from any other state, a SET pulse of low amplitude and long width is applied. However, to go to "00" state from any other state, it has to first transition to "11" state to avoid over programming. To go to "01" or "10" state, it first goes to "00" state and then to the final state after application of several short pulses. After each pulse, the read and verify method is applied to check whether the correct resistance value has been reached.

PRAM cell resistance is determined by the programming strategy and current profile. The Simulation Program with Integrated Circuit Emphasis (SPICE) parameters needed to simulate a PRAM cell are given in Table 19.1. These are used to obtain the initial resistance distributions of the four logical states of a 2-bit MLC PRAM. The variation in these parameters is used to calculate the shifts in these distributions, which again affect the error rate.

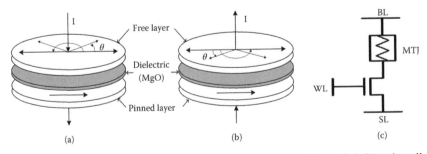

FIGURE 19.3 Spin-transfer torque magnetic memory structure: (a) parallel, (b) antiparallel, and (c) magnetic tunneling junction circuit structure.

TABLE 19.1

Phase-Change Memory Cell Parameter Values Used in SPICE Simulation for 45-nm Technology Node

	Parameter	Value ($\mu \pm \sigma$)
PRAM	CW	28 nm $\pm 2\%$
	d	49 nm $\pm 2\%$
	R_{SET}	29 kΩ
	R_{RESET}	2.3 MΩ
	R_{WRITE}	1 kΩ
CMOS	V_{dd}	1.0 V
	V_{th}	494 mV ± 45 mV
	Length	28 nm

The parameter values in Table 19.1 are used to calculate the energy and latency of read and write operations of a single cell PRAM. These values are then embedded into CACTI [4] to generate the energy and latency of a PRAM bank, as will be described in Section 19.3. Our approach is similar to that in Dong et al. (2009) and Dong et al. (2012) [5,6] where SLC PRAM parameters are embedded into CACTI to enable system-level study of energy consumption and latency [7,8].

19.2.2 SPIN TORQUE TRANSFER MAGNETIC RANDOM-ACCESS MEMORY

In STT-MRAM, the resistance of the magnetic tunneling junction (MTJ) determines the logical value of the data that is stored. MTJ consists of a thin layer of insulator (spacer-MgO) about ~1 nm thick sandwiched between two layers of ferromagnetic material [9]. Magnetic orientation of one layer is kept fixed and an external field is applied to change the orientation of the other layer. Direction of magnetization angle (parallel [P] or antiparallel [AP]) determines the resistance of MTJ, which is translated into storage value. Low resistance (parallel) state that is accomplished when magnetic orientation of both layers is in the same direction corresponds to storage of bit 0. By applying external field higher than critical field, magnetization angle of free layer is flipped by 180°, which leads to a high resistance state (antiparallel). This state corresponds to storage of bit 1. The difference between the resistance values of parallel and antiparallel states is called tunneling magnetoresistance (TMR) defined as TMR = $(R_{AP} - R_P) / R_P$, where R_{AP} and R_P are the resistance values at antiparallel and parallel states, respectively. Increasing the TMR ratio makes the separation between states wider and improves the reliability of the cell [10]. Figure 19.3a and b highlights the parallel and antiparallel states.

Figure 19.3c describes the cell structure of an STT-MRAM cell. It consists of an access transistor in series with the MTJ resistance. The access transistor is controlled through WL, and the voltage levels used in bit lines (BLs) and select lines (SLs) determine the current that is used to adjust the magnetic field. There are three modes of operation, read, write-0, and write-1.

For read operation, current (magnetic field) lower than critical current (magnetic field) is applied to MTJ to determine its resistance state. Low voltage (~0.1 V) is applied to BL, and SL is set to ground. When the access transistor is turned on, a small current passes through MTJ whose value is detected based on a conventional voltage sensing or self-referencing schemes [11].

During write operation, BL and SL are charged to opposite values depending on the bit value that is to be stored. During write-0, BL is high and SL is set to zero, whereas during write-1, BL is set to zero and SL is set to high. We distinguish between write-0 and write-1 because of the asymmetry in their operation. For instance, in 45-nm technology, write-1 needs 245 μA programming current while write-0 requires 200 μA.

A physical model of MTJ based on the energy interaction is presented [12]. Energies acting in MTJ are Zeeman, anisotropic, and damping energy [13] and the state change of an MTJ cell can be derived by combining these energy types:

$$\frac{d\overline{M}}{dt} = -\mu_0 \cdot M_s \cdot \overline{H} + \frac{\alpha}{M_s} \cdot \overline{M} \times \frac{d\overline{M}}{dt} + K \sin\theta \cos\theta$$

Here $\overline{M}$ is the magnetic moment, M_s is the saturation magnetization, μ_0 is the vacuum permeability, α is the damping constant, $\overline{H}$ is the magnetic field, θ is the magnetic angle of the free dielectric, and K is the anisotropic constant. Such an equation can be modeled using Verilog-A to simulate the circuit characteristics of STT-MRAM [12]. For instance, differential terms are modeled using capacitance, whereas Zeeman and damping energy are described by voltage-dependent current source.

The nominal values and variance of the device parameters are listed in Table 19.2. We consider 40 mV variation for random dopant fluctuation (RDF) when the width of 128 nm is equivalent to W/L = 4, and scaled it for different W/L ratios. The SPICE values have been used to calculate the energy and latency of a single cell during read and write operations and embedded into CACTI for system-level simulation. The parameter variations have been used to estimate the error rates as will be demonstrated in Section 19.4.

TABLE 19.2

Spin-Transfer Torque Magnetic Memory Parameter Values Used in SPICE Simulation for 45-nm Technology Node

	Parameter	Value
MTJ	R_p (Parallel)	$2.25 \pm 6\%$
	R_{AP} (Antiparallel)	$4. \pm 6\%$
	MTJ initial angle	$0 \pm 0.1\pi$
CMOS	W	$96\,nm \pm 6\%$
	L	$45\,nm \pm 6\%$
	V_{th}	$0.4\,V \pm 6\%$

19.2.3 COMPARISON OF ENERGY AND LATENCY IN SCALED TECHNOLOGIES

The parameter values of NVM memories are embedded into CACTI [4] and used to generate the latency and energy results presented in this section. Because PRAM and STT-MRAM are resistive memories, the equations for BL energy and latency had to be modified. The rest of the parameters are the same as the default parameters used in DRAM memory simulator with International Technology Roadmap for Semiconductors low operation power setting used for peripheral circuits. For 2-bit MLC PRAM, cell parameters were obtained using the setting in Table 19.1. A total of 256 cells corresponding to a 512-bit block were simulated for write/read operations. Note that the write latency and write energy of two intermediate states "01" and "10" are much higher than that of "11" or "00" states [14]. This is because the write operation of intermediate states requires multiple current pulses with a read and verify step after each current pulse.

The write latency and write energy simulation results of a 2M 8-way 512 bits/block cache for different memory technologies are presented in Figure 19.4. For SRAM and DRAM, we do not distinguish between read and write for both latency and energy while for PRAM and MRAM, they are quite distinct and are shown separately. The results show that PRAM and STT-MRAM has higher write energy than SRAM and DRAM. For 45-nm technology, the write energy of PRAM is 7.3 nJ, which is seven times higher than that of SRAM and DRAM. STT-MRAM has much lower write energy than PRAM, but it is still twice than that of SRAM and DRAM. Note that the read energy of PRAM and STT-MRAM is less than that of SRAM and DRAM due to their simple 1T1R structure without precharging during read. The asymmetry between write and read energy of NVMs such as PRAM and STT-MRAM has been exploited in some NVM-based heterogeneous memories as will be described in Section 19.3.2.

The latency comparison in Figure 19.4b shows that MLC PRAM has the highest write latency due to multistep programming in MLC. STT-MRAM has comparable read latency to SRAM and DRAM though the write latency is higher. Thus, STT-MRAM is being considered a suitable replacement of SRAM for high level caches.

Next, we describe the trends of these parameters for scaled technology nodes. The scaling rule in PRAM is the constant-voltage isotropic scaling rule, assuming that phase change material remains the same during scaling and the three dimensions are scaled by the same factor k. Thus, the resistance for both amorphous and crystalline states increases linearly. Because the voltage is constant, the current through the material decreases linearly by $1/k$ while the current density increases linearly by k. Assuming the melting temperature remains the same during scaling, the time of phase changing (thermal RC constants) decreases by $1/k^2$. Based on this scaling rule and the published data for PRAM at 90-nm [15] and 45-nm [16] technology nodes, the critical electrical parameters of PRAM such as resistance, programming current, and the RC constant are predicted down to 16 nm.

The scaling rule of STT-MRAM is a constant-write-time (10 nanoseconds) scaling [17] with the assumption that the material remains the same. If the dimensions l and w of the MTJ are scaled by a factor k, and d (the thickness) scales by k–1/2, the resistance gets scaled by k–3/2 and the critical switching current gets scaled by k3/2.

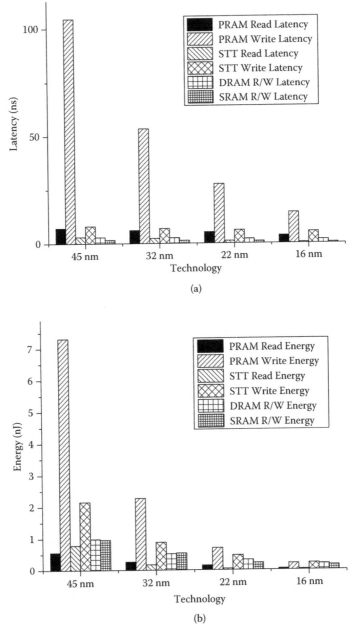

FIGURE 19.4   (a) Latency per access and (b) dynamic energy of different memory technologies at different technology nodes.

Based on this scaling rule and the published data of STT-MRAM in 45-nm technology node [18], the critical electrical parameters such as resistance and switching current are predicted down to 16 nm.

Figure 19.4 describes the effect of technology scaling on the dynamic energy and latency of different memory technologies. From Figure 19.4a, we see that SRAM has the lowest read latency at 45 nm and 32 nm, but for 22-nm and lower, the read latency of MRAM is comparable to that of SRAM. In terms of write latency, SRAM is the lowest for all technology nodes. The write latency for MRAM does not scale much and the write latency of PRAM scales according to the $1/k^2$ scaling rule. So, at lower technology nodes, the PRAM write latency is not that high compared to the MRAM write latency. However, they are both higher than SRAM and so are less suitable for L1 cache unless a write-back buffer is possibly used as in [19,20].

From Figure 19.4b, we also see that in all the technology nodes, PRAM has the highest write energy while the write energy of SRAM is the lowest. For read energy, PRAM and MRAM always have lower energy consumption than DRAM and SRAM. The energy of PRAM and MRAM scale very fast compared to SRAM and DRAM; PRAM write energy scales even faster than MRAM due to the $1/k^2$ scaling rule of its write latency. PRAM has the lowest read energy at 45 nm, but read energy of MRAM drops faster with technology scaling, making it the lowest one for 32 nm and lower technology nodes. Among all the technologies, the leakage power of SRAM is the highest. CACTI simulation results show that at 32-nm technology, 2M SRAM has 10 times more leakage than 2M PRAM and 2M MRAM. This is because PRAM and STT-MRAM cells have very low leakage compared to SRAM cells. All of this implies that the use of SRAM in L2 cache will increase the overall power consumption, as will be demonstrated in Section 19.3.

19.3   HETEROGENEOUS MEMORY DESIGN

Most multiple level caches, which are integrated on chip with CPU, consist of two or three levels of SRAM-based caches with gradually increasing size. The last or the highest cache level tends to occupy a large on-chip silicon area, and thus incurs a significant amount of leakage power consumption. Emerging NVMs, such as PRAM and STT-MRAM, which have high data storage density and low leakage power, have been considered as promising replacements of SRAM in higher levels of memory hierarchy. However, the access latency, especially the write latency of these memories, is much longer than that of SRAM. Thus, while using emerging memories in high cache level (L2 or L3) can dramatically reduce the leakage power consumption, the instructions per cycle (IPC) remains constant or decreases slightly.

19.3.1   CASE STUDY OF HETEROGENEOUS MEMORY

In this section, we analyze heterogeneous memory under the same area rule, that is, the competing memory technologies have the same area (instead of memory size or density). Accordingly, a 4 MB STT-MRAM has the same area as a 16 MB MLC PRAM or a 512 KB SRAM.

We consider three configurations with comparable total area: Config (i) 32 KB SRAM in L1 and 512 KB SRAM in L2, Config (ii) 32 KB SRAM in L1 and 4 MB MRAM in L2, and Config (iii) 32 KB SRAM in L1 and 16 MB MLC PRAM in L2. We use SimpleScalar 3.0e [21] and our extension of CACTI to generate the latency and energy of the different cache system configurations. All caches are 8-way with 64 bytes per block.

Figure 19.5 shows the average energy comparison of the three configurations in 45-nm technology for seven benchmarks, namely gcc, craft, and eon from SPEC2000 and sjeng, perlbench, gcc06, and prvray from SPEC2006. For each benchmark, the first bar corresponds to Config (i), the second bar corresponds to Config (ii), and the third bar corresponds to Config (iii). The energy values of Configs (ii) and (iii) are normalized with respect to that of Config (i). For all the benchmarks, Config (i) has the largest energy consumption. Config (iii) has the second largest energy consumption. Config (ii) has the lowest energy consumption, which is 70% and 33% lower than Config (i) and Config (ii). The leakage energy of Config (iii) is almost the same as that of Config (ii) because the leakage of STT-MRAM and PRAM are both very low. The high dynamic energy of Config (iii) results from the repeated read&verify process for programming intermediate states in MLC PRAM.

Figure 19.6 plots the normalized IPC of the three configurations for the same set of benchmarks. The IPC increase is modest; it increases by 8% for Config (ii) and by 11% for Config (iii). This increase is because for the same area constraint, large L2 cache reduces the L2 cache miss penalty. Thus from Figures 19.5 and 19.6, we conclude that Config (ii), which uses STT-MRAM, has the largest IPC improvement with the lowest energy consumption.

Next, we present an example of heterogeneous main memory design using a small-size DRAM as cache on top of large PRAM. The configurations used in GEM5 are listed in Table 19.3 [22]. Our workload includes the benchmarks of SPEC CPU INT 2006 and DaCapo-9.12. For the GEM5 simulations, the PRAM memory latency is obtained by CACTI and expressed in number of cycles corresponding to the processor frequency of 2 GHz.

Figure 19.7 shows the normalized IPC, lifetime, and energy of PRAM- and STT-MRAM-based heterogeneous main memory with different sized DRAM cache. The lifetime of PRAM is defined as the maximum number of program/erase (P/E)

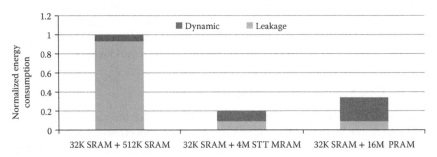

FIGURE 19.5 Normalized energy consumption of the three configurations for 45-nm technology. The normalization is with respect to Config (i).

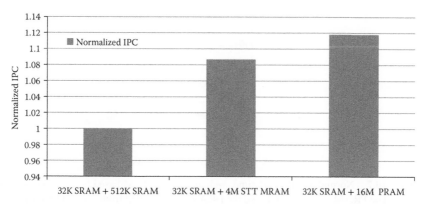

FIGURE 19.6 Normalized IPC (respect to Config (i)) of the three configurations for 45-nm technology. The normalization is with respect to Config (i).

TABLE 19.3
GEM5 System Evaluation Configuration

Processor	Single Core 2 GHz. Pipeline 16 Stages; Out-of-Order Execution
L1 cache(SRAM)	ICache and DCache 64 KB, each block is 64 bytes, 4-way. Latency is four cycles
L2 cache(SRAM)	L2 Cache 2 MB, each block is 64 bytes, 8-way. Latency is 16 cycles
Memory bank(PRAM)	8 GB PRAM memory + 256 KB/512 KB/1 M/2 M/4 M DRAM cache. Each block is 64 bytes. Read latency is 95, write latency corresponds to programming strategy
Benchmarks	Spec2006, DaCapo
Instruction fetch	Four instructions per cycle; fetch and at first predicted taken branch
Regs	Physical Integer Regs: 256; Physical Float Regs: 256
Execution engine	4-wide decode/rename/dispatch/issue/writeback; Load Queue: 64-entry; Store Queue: 64-entry
Branch predictor	4 KB-entry, 4-way BTB (LRU), 1-cycle prediction delay; 32-entry return address stack; 4096-entry GShare. 15-cycle min. branch misprediction penalty

Source: GEM5 simulator, available at http://www.m5sim.org/Main_Page, February 14, 2014.

cycles for which data stored in memory remains reliable according to a 10^{-8} block failure rate (BFR) constraint. As the DRAM size of the PRAM-based heterogeneous memory increases, the frequency of PRAM access decreases and the lifetime of PRAM, in terms of P/E cycles, increases. The energy consumption has a significant

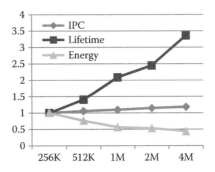

FIGURE 19.7   Normalized IPC, lifetime, and energy of PRAM-based heterogeneous main memory. The PRAM is of size 8 GB and the DRAM size is varied from 256 KB to 4 MB.

reduction (about 50%) when cache size increases from 256 KB to 1M. That is because the amount of access to PRAM, which is costly in terms of high energy consumption, is reduced. However, the total energy consumption becomes flat as the cache size keeps increasing. This is because the reduction in PRAM energy due to very few accesses is offset by the increase in leakage and refresh energy of a large DRAM cache. Similar to IPC in heterogeneous cache design, the IPC of PRAM-based main memory increases slowly with increase in the size of DRAM cache. In general, if we want to achieve a balance between latency and energy, 1 M DRAM cache seems to be an efficient configuration. If long lifetime is also required, then the cache size should be increased though this does not result in much gain in performance and energy.

For STT-MRAM-based heterogeneous memory, the normalized lifetime and energy have the same trend as PRAM-based heterogeneous memory. That is because the reduction of accesses to the main memory is the same. The IPC improvement due to increased DRAM size is even lower than that in PRAM-based heterogeneous memory because the access latency of STT-MRAM is lower than that of PRAM, as shown in Figure 19.8.

19.3.2   RELATED WORK

19.3.2.1   Heterogeneous Cache Architecture

In Chen et al. [23], STT-MRAM has been studied as the last-level (L2) cache to reduce cache leakage power. To address possible degradation in IPC performance, the NMOS transistor width of the access transistor is varied and the difference between read and write latency studied. Simulation results show that while some benchmarks favor the use of relatively small NMOS transistors in memory cells, others favor the use of relatively large NMOS transistors.

A relationship between retention time of STT-MRAM and its write latency has been established in Jog et al. [24] and used to calculate an efficient cache hierarchy. Compared to SRAM-based design, the proposed scheme lowers write latency and improves performance and energy consumption by 18% and 60%, respectively.

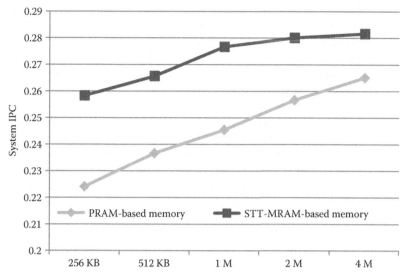

FIGURE 19.8 IPC of PRAM and STT-MRAM-based heterogeneous main memories. The DRAM size is varied from 256 KB to 4 MB.

Cache memories have also been designed where PRAM is used as L2 or the last level cache. In Joo et al. [25], design techniques are proposed to reduce the write energy consumption of a PRAM cache and to prolong its lifetime. This is done by using read-before-write, data inverting, and wear leveling. Simulation results show that compared with the baseline 4 MB PRAM L2 cache having less than an hour of lifetime, Joo et al. [25] achieved 8% of energy saving and 3.8 years of lifetime.

A novel architecture RHCA in which SRAM and PRAM/MRAM serve in the same cache level has been proposed in Wu et al. [26]. The PRAM/MRAM part is used to store data that is mostly used for reads and SRAM is used to store data that will be updated frequently. Because the read/write access ratio for a cache line changes over time, data is transferred from the fast SRAM region to the slow PRAM/MRAM region and vice versa dynamically.

19.3.2.2 Heterogeneous Architecture for Main Memory

Different memory combinations have been studied for the main memory level. Most of the work has been geared toward partially replacing DRAM, which has problems of large power consumption, especially due to refresh power and is not amenable to aggressive scaling. Unfortunately, PRAM has high programming energy and reliability problems. Thus, most approaches try to hide the write latency of PRAM and reduce the number of PRAM accesses by putting DRAM/Flash buffer or cache before PRAM-based main memory. The major differences among these architectures are memory organization, such as buffer/cache size or line size, and data control mechanisms.

An architecture where the DRAM is used as a buffer in a DRAM+PRAM architecture is proposed in Qureshi et al. [27]. The DRAM buffer is organized similar to a hardware cache that is not visible to the OS, and is managed by the DRAM controller. This organization successfully hides the slow write latency of PRAM, and also addresses the endurance problem of PRAM by limiting the number of writes to the PRAM memory. Evaluations showed that a small DRAM buffer (3% the size of PRAM storage) can bridge the latency gap between PRAM and DRAM. For a wide variety of workloads, the PRAM-based hybrid system provides an average speed up of 3X while requiring only 13% area overhead.

There are several DRAM+PRAM hybrid architectures where DRAM is used as cache. The architecture in Park et al. [28] uses DRAM as a last level cache. The DRAM can be bypassed when reading data from PRAM with the low cache hit rate. In this case, DRAM has lower refresh energy while the miss rate increases. Dirty data in DRAM is kept for a longer period than clean data to reduce the write backs to PRAM. This memory organization reduces the energy consumption by 23.5%–94.7% compared to DRAM-only main memory with very little performance overhead.

Another hybrid architecture has been proposed in Ferreira et al. [29], which focuses on the lifetime of PRAM. It achieves the same effect of lazy write by restricting CPU writes to DRAM and only writing to PRAM when a page is evicted. Such a scheme has been shown to have significant impact on endurance, performance, and energy consumption. Ferreira et al. [29] claim that the lifetime of PRAM can be increased to 8 years.

A PDRAM memory organization where data can be accessed and swapped between parallel DRAM and PRAM main memory banks has been proposed in Dhiman et al. [30]. To maintain reliability for PRAM (because of write endurance problem), Dhiman et al. [30] introduced a hardware-based book-keeping technique that stores the frequency of writes to PRAM at a page-level granularity. It also proposes an efficient operating-system-level page manager that utilizes the write frequency information provided by the hardware to perform uniform wear leveling across all the PRAM pages. Benchmarks show that this method can achieve as high as 37% energy savings at negligible performance overhead over comparable DRAM organizations.

19.4 RELIABILITY OF NONVOLATILE MEMORY

One major drawback of NVMs is that they suffer from reliability degradation due to process variations, structural limits, and material property shift. For instance, repeated use of high currents during RESET programming of PRAM results in S_b enrichment at the contact reducing the capability of heating the phase change material to full amorphous phase and results in hard errors. Process variations in the metal–oxide–semiconductor field-effect transistor current driver in STT-MRAM impact the programming current and lead to unsuccessful switch. In order for NVMs to be adopted as a main part of heterogeneous memory, it is important that the reliability concerns of these devices also be addressed.

19.4.1 RELIABILITY OF PHASE-CHANGE MEMORY

As described in Section 19.2, the logical value stored in PRAM is determined by the resistance of the phase change material in the memory cell. Assuming there is no variation in the phase change material characteristic and there is no sense amplifier mismatch, the primary cause of errors in PRAM is due to overlap of the resistance distributions of different logical states, as shown in Figure 19.9a. The resistance distributions of all the states shift from the initial position due to the change in the material characteristics such as structure relaxation or recrystallization [31,32]. There are three threshold resistances, $R_{th\,(11,10)}$, $R_{th\,(10,01)}$, and $R_{th\,(01,00)}$, to identify the boundaries between the four states. A memory failure occurs when the resistance distribution of one state crosses the threshold resistance; the error rate is proportional to the extent of overlap. Figure 19.9b shows the resistance distributions of state "00" obtained by using the parameters in Table 19.2 for running 10,000 point Monte-Carlo simulations. We see that the resistance distribution curve of state "00" has a long tail, which makes it more prone to error.

The reliability of a PRAM cell can be analyzed with respect to data retention, cycling endurance, and data disturbs [33]. For PRAM, data retention depends on the stability of the resistance in the crystalline and amorphous phases. While the crystalline phase is fairly stable with time and temperature, the amorphous phase suffers from resistance drift and spontaneous crystallization. Let *data storage time* (DST) be the time that the data is stored in memory between two consecutive writes. Then, the resistance drift caused due to DST can be modeled by

$$R_t = R_A \left(\frac{t}{t_0} \right)^v + R_e$$

where R_t is the resistance at time t, R_A is the effective resistance of amorphous part, R_e is the effective resistance of crystalline part, and v is the resistance drift coefficient, which is 0.026 for all the intermediate states. Note that R_A and R_e have different values for the different states. Soft errors occur when the resistance of a state increases and crosses the threshold resistance, demarcating its resistance state with the state with higher resistance.

Hard errors occur when the data value stored in one cell cannot be changed in the next programming cycle. There are two types of hard errors in SLC PRAM: stuck-RESET failure and stuck-SET failure [33]. Stuck-SET or stuck-RESET means the value of stored data in PRAM cell is stuck in "1" or "0" state no matter what value has been written into the cell. These errors increase as the number of programming cycles (NPC) increases.

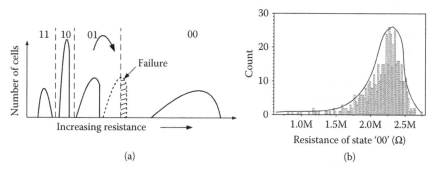

(a)   (b)

FIGURE 19.9   (a) An example of a failure caused by the "01" resistance shift. (b) Resistance distribution of the state "00."

Stuck-SET failure is due to repeated cycling that leads to Sb enrichment at the bottom electrode [34]. The resistance reduction is a power function of the NPC and is given by $\Delta R = a \times NPC^b$, where a equals 151,609 and b equals 0.16036 [35]. The stuck-SET failure due to the resistance drop of state "00" is the main source of hard errors in PRAM. This kind of errors occur if the resistance distribution of state "00" crosses $R_{th\,(00,01)}$.

For MLC PRAM, the failure characteristics due to NPC is similar to that in SLC PRAM, but the number of hard errors in MLC PRAM is larger than that in SLC PRAM. This is to be expected since the threshold resistance between state "00" and state "01" in MLC PRAM is higher than the threshold resistance between state "0" and state "1" in SLC PRAM. As a result, for the same NPC, the number of errors due to distribution of state "00" crossing $R_{th\,(00,01)}$ is higher.

In summary, hard error rate is determined by the resistance of state "00" decreasing and its distribution crossing $R_{th\,(01,00)}$ and soft error rate is primarily determined by the resistance of state "01" increasing and its distribution crossing $R_{th\,(01,00)}$. Thus, the threshold resistance $R_{th\,(01,00)}$ plays an important role in determining the total error rate. Because we can estimate the resistance shift quantitatively, we can control the error rate by tuning the threshold resistance [35].

19.4.1.1   Enhancing Reliability of Phase-Change Memory

Many architecture-level techniques have been proposed to enhance the reliability of PRAM. Techniques to reduce hard errors in SLC PRAM have been presented in Qureshi et al., Seong et al., Schechter et al., and Yoon et al. [27,36–38]. Wear leveling techniques and a hybrid memory architecture that reduce the number of write cycles in PRAM have been proposed in Qureshi et al. [27]. The schemes in Seong et al. [36] and Schechter et al. [37] can identify the locations of hard errors based on read-and-verify process. While additional storage area is needed to store the location addresses of hard errors in Schechter et al. [37], iterative error partitioning algorithm is proposed in Seong et al. [36] to guarantee that there is only one hard error in a subblock so that it can be corrected during read operation. For correcting soft errors in MLC PRAM, a time tag is used in Xu et al. [39] to record the retention time information for each memory block or page and this information is used to determine the threshold resistance that minimizes the soft error bit error rate (BER). However, tuning of threshold resistance for reducing only soft errors has an adverse effect on its hard error rate. A multitiered approach spanning device, circuit, architecture, and system levels has been proposed for improving PRAM reliability [35,40]. At the device level, tuning the programming current profile can affect both the memory reliability as well as programming energy and latency. At the circuit level, there is an optimal threshold resistance for a given data retention time and number of programming cycles that results in the lowest error rate (soft errors + hard errors). At the architecture level, Gray coding and 2-bit interleaving distribute the odd and even bits into an odd block that has very low BER and an even block that has comparatively high BER. This enables us to employ a simpler error control coding (ECC) such as Hamming on odd block and a combination of subblock flipping [14] and stronger ECC on even block. The multitiered scheme makes it possible to use simple ECC schemes and thus, has lower area and latency overhead.

19.4.2 Reliability of Spin-Transfer Torque Magnetic Memory

The primary causes of errors in STT-MRAM are due to variations in MTJ device parameters, variations in CMOS circuit parameters, and thermal fluctuation. Recently, many studies have been performed to analyze the impact of MTJ device parametric variability and the thermal fluctuation on the reliability of STT-MRAM operations. Li et al. [41] summarized the major MTJ parametric variations affecting the resistance switching and proposed a "2T1J" STT-MRAM design for yield enhancement. Nigam et al. [42] developed a thermal noise model to evaluate the thermal fluctuations during the MTJ resistance switching process. Joshi et al. [43] conducted a quantitative statistical analysis on the combined impacts of both CMOS/MTJ device variations and thermal fluctuations. In this section, we also present the effects of process variation and geometric variation.

Typically, there are two main types of failures that occur during the read operation: read disturb and false read. Read disturb is the result of the value stored in the MTJ being flipped because of large current during read. False read occurs when current of parallel (antiparallel states) crosses the threshold value of the antiparallel (parallel) state. In our analysis, we find that the false read errors are dominant during the read operation; thus, we focus on false reads in the error analysis.

During write operation, failures occur when the distribution of write latency crosses the predefined access time. Write-1 is more challenging for an STT-MRAM device due to the asymmetry of the write operation. During write-1, access transistor and MTJ pair behaves similar to a source follower that increases the voltage level at the source of the access transistor and reduces the driving write current. Such a behavior increases the time required for a safe write-1 operation.

The variation impacts of the different parameters are presented in Figure 19.10 for read and write operations. To generate these results, we changed each parameter one at a time and did Monte Carlo simulations to calculate the contribution of each variation on the overall error rate. We see that variation in access transistor size is very effective in shaping the overall reliability; it affects the read operation by 37% and write operation by 44% with the write-0 and write-1 having very similar values. The threshold voltage variation affects the write operation more than the read operation. Finally, the MTJ geometry variation is more important in determining the read error rate as illustrated in Figure 19.10 [35].

19.4.2.1 Enhancing Reliability of Spin-Transfer Torque Magnetic Memory

The reliability of an STT-MRAM cell has been investigated by several researchers. While Joshi et al. [43] studied the failure rate of a single STT-MRAM cell using basic models for transistor and MTJ resistance, process variation effects such as RDF and geometric variation were considered in Chatterjee et al. [10] and Zhang et al. [44]. A methodology of optimizing STT-MRAM cell design that chose the MTJ device parameters and the NMOS transistor sizes to minimize write errors was proposed in Zhang et al. [18]. In Li et al. [41], an architecture-aware cell sizing algorithm that reduces read failures and cell area at the expense of write failures was proposed. In Yang et al. [14], we presented a multitiered approach spanning circuit level and system level to improve STT-MRAM reliability. By

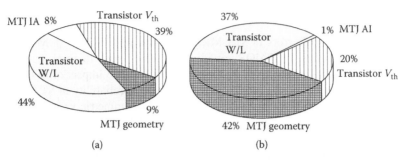

FIGURE 19.10   Effects of different variations on STT-MRAM. (a) Write operation. (b) Read operation.

using a combination of an increase in the W/L ratio of the access transistor, applying a higher voltage difference across the memory cell, and adjusting pulse width in write operation, the error rate was significantly reduced. As a result, a low-cost ECC was able to achieve very low BFR.

We are not aware of any work that explicitly addresses the reliability of NVM-based heterogeneous memory systems. Existing DRAM + PRAM systems improve the reliability of PRAM implicitly by reducing the number of hard errors due to reduced number of P/E cycles. However, such a system is likely to see an increase in the number of soft errors because the time between two consecutive writes in the one PRAM page is larger. In this case, stronger ECC will be required to deal with increased soft errors that result in long decoding latency in read operation.

For STT-MRAM, programming current amplitude and width affect the probability of a successful write. To meet programming latency and energy constraints, a homogeneous memory system may have high error rate that has to be compensated by strong ECC. However, DRAM or SRAM buffer can also reduce the number of accesses to STT-MRAM, which would reduce the energy of the hybrid memory. In that case, the STT-MRAM part can still use high current pulse amplitude and long pulse width to achieve low error rate without adversely affecting the energy and latency performance of the system.

19.5   CONCLUSION

To address the scaling problems of CMOS-based memory devices in sub-nano-technology and provide desirable performance, heterogeneous memory systems built with emerging memory technologies are needed. This chapter introduces a design methodology for such systems through device modeling, architecture level and system-level simulation, as well as reliability analysis. Compact device models of PRAM and STT-MRAM are built to facilitate study of memory latency, energy, and reliability. By embedding device model parameters into circuit and architecture simulators, system-level optimization can be achieved for different performance metrics. We also provide reliability analysis and methods to improve reliability for PRAM and STT-MRAM, because it affects memory energy, performance, and lifetime.

REFERENCES

1. R. Venkatesan, V. J. Kozhikkottu, C. Augustine, A. Raychowdhury, K. Roy, and A. Raghunathan, "TapeCache: A high density, energy efficient cache based on domain wall memory," *ACM/IEEE International Symposium on Low Power Electronics and Design*, pp. 185–190, 2012.

2. H. S. P. Wong, S. Raoux, S. Kim, J. Liang, J. P. Reifenberg, B. Rajendran, M Asheghi, and K. E. Goodson, "Phase change memory," *Proceedings of the IEEE*, pp. 2201–2227, December 2010.

3. G. W. Burr, M. J. Breitwisch, M. Franceschini, D. Garetto, K. Gopalakrishnan, B. Jackon, B. Kurdi et al., "Phase change memory technology," *Journal of Vacuum Science and Technology B*, 28(2): 223–262, April 2010.

4. S. Thoziyoor, N. Muralimanohar, J. H. Ahn, and N. P. Jouppi, "CACTI 5.1 technical report," HP Labs, Palo Alto, CA, Tech. Rep. HPL-2008-20, 2008.

5. X. Dong, N. Jouppi, and Y. Xie, "PCRAMsim: System-level performance, energy, and area modeling for phase-change RAM," *IEEE/ACM International Conference on Computer-Aided Design*, San Jose, CA, pp. 269–275, 2009.

6. X. Dong, C. Xu, Y. Xie, and N. P. Jouppi, "NVSim: A circuit-level performance, energy, and area model for emerging nonvolatile memory," *IEEE Transactions on Computer-Aided Design of Integrated Circuits and Systems*, 31(7): 994–1007, July 2012.

7. G. Sun, Y. Joo, Y. Chen, D. Niu, Y. Xie, Y. Chen, and H. Li, "A Hybrid solid-state storage architecture for the performance, energy consumption, and lifetime improvement," *IEEE 16th International Symposium on High Performance Computer Architecture*, pp. 1–12, January 2010.

8. G. Dhiman, R. Ayou, and T. Rosing, "PDRAM: A hybrid PRAM and DRAM main memory system," *IEEE Design Automation Conference*, San Francisco, CA, pp. 664–669, July 2009.

9. T. Kawahara, R. Takemura, K. Miura, J. Hayakawa, S. Ikeda, Y. Lee, R. Sasaki et al., "2 Mb SPRAM (spin-transfer torque RAM) with bit-by-bit bi-directional current write and parallelizing-direction current read," *IEEE Journal of Solid State Circuits*, 43(1): 109–120, January 2008.

10. S. Chatterjee, M. Rasquinha, S. Yalamanchili, and S. Mukhopadhyay, "A scalable design methodology for energy minimization of STTRAM: A circuit and architecture perspective," *IEEE Transactions on VLSI Systems*, 19(5): 809–817, May 2011.

11. Y. Chen, H. Li, X. Wang, W. Zhu, and T. Zhang, "A 130nm 1.2V/3.3V 16Kb spin-transfer torques random access memory with non-deterministic self-reference sensing scheme," *IEEE Journal of Solid State Circuits*, pp. 560–573, February 2012.

12. Z. Xu, K. B. Sutaria, C. Yang, C. Chakrabarti, and Y. Cao, "Compact modeling of STT-MTJ for SPICE simulation," *European Solid-State Device Research and Circuits Conference*, Bucharest, Romania, pp. 338–342, September 2013.

13. J. Kammerer, M. Madec, and L. Hébrard, "Compact modeling of a magnetic tunnel junction – Part I: dynamic magnetization model," *IEEE Transactions on Electron Devices*, 57(6): 1408–1415, June 2010.

14. C. Yang, Y. Emre, Z. Xu, H. Chen, Y. Cao, and C. Chakrabarti, "A low cost multi-tiered approach to improving the reliability of multi-level cell PRAM," *Journal of Signal Processing*, 1–21, DOI: 10.1007/s11265-013-0856-x, May 2013.

15. R. Annunziata, P. Zuliani, M. Borghi, G. De Sandre, L. Scotti, C. Prelini, M. Tosi, I. Tortorelli, and F. Pellizzer, "Phase change memory technology for embedded non volatile memory applications for 90nm and beyond," *IEEE International Electron Devices Meeting*, Baltimore, MD, pp. 1–4, Dec. 2009.

16. G. Servalli, "A 45nm generation phase change memory technology," *IEEE International Electron Devices Meeting*, Baltimore, MD, pp. 1–4, December 2009.

17. R. Dorrance, F. Ren, Y. Toriyama, A. Amin, C. K. Ken Yang, and D. Marković, "Scalability and design-space analysis of 1T-1MTJ memory cell," *IEEE/ACM International Symposium on Nanoscale Architectures*, pp. 32–36, 8–9, June 2011.

18. Y. Zhang, X. Wang, and Y. Chen, "STT-RAM cell design optimization for persistent and non-persistent error rate reduction: A statistical design view," *IEEE Transaction on Magnetics*, 47(10): 2962–2965, October 2011.

19. R. I. Bahar, G. Albera, and S. Manne, "Power and performance tradeoffs using various caching strategies," *International Symposium on Low Power Electronics and Design*, Monterey, CA, pp. 64–69, August 1998.

20. B. Lee, E. Ipek, O. Mutlu, and D. Burger, "Architecting phase change memory as a scalable DRAM alternative", *International Symposium on Computer Architecture*, pp. 2–13, June 2009.

21. Simplescalar 3.0e, available at http://www.simplescalar.com/, March 22, 2011.

22. GEM5 simulator, available at http://www.m5sim.org/Main_Page, February 14, 2014.

23. Y. Chen, X. Wang, H. Li, H. Xi, W. Zhu, and Y. Yan, "Design margin exploration of spin-transfer torque RAM (STT-RAM) in scaled technologies", *IEEE Transactions on Very Large Scale Integration Systems*, 18(12): 1724–1734. December 2010.

24. A. Jog, A. Mishra, Cong Xu, Y. Xie, V. Narayanan, R. Iyer, and C. Das, "Cache revive: Architecting volatile STT-RAM caches for enhanced performance in CMPs," *Design Automation Conference,* San Francisco, CA, pp. 243–252, June 2012.

25. Y. Joo, D. Niu, X. Dong, G. Sun, N. Chang, and Y. Xie, "Energy- and endurance-aware design of phase change memory caches," *Design, Automation and Test in Europe Conference and Exhibition*, Dresden, Germany, pp. 136–141, March 2010.

26. X. Wu, J. Li, L. Zhang, E. Speight, and Y. Xie, "Power and performance of read-write aware Hybrid Caches with non-volatile memories," *Design, Automation & Test in Europe Conference & Exhibition,* Nice, France, pp. 737–742, April 2009.

27. M. K. Qureshi, V. Srinivasan, and J. Rivers, "Scalable high performance main memory system using phase-change memory technology," *International Symposium on Computer Architecture*, pp. 1–10, June 2009.

28. H. Park, S. Yoo, and S. Lee, "Power management of hybrid DRAM/PRAM-based main memory," *Design Automation Conference*, New York, pp. 59–64, June 2011.

29. A. Ferreira, M. Zhou, S. Bock, B. Childers, R. Melhem, and D. Mosse, "Increasing PCM main memory lifetime," *Design, Automation & Test in Europe Conference & Exhibition*, Dresden, Germany, pp. 914–919, March 2010.

30. G. Dhiman, R. Ayoub, and T. Rosing, "PDRAM: A hybrid PRAM and DRAM main memory system," *Design Automation Conference,* San Francisco, CA, pp. 664–669, July 2009.

31. S. Lavizzari, D. Ielmini, D. Sharma, and A. L. Lacaita, "Reliability impact of chalcogenide-structure relaxation in phase-change memory (PCM) cells—Part II: physics-based modeling," *IEEE Transactions on Electron Devices*, 56(5): 1078–1085, March 2009.

32. D. Ielmini, A. L. Lacaita, and D. Mantegazza, "Recovery and drift dynamics of resistance and threshold voltages in phase-change memories," *IEEE Trans. Electron Devices*, 54(2): 308–315, 2007.

33. K. Kim and S. Ahn, "Reliability investigation for manufacturable high density PRAM," *IEEE 43rd Annual International Reliability Physics Symposium*, pp. 157–162, 2005.

34. S. Lavizzari, D. Ielmini, D. Sharma, and A. L. Lacaita, "Reliability impact of chalcogenide-structure relaxation in phase-change memory (PCM) cells—Part I: Experimental Study," *IEEE Transactions on Electron Devices*, pp. 1070–1077, May 2009.

35. C. Yang, Y. Emre, Y. Cao, and C. Chakrabarti, "Improving reliability of non-volatile memory technologies through circuit level techniques and error control coding," *EURASIP Journal on Advances in Signal Processing*, 2012: 211, 2012.

36. N. H. Seong, D. H. Woo, V. Srinivasan, J. A. Rivers, and H. H. S. Lee, "SAFFER: Stuck-At-Fault Error Recovery for Memories," Annual *IEEE/ACM International Symposium on Microarchitecture,* Atlanta, GA, pp. 115–124, 2009.

37. S. Schechter, G. H. Loh, K. Strauss, and D. Burger, "Use ECP, not ECC, for hard failures in resistive memories," *International Symposium on Computer Architectures,* pp. 141–152, June 19–23, 2010.

38. D. H. Yoon, N Muralimanohar, J. Chang, P. Ranganathan, N. P. Jouppi, and M. Erez, "FREE-p: Protecting non-volatile memory against both hard and soft errors," *IEEE 17th International Symposium on High Performance Computer Architecture*, pp. 466–477, 2009.

39. W. Xu and T. Zhang, "A time-aware fault tolerance scheme to improve reliability of multi-level phase-change memory in the presence of significant resistance drift," *IEEE Transactions on Very Large Scale Integration Systems*, 19(8): 1357–1367, June 2011.

40. C. Yang, Y. Emre, Y. Cao, and C. Chakrabarti, "Multi-tiered approach to improving the reliability of multi-level cell PRAM," *IEEE Workshop on Signal Processing Systems*, Quebec City, pp. 114–119, October 2012.

41. J. Li, P. Ndai, A. Goel, S. Salahuddin, and K. Roy, "Design paradigm for robust spin-torque transfer magnetic RAM (STT MRAM) from circuit/architecture perspective," *IEEE Transactions on Very Large Scale Integration Systems*, 18(12): 1710–1723, 2010.

42. A. Nigam, C. Smullen, V. Mohan, E. Chen, S. Gurumurthi, and W. Stan, "Delivering on the promise of universal memory for spin-transfer torque RAM (STT-RAM)," *International Symposium on Low Power Electronics and Design*, Fukuoka, Japan, pp. 121–126, August 2011.

43. R. Joshi, R. Kanj, P. Wang, and H. Li, "Universal statistical cure for predicting memory loss," *IEEE/ACM International Conference on Computer-Aided Design,* San Jose, CA, pp. 236–239, November 2011.

44. Y. Zhang, Y. Li, A. Jones, and Y. Chen. "Asymmetry of MTJ switching and its implication to STT-RAM designs," *In Proceedings of the Design, Automation, and Test in Europe*, Dresden, Germany, pp. 1313–1318, March 2012.

20 Soft-Error Resilient Circuit Design

Chia-Hsiang Chen, Phil Knag, and Zhengya Zhang

CONTENTS

20.1 INTRODUCTION

Soft errors are nondestructive, nonpermanent, and nonrecurring errors. They were first observed in dynamic random-access memory due to α particles emitted by lead-based package in the 1970s [1]. Neutrons in cosmic rays were found as another important source of soft errors [2–4]. These energetic particles travel through the silicon substrate and create minority carriers. When enough minority carriers are collected by a nearby transistor's drain diffusion node, it will result in a potential disruption of the stored 0 or 1 state, or a voltage transient, resulting in soft errors [5–7]. Soft errors belong to the broader class of single-event effects, defined as any measurable or observable change in state of performance of a device resulting from a single energetic particle strike [8]. Soft errors include single-event upset (SEU), that is, a soft error caused by a single energetic particle strike [5,6], and single-event transient (SET), that is, a momentary voltage spike at a circuit node caused by a single energy particle strike [9].

Soft-error occurrence depends on two factors: the charge collection volume near the sensitive node, and the critical charge necessary to change a stored state of a diffusion node, known as Q_{crit}. With technology scaling, scaled device size and supply voltage decrease the critical charge, but the scaled device size also reduces the charge collection volume [10–12]. Recent results point to a relatively constant device- and cell-level soft-error rate (SER) in newer technology nodes [13,14]. However, technology scaling has increased the number of devices per chip, thus the chip-level SER has worsened. Soft errors have become a major concern for robust systems, especially enterprise systems that require high reliability [12]. Soft errors induced by high-energy heavy ions also present major challenges to the design of flight [15] and space systems [14].

In the last twenty years, a variety of soft-error hardening techniques have been developed for memory cells, latches, and combinational circuits. The techniques belong to one of two broad classes: spatial redundancy and temporal redundancy. In the following, some of the most well-known techniques are reviewed. Some of these techniques have become key enablers of the latest commercial microprocessors.

20.2   ENGINEERING OF CIRCUIT NODES

Soft-error resiliency through circuit node engineering aims at increasing the critical charge, or decreasing the charge collection volume, or both, at selective nodes that are most sensitive to soft errors with the smallest impact on performance, power, and area [16,17]. Figure 20.1a shows stack tapering and explicit capacitance addition to the node fb in the latch design [16]. Stack tapering reduces the diffusion area of the transistor that is connected to fb to reduce its charge collection. The extra capacitance added to fb increases its critical charge. The two approaches provide 2 times improvement in SER under neutron beam testing. The higher capacitance and transistor stack increase the setup time by 10% and power by 4% [16].

The extra capacitor can be implemented using transistor gate capacitor. However, the gate capacitance is voltage dependent. When devices turn off, the gate capacitance is also lower due to the disappearing gate-to-channel capacitor. To maintain a relatively constant and high capacitance, a pair of n-type metal–oxide–semiconductor (NMOS) and p-type metal–oxide–semiconductor (PMOS) gate capacitors is added to a latch to provide a relatively flat capacitance across the entire voltage range [17]. The additional capacitance improves the SER of the latch by 2 times. The extra capacitance does not affect the D-to-Q time, but increases the setup time. To minimize the timing penalty, a switched capacitance design, shown in Figure 20.1b [17], can be used to add the extra capacitor only when the latch is holding. The switched capacitance is disconnected when the latch is sampling, so that the timing is not affected. Because the transmission gate switch adds resistance, upsizing the transmission gate reduces the resistance for a better SER. Another alternative is a stack node design shown in Figure 20.1c [17], where the extra capacitors are switched on by the clocked devices when the latch is holding.

20.3   DUAL INTERLOCKED STORAGE CELL

The dual interlocked storage cell (DICE) [18] was invented as a hardened storage to prevent SEUs. DICE is a dual-redundancy, circuit-level hardening technique. DICE is constructed as in Figure 20.2a [18], where inverters are implemented using single NMOS or PMOS devices (the first letter in the label indicates whether it is an NMOS or PMOS device, for example, N0 is an NMOS device and P1 is a PMOS device). DICE contains four storage nodes: X0, X1, X2, and X3, connected in a ring. Two of

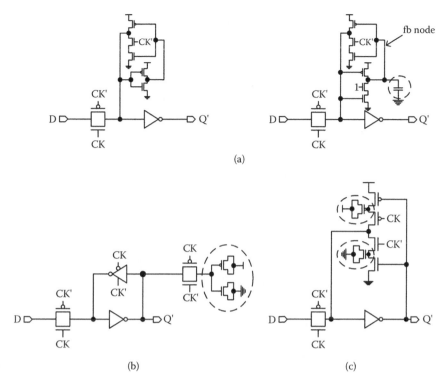

FIGURE 20.1 Node engineering approaches: (a) standard latch design (left) and stack tapering and explicit capacitance addition (right), (b) using a pair of n-type metal–oxide–semiconductor and p-type metal–oxide–semiconductor as extra capacitance, and (c) switched capacitance as extra capacitance. (T. Karnik et al., "Selective node engineering for chip-level soft error rate improvement," *IEEE Symp. VLSI Circuits Dig. Tech. Papers*, © 2002 IEEE.)

the nodes are redundant copies. An upset on one node affects one neighboring node, but the upset cannot propagate to the two remaining nodes.

The DICE design can be understood with an example. Suppose DICE is storing 0 at X0 and X2, and storing 1 at X1 and X3. X0 and X2 are independently controlled, and so are X1 and X3. If one node is upset, for example, X0 changes from 0 to 1, it will turn on N3, changing X3 to 0; turn off P1, leaving X1 floating at 1. X3 will then turn off N2, leaving X2 floating at 0. Therefore, a single-node upset will propagate to only one neighboring node, leaving the remaining two nodes unaffected. The unaffected nodes, X1 and X2, will eventually overcome the upsets by driving X0 low through N0, and driving X3 high through P3.

DICE protects against a single storage node upset. If two nonadjacent nodes are upset, DICE will not be able to recover from it. In the layout of the DICE cell, it is important to keep nonadjacent storage nodes physically separate, and the transistors that are responsible for recovering a storage node should be placed away from the storage node to minimize the impact of multiple-node upsets [19]. Because the error recovery in DICE relies on two floating nodes that can be affected by leakage, it is important to test the DICE circuitry for high leakage that could render the design ineffective.

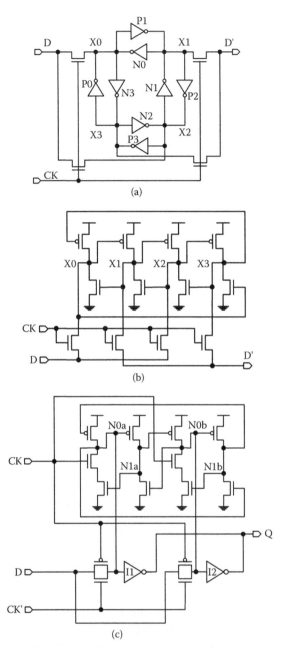

FIGURE 20.2 Dual interlocked storage cell designs. (a) Dual interlocked storage cell struc-
ture, (b) dual interlocked storage cell memory cell, and (c) dual interlocked storage cell latch.
(T. Calin et al., "Upset hardened memory design for submicron CMOS technology," *IEEE
Trans. Nucl. Sci.*, 43(6), © 1996 IEEE; P. Hazucha et al., "Measurements and analysis of
SER-tolerant latch in a 90-nm dual-Vt CMOS process," *IEEE J. Solid-State Circuits*, 39(9),
© 2004 IEEE.)

A higher supply voltage reduces recovery time, but it also increases the leakage current that can potentially damage the floating storage nodes during recovery [19]. The choice of the device threshold is also faced with the same trade-off between recovery time and reliability of floating nodes during recovery.

The circuit schematic for a DICE memory cell uses only eight transistors, and two or four transistors for access, as shown in Figure 20.2b [18]. A DICE latch can be created similarly as shown in Figure 20.2c [19]. In this design, two output buffers I1 and I2 are attached to the primary and the redundant storage, respectively, to reduce the possible output glitch. When one storage node, for example, N0a, is upset, the resulting glitch in the output of I1 can be balanced by the stable output of I2. Clever layout of this DICE latch results in only a 44% area overhead and a 34% energy overhead for achieving the same timing performance as a standard latch. The DICE latch exhibited 29 times lower SER compared to a standard latch [19].

20.4 SOFT-ERROR HARDENED LATCH

A particle strike generates electrons and holes. When enough electrons are collected by an n-diffusion, a stored 1 is changed to 0; conversely, when enough holes are collected by a p-diffusion, a stored 0 is changed to 1. The upset is directional: n-diffusion upset is from 1 to 0 only, and p-diffusion upset is from 0 to 1 only. Taking advantage of this fact, a soft-error hardened (SEH) latch is designed to have only three storage nodes for a smaller design [20] as shown in Figure 20.3a. There are three storage nodes: DH, as labeled in Figure 20.3a, stores the inverted D, and PDH and NDH, as labeled in Figure 20.3a, both store D of the same polarity. PDH is tied to p-diffusion only, and NDH is tied to n-diffusion only. When PDH and NDH are storing 0, a particle strike causes PDH to be charged to 1, but NDH remains at 0; when PDH and NDH are storing 1, a particle strike causes NDH to be discharged to 0, but PDH remains at 1. PDH and NDH cannot be upset at the same time.

SEH latch protects against a single storage node upset. Suppose DH stores 1, and PDH and NDH both store 0 (NDH is floating). If PDH is upset and charged to 1, it will turn off P1, leaving DH floating at 1. Q still stays at 0, which keeps N3 off, so that NDH is still floating at 0. NDH and Q keep P2 and P3 on to discharge PDH for error recovery. Suppose DH is upset and discharged to 0, then Q will change to 1. However, it will have no effect on PDH and NDH, except that both PDH and NDH are floating at 0. PDH keeps P1 on to charge DH for error recovery. The error shows up at Q as a 1-glitch. The DH upset recovery time depends on PDH to drive P1 to recover from a 1-to-0 upset. However, PDH stores V_t, a weak 0, which slows down the recovery. Similarly, DH depends on NDH to drive N1 to recover from a 0-to-1 upset, but NDH stores $V_{DD} - V_t$, a weak 1, that also slows down the recovery.

SEH latch is unable to recover from multiple node upset. However, PDH and NDH cannot be upset simultaneously. PDH and DH can fail at the same time (PDH changes from 0 to 1, and DH changes from 1 to 0), and NDH and DH can also fail at the same time (NDH changes from 1 to 0, and DH changes from 0 to 1). To minimize such possibilities, it is important to keep n-diffusion of DH away from PDH, and p-diffusion of DH away from NDH [20].

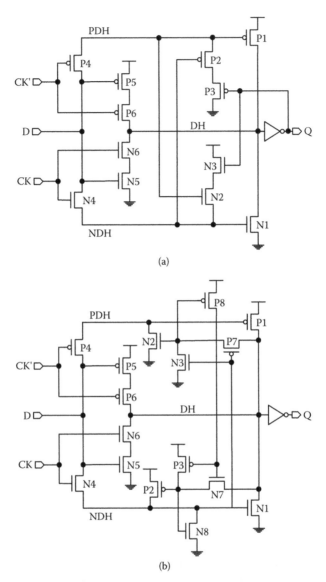

FIGURE 20.3 Soft-error hardened latch designs. (a) Soft-error hardened latch design, and (b) Type-B soft-error hardened latch design. (Y. Arima et al., "Cosmic-ray immune latch circuit for 90nm technology and beyond," *IEEE Int. Solid-State Circuits Conf. Dig. Tech. Papers*, © 2004 IEEE; Y. Komatsu et al., "A soft-error hardened latch scheme for SoC in a 90 nm technology and beyond," *Proc. IEEE Custom Integrated Circuits Conf.*, © 2004 IEEE.)

The incomplete swing of PDH and NDH in SEH latch results in a long recovery time. During steady state, either PDH (when storing 1) or NDH (when storing 0) is floating, and sensitive to leakage and noise coupling. A Type-B SEH latch is designed, as shown in Figure 20.3b [21], to overcome these problems by keeping both PDH and NDH static at full swing. In steady state, N2 and P8 keep PDH static, and N8 and P2 keep NDH static. An upset in P8 can be cancelled by N2, and vice versa. Similarly, an upset in N8 is cancelled by P2, and vice versa. The cancellation lowers the PDH and NDH upset probability. When DH is upset, one of PDH and NDH is floating. However, at full swing, the floating PDH or NDH will drive P1 or N1 to recover DH quickly and a glitch on Q is shortened.

The original SEH latch can be implemented with small area penalty of 15% with almost no increase in power, while its timing performance is 33% worse than a standard latch. It provides 197 times improvement in α-particle immunity and 34 times improvement in neutron immunity, but the recovery time is as long as 3.1 ns [21]. The Type-B SEH latch's area penalty is higher at 33%. It loses the single error polarity in storage nodes PDH and NDH, so its upset immunity is slightly worse than the original SEH latch: 99 times improvement in α-particle immunity and 25 times improvement in neutron immunity. By keeping storage nodes static and full swing, it allows for a fast recovery time of only 50 ps [21].

20.5 BUILT-IN SOFT-ERROR RESILIENCE

Replacing standard memory cells and latches with custom DICE and SEH cells may be costly. Built-in soft-error resilience (BISER) takes advantage of the built-in scan latches and flip-flops that are already part of many microprocessors, as shown in Figure 20.4a [22,23]. The scan flip-flop is enabled in testing by the two-phase clock SCA and SCB to sample and propagate the test input SI through the scan chain. The test input from the scan flip-flop is loaded to the primary flip-flop by asserting UPDATE. The output of the pipeline stage can be sampled by raising CAPTURE.

BISER makes use of the scan flip-flop as a redundant storage by connecting its data input and clock together with the primary flip-flop. The outputs of the primary and scan flip-flop O3 and O4 are fed to a checker, or C-element, as shown in Figure 20.4b (similarly, the master latch outputs O1 and O2 are also connected to a C-element). The C-element acts as a buffer when O3 and O4 agree, otherwise it blocks O3 and O4, allowing its output Q to retain its previous value. When CK is low, the slave latches are holding data that are subject to upsets. When CK is high, the master latches are holding and are subject to upsets. BISER protects against single latch upset, as the C-element blocks the error propagation. However, BISER fails when two latches, one in the primary flip-flop and one in the scan flip-flop, are upset, as the C-element will pass the error to the output. BISER also prevents the upset of the keeper, because O3 and O4 will drive the C-element to overcome the upset. The C-element needs to be sized stronger than the keeper for the error recovery.

The two flip-flops in BISER are not as tightly coupled as the redundant storage nodes in DICE or SEH latches, and the scan flip-flop can be turned off to save power if the protection is not needed [23]. The scan flip-flop needs to be designed to match the speed of the primary flip-flop, as a slow scan flip-flop will slow down the checking in the C-element and degrade the performance. BISER makes use of intrinsic redundancy, so the area penalty is only 17%. BISER dissipates 2.43 times power for a 20 times improvement in cell-level SER [23].

A potential weakness of BISER is that the combinational C-element is unprotected and vulnerable to SET [24]. When an SET is sampled by latches, it turns into upsets.

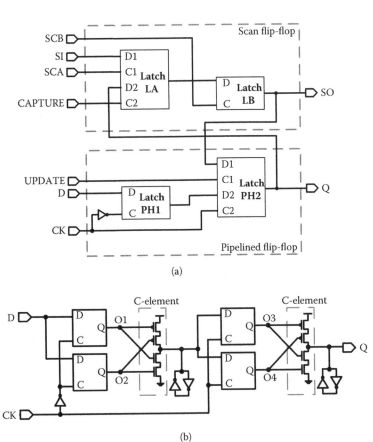

(a)

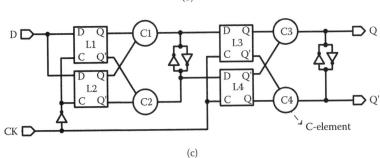

(b)

(c)

FIGURE 20.4 Built-in soft-error resilience (S. Mitra et al., "Robust system design with built-in soft-error resilience," *Computer*, 38(2), © 2005 IEEE; M. Zhang et al., "Sequential element design with built-in soft error resilience," *IEEE Trans. Very Large Scale Integr. (VLSI) Syst.*, 14(12), © 2006 IEEE.) and bistable cross-coupled dual modular redundancy designs. (J. Furuta et al., "A 65 nm bistable cross-coupled dual modular redundancy flip-flop capable of protecting soft errors on the C-element," *IEEE Symp. VLSI Circuits Dig. Tech. Papers*, © 2010 IEEE.) (a) Primary flip-flop and scan flip-flop, (b) built-in soft-error resilience latch design, and (c) bistable cross-coupled dual modular redundancy latch design.

At a low clock frequency, upsets in latches dominate and BISER remains effective. However, more frequently sampling at a higher clock frequency will likely register more SETs as upsets. If a BISER latch is followed by another BISER latch, as shown in Figure 20.4b [24], the SETs incurred by the C-element of the first BISER latch will be sampled by a pair of latches in the second BISER latch, rendering it completely ineffective. The weakness of BISER is evidenced in particle testing, where it is shown to lose its effectiveness at a higher clock frequency.

To enhance the reliability of BISER at a high clock frequency, bistable cross-coupled dual modular redundancy (BCDMR) has been invented to harden the C-element using a pair of cross-coupled C-elements, as shown in Figure 20.4c [24]. The cross-coupled C-elements offer a better protection against SET. For example, an SET in C1 will be sampled by L3, but C2 will keep its output stable, so L4 still retains the correct data. In this way, BCDMR improves the SET immunity. It has demonstrated 150 times better SER at a 160 MHz clock frequency than the BISER flip-flop in α-particle testing [24]. The C-elements in BCDMR can be downsized, as the differential drive easily overcomes the keeper. Therefore, the area and power dissipation of BCDMR are comparable to BISER, and the speed of BCDMR is slightly better than BISER.

20.6 TEMPORAL REDUNDANCY TECHNIQUES

DICE, SEH, BISER, and BCDMR are based on spatial redundancy: when one copy is perturbed by a particle strike, the redundant copies help it recover. The counterpart of spatial redundancy is temporal redundancy, that is, to record redundant (multiple) samples in time. Because soft errors often last for a short duration, the multiple samples in time enable the detection of transient errors. Error detection is followed by error correction through either re-execution or rollback to the last checkpoint.

The clock-shift temporal redundancy is shown in Figure 20.5a [25], where the output of the combinational circuit is sampled by latch1 that is clocked by CK, and a parallel latch2 that is clocked by a delayed clock CK + δ. To be able to detect an SET pulse of duration up to T_{tr}, latch2 sampling clock offset δ is set to $T_{tr} + T_{su}$. A SET pulse of T_{tr} or shorter is captured by at most one latch, but not both, so the error can be detected by the comparator. The clock-shift design also uncovers SEU occurrence in one of the latches. If $\delta > 0$, the clock-shift design incurs no speed penalty, as latch1 still runs at the same speed using the same clock signal as the unprotected case; whereas latch2 sampling at CK + δ increases the hold time of the combinational circuit by δ. If the fast path in the combinational circuit is $t_{min} < \delta$, the sampling by latch2 will result in a race condition. A solution to the problem is to pad fast paths by extra delays to ensure that $t_{min} > \delta$. The clock-shift design requires a delayed clock CK + δ to be distributed, which may be expensive.

An alternative time-shift temporal redundancy is shown in Figure 20.5b [25] to eliminate the delayed clock. In the time-shift design, both latch1 and latch2 sample at CK. The output of the combinational circuit is sampled by latch1, and a δ-delayed output is sampled by latch2. Similar to the previous approach, this design captures two time samples that are spaced by δ. However, the clock period needs to be increased by δ to allow latch2 to sample the slow paths correctly. The time-shift design incurs a speed penalty, but it eliminates race conditions. Time shift is

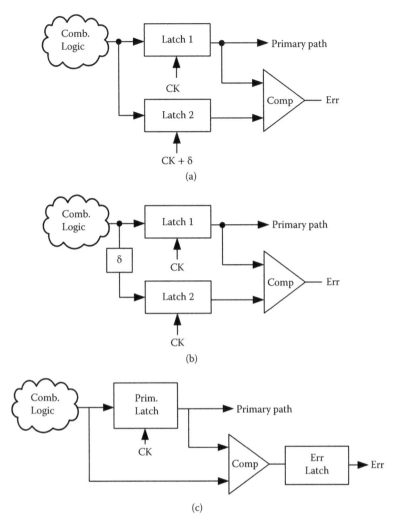

FIGURE 20.5 Temporal redundancy design approaches. (a) Clock-shift temporal redundancy, (b) time-shift temporal redundancy, and (c) transition-detection temporal redundancy. (From M. Nicolaidis, *Proc. VLSI Test Symp.*, 86–94, 1999. With permission; L. Anghel and M. Nicolaidis, "Cost reduction and evaluation of a temporary faults detecting technique," *Proc. Design, Automation and Test in Europe Conf. and Exhibition*, 591–598, © 2000 IEEE.)

incorporated in the improved BISER [26] and BCDMR designs [24] to improve the protection against SET.

Both clock-shift and time-shift temporal redundancy designs use two latches and a comparator. A low-cost variation of temporal redundancy, illustrated in Figure 20.5c [27], uses only one latch by comparing the output of latch1 (O1) with its input (O2). The design detects transitions at O2 by comparing it with the latched sample O1. Note that the error flag is only valid during a small window after O1 is sampled and before O2 is overwritten by a fast path, that is, between CK and CK + t_{min}, so the error flag has to be sampled during this window. To enable detection of an SET pulse of longer than δ, fast paths have to be padded such that $t_{min} > \delta$. The SEU detection in the transition-detection design is weaker compared to clock-shift or time-shift approaches, because it is only valid between CK and CK $+ t_{min}$.

20.7 RAZOR: DOUBLE SAMPLING SEQUENTIAL CIRCUITS

Razor was invented to detect and correct timing errors caused by supply voltage scaling [28]. With the timing errors corrected by Razor, the supply voltage can be reduced to save power. The circuit schematic of Razor is shown in Figure 20.6a [28], made of a master-slave main flip-flop and a parallel shadow latch. The main flip-flop samples at CK, and the shadow latch samples at a delayed clock CK $+ \delta$. An XOR gate compares the samples from the main flip-flop and the shadow latch, and an error is flagged if a mismatch is detected. A Razor flip-flop consumes 66% more energy than a standard flip-flop in an error-free operation at a 10% activity factor. Detecting and correcting an error consumes another 210% more energy [28]. The extra energy in error detection and correction is usually negligible as the error rate is kept low.

Razor's error detection mechanism follows the clock-shift temporal redundancy approach [25], but Razor was designed to detect timing errors, not soft errors. In the case of timing errors, the worst-case slow path may not meet the timing of the main flip-flop, but it will be correctly sampled by the shadow latch (by design). The correct sample in the shadow latch can be used to correct timing errors. However, in the case of soft errors, soft errors can occur in the main flip-flop or the shadow latch, and there is no guarantee that the shadow latch holds the correct sample. Therefore, Razor can be used for soft-error detection, but not correction.

Razor also incurs possible race conditions due to the delayed sampling by the shadow latch. A longer δ allows more timing errors to be detected and corrected and the supply voltage to be reduced further, but it increases the t_{min} constraint and delay padding of short paths. Razor is well-suited for critical flip-flops for critical slow paths. In the design demonstrated by D. Ernst et al. [28], Razor was used in 192 critical flip-flops out of a total of 2408 flip-flops, or 8% of the flip-flops, to reduce the cost and minimize the delay padding. However, this approach does not apply if Razor is used for soft-error detection, because soft errors occur in both critical and noncritical paths. Razor needs to be added to all flip-flops for the complete soft-error coverage.

The design of Razor reveals an important metastability issue of temporal redundancy techniques. Because soft errors can occur anytime, it is possible that the data at the input of a latch transitions at the same time as the clock transitions, resulting in a possible metastability [28], where the output is undetermined as it does not settle to either a high or a low voltage level. Metastability gives rise to undetected errors and misdetected errors. Razor incorporates a metastability detector using two parallel inverters [28], inv_n and inv_p shown in Figure 20.6a, whose switching thresholds are skewed low and high, respectively. The two inverters will sample the metastable input and give two different outputs, allowing metastability to be detected. Alternatively, the metastable output can be latched a second time, and the probability

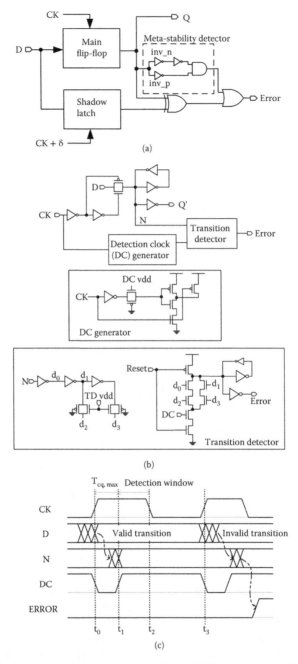

FIGURE 20.6 Razor doubling sampling techniques. (a) Razor flip-flop, (b) Razor II latch, and (c) Razor II timing diagram. (D. Ernst et al., "Razor: a low-power pipeline based on circuit-level timing speculation," *Proc. IEEE/ACM Int. Symp. Microarchitecture*, © 2003 IEEE; S. Das et al., "RazorII: in situ error detection and correction for PVT and SER toler- ance," *IEEE J. Solid-State Circuits*, 44(1), © 2009 IEEE.)

that the output of the second latch is metastable becomes negligible. Resolving metastability introduces area and power penalties.

To reduce the area and power of Razor, a low-cost Razor II was introduced [29]. Razor II performs error detection using transition detection as described previously. Razor II removes circuit-level timing error correction, and instead relies on architectural replay for error correction. As the error rate will be kept very low, the cost of architectural replay will be negligible. The schematic of the Razor II latch is shown in Figure 20.6b [29]. The transition detector is attached to the storage node N. The detector creates four copies of N: d_0, d_1, d_2, d_3 that are phased apart by an inverter or transmission gate delay (d_1, d_2 = N; d_0, d_3 = ~N, the inverted N). The four delayed copies are fed to a dynamic comparator. The comparator is reset high and the error flags false. A transition on node N creates a momentary glitch on d_0 first, so that $d_0 = d_2$ for a moment; and then the glitch propagates to d_1, so that $d_1 = d_3$ for a moment, which trigger the discharge of the dynamic comparator and setting the error flag to true.

The transition detection is only enabled after the output of the combinational circuit is correctly sampled by the latch, which occurs by t_1, that is, $T_{cq,max}$ after the rising edge of the clock t_0, as illustrated in Figure 20.6c [29]. A detection clock DC disables the transition detector from t_0 to t_1, so that any errant transition on N for the rest of the clock period will be detected as errors. Razor II latch is able to detect timing errors between t_1 and t_2, the clock falling edge. The timing error detection window can be adjusted by duty cycling the clock. The metastability detection is removed in this design by ensuring that the latch input never transitions near the clock falling edge. Razor II latch detects soft errors due to SET between t_1 and t_2, and SEU between t_1 and t_3, the next clock rising edge. Note that soft errors can cause metastability, which is not detected in this design. The simplified Razor II latch consumes 28.5% more power than a standard latch at a 10% activity factor [29].

20.8 CONFIDENCE-DRIVEN COMPUTING

The error detection window of a clock-shift temporal redundancy design, for example, Razor, is limited to δ after the sampling edge, or δ before the sampling edge in a time-shift design. The length of δ determines the error coverage, but it is limited by short paths in a clock-shift design and clock speed in a time-shift design. Using transition detection, for example, in Razor II, the error coverage is widened, but still limited to a fraction of a clock cycle. The error coverage is also fixed in design time and it is difficult to adapt in runtime. Confidence-driven computing (CDC) provides protection against nondeterministic errors over a wide range of rate and duration [30,31]. The concept of CDC is to employ a tunable confidence threshold for an adjustable reliability.

CDC is implemented by placing a confidence estimator (CE) before a pipeline flip-flop, as illustrated in Figure 20.7 [30,31]. CE consists of a flip-flop to store the output of the combinational circuit, a checker to compare the output with the previous sample stored in the flip-flop, and a counter to keep track of the confidence level. The confidence level is raised on an agreement and reset to zero on a disagreement. A controller ensures the confidence level reaches a required threshold before sending a request to the next pipeline stage. When the next pipeline stage becomes ready to

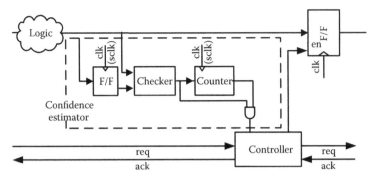

FIGURE 20.7 Design of confidence estimator in confidence-driven computing. (From C. H. Chen et al., "Design and Evaluation of Confidence-Driven Error-Resilient Systems," *IEEE Trans. Very Large Scale Integr. (VLSI) Syst., to be published, © 2013 IEEE.*

accept a new input, an acknowledgement is sent back and the controller enables the pipeline flip-flop to pass the input to the next stage.

CDC looks for consecutive agreements in time for error detection and correction. A confidence threshold of n requires n agreements. When an error is detected, the confidence level is reset to 0 and the confidence accumulation starts all over again. The error detection window and the error protection level are adjustable by the confidence threshold, but the throughput is also reduced proportionally. A confidence threshold of n decreases the throughput by n in an error-free operation. Errors decrease the throughput further. However, at a lower error rate, the throughput degradation due to errors is negligible. Experiments show that the error protection can be improved by orders of magnitude with each increment of the confidence threshold by effectively trading off the throughput. The flexibility of CDC comes at a cost of extra area and energy. With a careful design, the area and energy penalty can be limited to as low as 12% and 19%, respectively [30,31].

20.9 ERROR RESILIENT SYSTEM ARCHITECTURE FOR PROBABILISTIC APPLICATIONS

A soft-error resilient system can be constructed bottom-up using hardened circuits as described previously. It is, however, more efficient sometimes to construct a resilient system top-down. For emerging applications, such as estimation, learning, and inference for communications, multimedia, and machine learning, the top-down approach can be especially advantageous. These emerging applications are often computationally intensive and data intensive that require massively parallel architectures. They often deal with noisy inputs from the physical world, and the outputs of these applications have a soft quality metric, for example., in the form of probability, confidence, and bit error rate. The computation is often probability based and the outputs are obtained iteratively. As a result, these emerging applications are intrinsically error resilient.

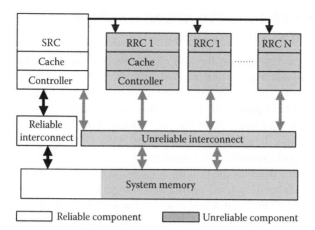

FIGURE 20.8 Error resilient system architecture for probabilistic applications. (H. Cho et al.,"ERSA: Error resilient system architecture for probabilistic applications," *IEEE Trans. Comput.-Aided Des. Integr. Circuits Syst,.* 31(4), © 2012 IEEE.)

Error resilient system architecture (ERSA), illustrated in Figure 20.8 [32], is a good example of an efficient system architecture that exploits an application's intrinsic error resilience. ERSA is a multicore architecture that is designed with asymmetric reliability that consists of a small number of highly reliable cores (SRCs) implemented using error hardening techniques, and a large number of less reliable, relaxed reliability cores (RRCs) implemented with minimal error hardening. SRC assigns computation tasks to RRCs and monitor their executions. The overhead of a small number SRC is amortized by a large number RRCs, so the overall overhead remains low. The area overhead of ERSA compared to an unprotected system is only 0.7%, and the power penalty is 1.9% [32].

ERSA uses lightweight execution monitoring, including watchdog timers, memory access boundary monitors, and exception monitors to ensure that the RRCs are responsive, accessing their intended memory block, and reporting unexpected exceptions. When an RRC is stuck and runs into exceptions, the RRC is restarted by the SRC. The monitoring is light weight, as opposed to heavy error detection and correction. The execution monitoring prevents crashes. To ensure the quality of the output, a set of software rules are used to check the validity of the output, damping the fluctuations, and filter out invalid results based on application-specific criteria. These software rules are also light weight and inexpensive to implement.

Experiments have demonstrated that the enhanced ERSA using both execution monitoring and software rule checking achieves significantly better computation accuracies to deliver acceptable outputs. Even as the error rate increases, the execution time on the enhanced ERSA remains low [32]. The cross-layer resilient architecture takes advantage of the application's error resilience in designing inexpensive hardware. The approach can be a highly cost-effective solution toward the increasingly more complex system design and resiliency problems.

20.10   COMMERCIAL APPLICATIONS OF ERROR RESILIENT CIRCUIT DESIGNS

The design of the Intel 65-nm Itanium quad-core processor is faced with a reliability challenge due to the triple increase in logic circuits over the previous generation of products, and the requirement of maintaining the same SER [33]. To meet the reliability requirement, the design chose to harden 99% of the latches in the system interface and 33% of the core latches. The DICE structure is incorporated in the pulse latch as shown in Figure 20.9a [33]. The pulse clock PCK enables sampling of the input over a short pulse window. When D = 0, in1 pulls up storage node B and disables the pull up of C, while q1 pulls down C and disables pull down of B. When D = 1, in1 pulls down storage node B and disables the pull down of A, while q2 pulls up A and disables the pull up of B. With the feedback removed, writing is faster than the standard DICE latch by up to 30%, and the clock-to-Q delay matches a standard pulse latch. By separating the diffusions of storage nodes in layout, the DICE pulse latch has demonstrated 100 times lower SER over an unprotected latch with a 34% area penalty and 25% higher active power [33].

A DICE register file design is shown in Figure 20.9b [33]. When bitw = 1, the bit line tries to pull up q and q1, which is facilitated by disabling the pull downs through nq and nq1, making it easier to write 1. When bitw = 0, the design is similar to the standard DICE latch. Therefore, writing 0 is slower than writing 1. The loading on the bit line worsens the write speed. However, write timing is generally not critical for a register file. The DICE register file cell achieves an 80 times lower SER over an unprotected cell with a 25% area penalty and 24% higher standby power [33].

Compared to devices in a bulk CMOS process, silicon-on-insulator (SOI) devices have a thin body. Due to the much smaller charge collection area in the thin body, the upset rate of an SOI device is much lower than bulk devices [34]. To further reduce the upset rate, SOI devices can be electrically connected in series (using separate diffusions that are electrically connected) [35,36]. An upset turns on one device, but it is unlikely for all devices in series to turn on at the same time, thereby preventing an upset of a logic gate.

A stacked transistor approach is adopted in the design of the pulse latch in IBM POWER7™ microprocessor, as shown in Figure 20.9c [37]. Each device in the pulse latch is divided to two devices in a stack with their gates tied together, so it is logically equivalent but provides better upset protection. The transmission gate is not stacked because both sides of the transmission gate are equal for most of the time. An upset turns on the gate without inducing an error. The stacked transistor approach increases the area by less than 10% with only small impact on timing and power. It provides 6 times improvement in SER compared to unprotected bulk devices [37]. The SER of the stacked transistor SOI latch design is comparable to DICE, but its cost is lower than DICE.

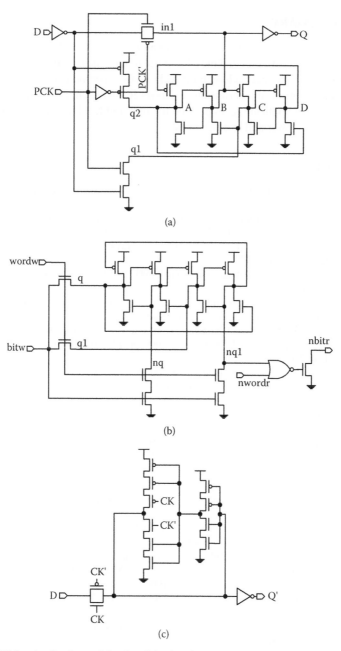

FIGURE 20.9 Applications of hardened latches in commercial microprocessors. (a) Dual interlocked storage cell pulse latch, (b) dual interlocked storage cell register file, and (c) stacked silicon-on-insulator device. (D. Krueger et al. "Circuit design for voltage scaling and SER immunity on a quad-core Itanium processor," *IEEE Int. Solid-State Circuits Conf. Dig. Tech. Papers*, © 2008 IEEE; J. Warnock et al., "POWER7 local clocking and clocked storage elements," *IEEE Int. Solid-State Circuits Conf. Dig. Tech. Papers*, © 2010 IEEE.)

20.11 SUMMARY

In this chapter, we review circuit design techniques to improve the soft error immunity by leveraging spatial redundancy, temporal redundancy, or architectural adaptation. The redundancy and architectural techniques consume extra area and power, and sometimes degrade the performance. However, each technique can be tailored to specific classes of applications to provide the required level of soft-error protection with the optimal power and overhead cost.

REFERENCES

1. T. C. May and M. H. Woods, "Alpha-particle-induced soft errors in dynamic memories," *IEEE Trans. Electron Devices*, 26(1): 2–9, January 1979.
2. D. Binder, E. C. Smith, and A. B. Holman, "Satellite anomalies from galactic cosmic rays," *IEEE Trans. Nucl. Sci.*, 22(6): 2675–2680, December 1975.
3. J. F. Ziegler and W. A. Lanford, "Effect of cosmic rays on computer memories," *Science*, 206(4420): 776–788, 1979.
4. E. Normand, "Single event upset at ground level," *IEEE Trans. Nucl. Sci.*, 43(6): 2742–2750, December 1996.
5. P. E. Dodd and L. W. Massengill, "Basic mechanisms and modeling of single-event upset in digital microelectronics," *IEEE Trans. Nucl. Sci.*, 50(3): 583–602, June 2003.
6. T. Karnik and P. Hazucha, "Characterization of soft errors caused by single event upsets in CMOS processes," *IEEE Trans. Dependable and Secure Computing*, 1(2): 128–143, April-June 2004.
7. R. C. Baumann, "Radiation-induced soft errors in advanced semiconductor technologies," *IEEE Trans. Dev. Mat. Rel.*, 5(3): 305–316, September 2005.
8. JESD. Measurement and reporting of alpha particle and terrestrial cosmic ray-induced soft errors in semiconductor devices, *JESD89A*, August 2001.
9. P. E. Dodd et al., "Production and propagation of single-event transients in high-speed digital logic ICs," *IEEE Trans. Nucl. Sci.*, 51(6): 3278–3284, December 2004.
10. P. Hazucha and C. Svensson, "Impact of CMOS technology scaling on the atmospheric neutron soft error rate," *IEEE Trans. Nucl. Sci.*, 47(6): 2586–2594, December 2000.
11. P. Shivakumar et al., "Modeling the effect of technology trends on the soft error rate of combinational logic," *Proc. Int. Conf. Dependable Syst. and Networks*, pp. 389–398, 2002.
12. R. Baumann, "Soft errors in advanced computer systems," *IEEE Design and Test of Comput.*, 22(3): 258–266, May–June 2005.
13. N. Seifert et al., "Radiation-induced soft error rates of advanced CMOS bulk devices," *Proc. IEEE. Int. Rel. Physics Symp.*, pp. 217–225, March 2006.
14. P. Roche et al., "A commercial 65 nm CMOS technology for space applications: Heavy ion, proton and gamma test results and modeling," *IEEE. Trans. Nucl. Sci.*, 57(4): 2079–2088, August 2010.
15. A. Taber and E. Normand, "Single event upset in avionics," *IEEE Trans. Nucl. Sci.*, 40(2): 120–126, April 1993.
16. T. Karnik et al., "Scaling trends of cosmic ray induced soft errors in static latches beyond 0.18μ," *IEEE Symp. VLSI Circuits Dig. Tech. Papers*, pp. 61–62, June 2001.
17. T. Karnik et al., "Selective node engineering for chip-level soft error rate improvement," *IEEE Symp. VLSI Circuits Dig. Tech. Papers*, pp. 204–205, Honolulu, HI, June 2002.
18. T. Calin, M. Nicolaidis, and R. Velazco, "Upset hardened memory design for submicron CMOS technology," *IEEE Trans. Nucl. Sci.*, 43(6): 2874–2878, December 1996.

19. P. Hazucha et al., "Measurements and analysis of SER-tolerant latch in a 90-nm dual-Vt CMOS process," *IEEE J. Solid-State Circuits*, 39(9): 1536–1543, September 2004.

20. Y. Arima et al., "Cosmic-ray immune latch circuit for 90nm technology and beyond," *IEEE Int. Solid-State Circuits Conf. Dig. Tech. Papers*, pp. 492–493, February 2004.

21. Y. Komatsu et al., "A soft-error hardened latch scheme for SoC in a 90 nm technology and beyond," *Proc. IEEE Custom Integrated Circuits Conf.*, pp. 329–332, October 2004.

22. S. Mitra et al., "Robust system design with built-in soft-error resilience," *Computer*, 38(2): 43–52, February 2005.

23. M. Zhang et al., "Sequential element design with built-in soft error resilience," *IEEE Trans. Very Large Scale Integr. (VLSI) Syst.*, 14(12): 1368–1378, December 2006.

24. J. Furuta et al., "A 65nm bistable cross-coupled dual modular redundancy flip-flop capable of protecting soft errors on the C-element," *IEEE Symp. VLSI Circuits Dig. Tech. Papers*, pp. 123–124, June 2010.

25. M. Nicolaidis, "Time redundancy based soft-error tolerance to rescue nanometer technologies," *Proc. VLSI Test Symp.*, pp. 86–94, 1999.

26. S. Mitra et al., "Combinational Logic Soft Error Correction," *Proc. IEEE Int. Test Conf.*, pp. 1–9, October 2006.

27. L. Anghel and M. Nicolaidis, "Cost reduction and evaluation of a temporary faults detecting technique," *Proc. Design, Automation and Test in Europe Conf. and Exhibition*, pp. 591–598, 2000.

28. D. Ernst et al., "Razor: a low-power pipeline based on circuit-level timing speculation," *Proc. IEEE/ACM Int. Symp. Microarchitecture*, pp. 7–18, December 2003.

29. S. Das et al., "RazorII: in situ error detection and correction for PVT and SER tolerance," *IEEE J. Solid-State Circuits*, 44(1): 32–48, January 2009.

30. C.-H. Chen et al., "A confidence-driven model for error-resilient computing," *Proc. Design, Automation and Test in Europe Conf. and Exhibition*, pp. 1–6, March 2011.

31. C.-H. Chen et al., "Design and Evaluation of Confidence-Driven Error-Resilient Systems," *IEEE Trans. Very Large Scale Integr. (VLSI) Syst.*, to be published.

32. H. Cho, L. Leem, and S. Mitra, "ERSA: Error resilient system architecture for probabilistic applications," *IEEE Trans. Comput.-Aided Des. Integr. Circuits Syst.*, 31(4): 546–558, April 2012.

33. D. Krueger, E. Francom, and J. Langsdorf, "Circuit design for voltage scaling and SER immunity on a quad-core Itanium processor," *IEEE Int. Solid-State Circuits Conf. Dig. Tech. Papers*, pp. 94–95, February 2008.

34. A. KleinOsowski et al., "Circuit design and modeling for soft errors," *IBM J. Research and Develop.*, 52(3): 255–263, May 2008.

35. J. Cai et al., "SOI series MOSFET for embedded high voltage applications and soft-error immunity," *Proc. IEEE Int. SOI Conf.*, pp. 21–22, October 2008.

36. C. Johnson et al., "A wire-speed power processor: 2.3GHz 45nm SOI with 16 cores and 64 threads," *IEEE Int. Solid-State Circuits Conf. Dig. Tech. Papers*, pp.104–105, February 2010.

37. J. Warnock et al., "POWER7 local clocking and clocked storage elements," *IEEE Int. Solid-State Circuits Conf. Dig. Tech. Papers*, pp. 178–179, February 2010.

Index